"Our economic system – particularly busine
(yet) include a mechanism for recognizing tl
nature. True Cost Accounting provides ɛ
values in business internal decision-making
reporting to financial markets. In today's oj
more than ever needs to demonstrate a ɭ⎯ɭ
financial value creation for shareholders. This purpose should include
generating social, human, and natural capital value for all stakeholders."
— *Peter Bakker, President and CEO, World Business Council for*
Sustainable Development

"*True Cost Accounting for Food* provides in-depth analysis of the environmental, health, and social costs of food systems that are rarely captured in
the price of food today. This valuable collection breaks new ground and
offers important insights into how these externalities can be better accounted for in ways that contribute toward positive change in food systems."
— *Jennifer Clapp, Professor and Canada Research Chair,*
University of Waterloo

"True Cost Accounting is the starting point for any serious conversation
about reforming food systems. This is no coincidence: prices shall continue
to lie, until social costs are incorporated and set the right incentives to guide
the choices of both producers and consumers. This book provides therefore
more than a state of play: it's an essential tool for future advocacy efforts."
— *Olivier De Schutter, UN Special Rapporteur on extreme poverty and human*
rights and Co-chair, International Panel of Experts on Sustainable
Food Systems (IPES-Food)

"Those who wish to help move the world towards a sustainable planet and
offer the next generation a chance to enjoy the wonders of its biodiversity
and benefit from its ecosystem services should embrace the tools and lessons
from this book to help transform our food systems from farm to plate."
— *Braulio Dias, Associate Professor of Ecology at the University of Brasilia and*
Former Executive Secretary of the UN Convention on Biological Diversity

"This book addresses a critically important topic: the need to open up
the 'hidden costs' of current food systems. This compelling need is not
being adequately considered by dominant policy makers or mainstream
thinkers in the agriculture sector and food industry. The present volume
serves to illuminate the issue from multiple perspectives and can serve to
inform international negotiations on food systems, agroecology and
biodiversity for more comprehensive policies."
— *Mohammad Hossein Emadi, PhD, Former Ambassador and Permanent*
Representative of Iran to FAO, WFP and IFAD and Chair of the
UN Committee on Agriculture

"The vision and core tenets of the Land and Justice Party I founded in Vanuatu in 2010 are in the principles behind True Cost Accounting in Food. This book will make a significant contribution to the international discourse needed to advance a just transformation of our food system, which reaches beyond monetary dimensions to include measures of progress for the social, cultural and environmental dimensions of life."

— *Ralph Regenvanu, Member of Parliament and President of the Land and Justice Party, Vanuatu*

"Since 2012, we have been committed to advancing True Cost Accounting (TCA) as a powerful tool for food systems transformation. In that time, TCA has evolved from a radical concept to a scientifically validated approach, driving policies and practices that create and sustain healthy, equitable, and resilient food systems. With this new book in hand, governments, farmers, corporations, investors, and others, can break away from the status quo and use TCA to make better decisions about the future of food."

— *Ruth Richardson, Executive Director, Global Alliance for the Future of Food*

"It is more than time that we rebalance the scales, making the cultivation of biodiversity and the consequent health of soils, plants, and people as the organizing principle of our food systems. This book offers many perspectives on how a genuinely honest valuation of the contribution of people and nature can bring about such changes."

— *Vandana Shiva, Navdanya International*

"Our food systems are not delivering the social, environment and nutritional outcomes we need to enable human prosperity. True Cost Accounting is one of the critical capabilities we need to build food systems that are equitable, healthy, and sustainable. By developing this capability our policy and legislative decision makers can help create the food system we all desire and deserve. This book creates the foundation for that work and is an important first step on that path."

— *Dr. Roy Steiner, Senior Vice President of Food at the Rockefeller Foundation*

"This book is a must-read for anyone interested in how to fix our food systems. We can only manage what we measure and this inspiring and instructive collection of examples informs how we, as a global community, can advance this crucial work."

— *Pavan Sukhdev, lead author of the United Nations report on TEEB (The Economics of Ecosystems & Biodiversity), CEO of GIST Impact, and President of WWF International*

True Cost Accounting for Food

This book explains how True Cost Accounting is an effective tool that we can use to address the pervasive imbalance in our food system.

Calls are coming from all quarters that the food system is broken and needs a radical transformation. A system that feeds many yet continues to create both extreme hunger and diet-related diseases, and one that has significant environmental impacts, is not serving the world adequately. This volume argues that True Cost Accounting in our food system can create a framework for a systemic shift. What sounds on the surface like a practice relegated to accountants is ultimately a call for a new lens on the valuation of food and a new relationship with the food that we eat, starting with the reform of a system out of balance. From the true cost of corn, and water, to incentives for soil health, the chapters economically compare conventional and regenerative, more equitable farming practices in food system structures, including taking an unflinching look at the true cost of cheap labour. Overall, this volume points towards the potential for our food system to be more human-centred than profit-centered and one that has a more respectful relationship to the planet. It sets forth a path forward based on True Cost Accounting for food. This path seeks to fix our current food metrics, in policy and in practice, by applying a holistic lens that evaluates the actual costs and benefits of different food systems, and the impacts and dependencies between natural systems, human systems, and agriculture and food systems.

This volume is essential reading for professionals and policymakers involved in developing and reforming the food system, as well as students and scholars working on food policy, food systems, and sustainability.

Barbara Gemmill-Herren, until she retired in 2015, was Delivery Manager for the Major Area of Work on Ecosystem Services and Biodiversity at the UN Food and Agriculture Organization (FAO). She was previously Executive Director of Environment Liaison Centre International, an international environmental non-governmental organization based in Nairobi, Kenya. She is an Associate Faculty in the Masters of Sustainable Food Systems Program at Prescott College in Arizona, USA, and a Senior Associate of the World Agroforestry Centre, Kenya.

Lauren E. Baker is Senior Director of Programs with the Global Alliance for the Future of Food, where she focuses on the intersections between food systems and health, climate change, agroecology, and TCA. Previously, Lauren led the Toronto Food Policy Council and was the Founding Director of Sustain Ontario. Lauren teaches in the Global Food Equity program at the University of Toronto, and at Ryerson University, Canada.

Paula A. Daniels is Co-founder and Chair of the Center for Good Food Purchasing. She is a lawyer and public policy leader in environmental, food, and water policy and has extensive experience in government through several appointed positions, including as Senior Advisor on Food Policy to Mayor Villairagosa of Los Angeles. She has been a faculty member at UCLA, Vermont Law School, and UC Berkeley.

Routledge Studies in Food, Society and the Environment

For more information about this series, please visit: www.routledge.com/books/series/RSFSE

True Cost Accounting for Food

Balancing the Scale

Edited by
**Barbara Gemmill-Herren,
Lauren E. Baker and Paula A. Daniels**

Routledge
Taylor & Francis Group
LONDON AND NEW YORK

First published 2021
by Routledge
2 Park Square, Milton Park, Abingdon, Oxon OX14 4RN

and by Routledge
605 Third Avenue, New York, NY 10158

Routledge is an imprint of the Taylor & Francis Group, an informa business

British Library Cataloguing-in-Publication Data
A catalogue record for this book is available from the British Library

Library of Congress Cataloging-in-Publication Data
Names: Gemmill-Herren, Barbara, editor. | Baker, Lauren E., editor. |
Daniels, Paula A., editor.
Title: True cost accounting for food : balancing the scale / edited
by Barbara Gemmill-Herren, Lauren E. Baker and Paula A. Daniels.
Description: New York : Routledge, 2021. |
Series: Routledge studies in food, society and the environment |
Includes bibliographical references and index. |
Identifiers: LCCN 2020053789 (print) | LCCN 2020053790 (ebook) |
ISBN 9780367506858 (paperback) | ISBN 9780367506896 (hardback) |
ISBN 9781003050803 (ebook)
Subjects: LCSH: Cost accounting. | Food prices. |
Food prices--Law and legislation.
Classification: LCC HF5686.C8 T718 2021 (print) |
LCC HF5686.C8 (ebook) | DDC 338.1/9--dc23
LC record available at https://lccn.loc.gov/2020053789
LC ebook record available at https://lccn.loc.gov/2020053790

ISBN: 978-0-367-50689-6 (hbk)
ISBN: 978-0-367-50685-8 (pbk)
ISBN: 978-1-003-05080-3 (ebk)

Typeset in Bembo
by Taylor & Francis Books

Contents

Illustrations

Figures

Tables

Boxes

Contributors

Helmy Abouleish was born 1961 in Graz, Austria. He is the CEO of the SEKEM Initiative in Egypt, founded by his father Ibrahim Abouleish in 1977. SEKEM promotes sustainable development in ecology, economy, societal, and cultural life. The SEKEM Holding produces, processes, and markets organic and biodynamic foodstuffs, textiles, and herbal medicine in Egypt and internationally and runs several educational institutions. SEKEM is regarded as Egypt's pioneer in organic farming, and was awarded the Right Livelihood Award ("Alternative Nobel Prize") in 2003. Helmy Abouleish studied economics and marketing in Cairo and has campaigned for many years in national and international politics to promote responsible competitiveness, social entrepreneurship, and tackling the greatest challenges of the 21st century, such as climate change and food security. He is also member of a number of international organizations and councils, such as Fair Trade USA. In June 2018 he was elected as president of Demeter International.

Jennifer Astone, PhD, Principal of Integrated Capital Investing, is a financial activist working to transform philanthropic investing. She founded Integrated Capital Investing in 2019 to catalyze foundations and investors to move their endowments, grants, and leadership into regenerative economies and healthy food systems. She served as Executive Director at the Swift Foundation, guiding its transitional investing portfolio over eight years. A researcher, advocate, writer, and coalition builder focused on community-led, movement solutions, Jennifer helped to launch the Agroecology Fund and the Transformational Investing in Food Systems Initiative. She is an RSF Social Finance Integrated Capital Fellow and received her doctorate in anthropology based on gender and agriculture fieldwork in Guinea, West Africa.

Lauren E. Baker, PhD, Senior Director of Programs with the Global Alliance for the Future of Food, has more than 20 years of experience facilitating cross-sectoral research, policy, and advocacy for sustainable food systems in non-profit, academic, business, policy, and philanthropic contexts. Lauren's expertise ranges from researching agricultural biodiversity in Mexico to negotiating and developing municipal food policy and programs. At the Global Alliance, Lauren's work is focused on the intersections between food systems and health, climate change, agroecology, and true cost accounting. Previously, Lauren led the Toronto Food Policy Council, a citizen advisory group embedded within the City of Toronto's Public Health Division, and was the Founding Director of Sustain Ontario—the Alliance for Healthy Food and Farming. Lauren teaches in the Global Food Equity program at the University of Toronto and at Ryerson University.

Tobias Bandel, Managing Director of Soil & More Impacts, spent several years as head of fruit and vegetable cultivation and sales at SEKEM in Egypt after studying agricultural sciences with a focus on soil science at the University of Hohenheim. In 2007 he co- founded the sustainability consultancy Soil & More Impacts. Today, his team of more than 20 colleagues develops sustainable business strategies, globally advises on soil fertility, and provides digital services for supplier management and the assessment of procurement risks. Soil & More Impacts is focussing its consulting services mainly on companies in the agricultural and food sector.

Mauricio Bellon, PhD, is an independent scientist and Honorary Fellow at the National Commission for the Use and Knowledge of Biodiversity (CONABIO), Mexico. He received his MS and PhD in ecology from the University of California, Davis and his undergraduate degree in agronomy from the Universidad Autónoma Metropolitana, Mexico. His research focuses on the reasons, incentives, and dynamics of crop diversity in agricultural systems—both at the inter-specific and infra-specific levels—in the developing world. He is interested in the links of crop diversity with crop evolution, human diets, seed systems, livelihoods, climate change, and agricultural technology. Previously he was Coordinator of Studies in Agrobiodiversity at CONABIO. Before this he was Principal Scientist and Programme Director, Diversity for Livelihoods Programme, at Bioversity International. He also has worked for the International Maize and Wheat Improvement Center, the International Rice Research Institute, and the National Autonomous University of Mexico. He is a member of the Mexican Academy of Sciences and Mexican National System of Researchers.

Rebecca Boehm, PhD, is an Economist for the Food and Environment program at the Union of Concerned Scientists. Prior to joining UCS, Dr. Boehm was a postdoctoral fellow at the Rudd Center for Food Policy and Obesity at the University of Connecticut, with a joint appointment in the UConn Agricultural and Resources Economics Department's Zwick Center for Food and Resource Policy. Her research has focused on understanding the implications of food choices for climate mitigation and adaptation, evaluating federal nutrition programs including the Food Insecurity Nutrition Incentive program, and assessing public health interventions to encourage healthy eating among children. Dr. Boehm has a BA in ecology and evolutionary biology from Princeton University and an MSc and PhD from the Tufts University Friedman School of Nutrition Science and the Agriculture, Food, and Environment program.

Christopher Bonnet holds an MSc in Spatial Planning and is currently Head of Environmental, Social and Governance (ESG) Business Services for P&C business Allianz Global Corporate & Specialty SE, the Allianz company for global business insurance and large corporate and specialty risks. The department serves as a Group center of competence for ESG integration in Allianz-wide P&C business. Before joining Allianz in 2014, Chris worked as a Senior Consultant Sustainability Services at KPMG Netherlands from 2009, where he provided audit and advisory services on sustainability for financial industry clients. After joining Allianz, Chris focused on developing the center of competence for ESG integration in Allianz-wide P&C business. In his position, he is focused on developing and implementing the Allianz ESG in Underwriting approach in the P&C portfolios within Allianz Group. Since August 2020 he has also served as a board member to the UNEP FI Principles of Sustainable Insurance Board and represents Allianz to drive sustainable insurance across the industry, together with dedicated peers.

Caroline Burgeff, PhD, has a Master's degree in Agricultural Sciences (Ingénieur Agronome) from the Faculté des Sciences Agronomiques de Gembloux (Belgium) and a PhD from the Universidad Nacional Autónoma de México. She has carried out research and academic activities on phytopathology, plant molecular biology, and development issues while working in academic institutions in Mexico and Belgium. In 2010 she joined the National Commission for the Use and Knowledge of Biodiversity (CONABIO), where she has worked mainly in the fields of GMO biosafety and agrobiodiversity.

Guillermo Castilleja, PhD, is a Senior Advisor at the Global Alliance for the Future of Food. Guillermo's time has been generously allocated by his foundation, the Gordon and Betty Moore Foundation, to help the Global Alliance further its work on true cost accounting. Before being seconded to the Global Alliance, Guillermo, in his position as senior fellow at the Moore Foundation, advised the foundation's president on programmatic strategy. Guillermo served six years at the helm of the foundation's Environmental Conservation Program, leading its efforts to protect critical ecosystems and balance long-term conservation with sustainable use. Before joining the foundation in 2010, Guillermo worked for WWF for 18 years, most recently as executive director for conservation at WWF International. In that capacity, he directed and coordinated its global conservation efforts, leading the development of place-based and policy priorities for the global network. At the beginning of his career at the WWF, he was country representative in Mexico and later vice-president for the Latin America and the Caribbean program. Through speaking engagements around the world, research published in peer-reviewed journals and the media, Guillermo has addressed the relationship between protected areas and food security, strategies for long-term financing of large-scale land conservation, and the social opportunities and challenges of biodiversity conservation. Guillermo graduated from the National University of Mexico and received a Master's degree in forestry, a Master's degree in philosophy, and a doctorate in forest ecology from Yale University.

Tim Crosby is Principal of the Thread Fund, which focuses on investing multiple forms of capital to generate social and environmental returns alongside financial returns. In addition, Tim is a Steering Committee Member of the Global Alliance for the Future of Food, Member Agroecology Fund, Chair Transformational Investing in Food Systems Initiative, Member Seattle Impact Investing Group, and Board Member of Center for Inclusive Entrepreneurship. Tim's previous work includes Co-Chair for the Sustainable Agriculture and Food System Funders, Director of Slow Money Northwest, and 15 years as a professional photographer and graphic designer. Tim holds an MBA in Sustainable Business from Presidio University and a BA in Anthropology from Kenyon College.

Paula A. Daniels is Co-founder and Chair of the Center for Good Food Purchasing, a social enterprise non-profit founded in July of 2015 as a national spin-off from the Los Angeles Food Policy Council, which Paula founded in 2011. She is a lawyer and public policy leader in environmental food and water policy, with recognition through academic appointments and other awards, including the Ashoka Fellowship (2018), the Resident Fellowship at the Bellagio Center of the Rockefeller Foundation (2016), the Stanton Fellowship of the Durfee Foundation (2012–13), the Pritzker Environment and Sustainability Education Fellow at the UCLA Institute of the Environment and Sustainability (2015); the Lee Chair in Real Estate Law and Urban Planning (2013). She has extensive experience in government. Among her appointed positions is: Senior Advisor to Mayor Villaraigosa of Los Angeles; Los Angeles Public Works Commissioner (a full-time executive position); commissioner with the California Coastal Commission; and the governing board of the California Bay-Delta Authority which oversaw the California State Water Project.

Adrian de Groot Ruiz, PhD, is Executive Director of Impact Institute, a social enterprise that aims to empower organizations and individuals to realize the impact economy by providing them with the tools to measure, report and manage impact. He is also co-founder and director of True Price, a social enterprise that has the mission to realize a sustainable economy based on true prices. Adrian is board member of the Impact Economy Foundation, which aims to create a global community of impact professionals to accelerate the transition to the impact economy. He is also co-initiator and board member of the foundation SDG Nederland, bringing together over 600 organizations around multi-stakeholder partnerships to realize the

Sustainable Development Goals. Adrian has overseen over 100 projects in impact measurement and valuation, including landmark projects in the banking, corporate, non-profit, and public sectors across four continents. He works on open source methods in the area of impact. He has contributed to the development of the Natural Capital Protocol and is an author of, among other publications, the *Framework for Impact Statements*, the *Integrated Profit & Loss Assessment Methodology* and the *True Price Principles*. He was recognized as a Global Shaper by the World Economic Forum. He is an often invited speaker on Natural and Social Capital Accounting in Europe, North America, and Latin America and has blogged in the Huffington Post. He holds an MSc in econometrics, a PhD in Economics, and was previously assistant professor of Finance at Radboud University. He has multiple publications in a global top 10 economics journals. Adrian is a joint Dutch and Mexican national.

Marta Echavarria, Director of EcoDecision has over 25 years of experience working in the conservation of natural landscapes, with a special interest in recognizing the economic value of water source conservation as a strategy for nature valuation. Working with different water users, Marta has developed innovating approaches to finance activities at different landscapes levels. She has been recognized as an international social entrepreneur and is considered a pioneer in the development of ecosystems services markets.

Frank Eyhorn, CEO of Biovision Foundation, is a sustainable agriculture and food systems expert with more than 20 years of experience in international development cooperation. He conducted comprehensive research on the impact of organic farming on the livelihoods of smallholders in Asia and Africa and has published various books, training manuals, and scientific papers. From 2000 to 2005 Frank coordinated organic farming projects in Asia at the Research Institute of Organic Farming. In 2006 he joined Helvetas, where he led the Organic and Fairtrade Competence Centre and headed the Rural Economy Team. From 2011 to 2019 he served on the World Board of IFOAM–Organics International, from 2014 as Vice-President. He is a member of the Steering Group of the Transformational Investing in Food Systems Initiative of the Global Alliance for the Future of Food. In 2020 he joined Biovision as CEO where he focuses on scaling up agroecological innovations in Sub-Saharan Africa and globally through multi-stakeholder processes, conducive policies, awareness raising, and targeted investments. Frank holds a PhD in Environmental Sciences and CAS in Development Cooperation and in International Organizations Management. He has authored various publications related to organic agriculture and its role to improve livelihoods of smallholders. Together with Adrian Muller, he is the lead author of the policy comment paper "Driving sustainability of global agriculture with organic farming" (Nature Sustainability, 2019).

Gábor Figeczky is Head of Global Policy at IFOAM–Organics International, leading the planning and implementation of the organization's global advocacy activities on sustainable agriculture and food systems, nutrition, biodiversity, and climate change, with a particular focus on UN processes and institutions. With a background in agriculture and ecology, he has worked in sustainable agriculture and nature conservation at senior positions in the civil society as well as the state sector including as CEO of WWF Hungary.

Emile Frison, PhD, is a member of the International Panel of Experts on Sustainable Food Systems. A Belgian national, he has spent his entire career in international agricultural research for development, including six years in Africa in Nigeria and Mauritania. In 2003 he became Director General of Bioversity International and developed a strategy entitled "Diversity for Well-being" focusing on the contribution of agricultural biodiversity to the nutritional quality of diets and to the sustainability, resilience, and productivity of smallholder agriculture. Dr. Frison is the lead author of the IPES-Food report "From Uniformity to Diversity: a paradigm shift from industrial agriculture to diversified agroecological systems". He is the Chair of the Board of Directors of Ecoagriculture Partners and a member of the Mission Board on Soil Health and Food of the European Commission.

Francisca Acevedo Gasman, PhD, obtained her biology degree from Grinnell College (Iowa, USA) and then went on to specialize in Plant Molecular Biology for her MSc (Colegio de Postgraduados, México) and PhD (Universidad Politécnica de Madrid, Spain). She has been working in México for the National Commission on the Knowledge and Use of Biodiversity (CONABIO) since the end of 2002 in biosafety and agrobiodiversity issues. She is currently the Coordinator of Agrobiodiversity of the General Coordination of Agrobiodiversity and Biological Resources at CONABIO.

Barbara Gemmill-Herren, PhD, was Delivery Manager, for the Major Area of Work on Ecosystem Services and Biodiversity at the UN Food and Agriculture Organization (FAO), until she retired in 2015. She was previously Executive Director of Environment Liaison Centre International, an international environmental non-governmental organization based in Nairobi, Kenya. At the FAO, she built and coordinated a global project on Pollination Services, implemented in Brazil, Ghana, Kenya, South Africa, India, Pakistan, and Nepal. In her last five years at FAO, she was responsible for the FAO's work on Ecosystem Services in Agricultural Production and was central to the FAO's new focus on Agroecology. Since leaving the FAO, she has been a contributor to the UN initiative on "The Economics of Ecosystems and Biodiversity for Agriculture and Food," exploring True Cost Accounting in Agriculture, and has led the "Beacons of Hope" initiative of the Global Alliance for the Future of Food. This initiative seeks to develop a framework bringing together evidence and stories of transitions towards more sustainable food and agriculture systems. She is currently a senior advisor to the FAO on Biodiversity Mainstreaming, a Senior

Associate at the World Agroforestry Centre and an Associate Faculty in the Masters of Sustainable Food Systems program at Prescott College, Prescott, Arizona.

Nadine Greiss graduated from Faculty of Arts, Ain Shams University, Egypt in 2012. She has more than ten years' work experience in the fields of creative writing, storytelling, and relations. She contributed to several publications (books, reports, and studies) about the SEKEM Initiative, sustainable development, biodynamic agriculture, anthroposophy, and social and cultural development. Nadine is currently responsible for SEKEM's digital communication, and she has published more than 100 articles.

Patrick Holden is the founder and chief executive of the Sustainable Food Trust, an organization founded in 2012 and working internationally to accelerate the transition to more sustainable food systems. Prior to this he was director of the Soil Association (until 2010) during which time he led high-profile campaigns on the misuse of antibiotics, genetic engineering, and the case for vaccination in the 2001 foot-and-mouth outbreak. He farms some 300 acres in West Wales, now the longest established organic dairy farm in the principality, where he produces Hafod, a raw milk cheddar from the milk of his 80 Ayrshire cows. He was awarded a CBE for services to organic farming in 2005, is Patron of the UK Biodynamic Farming Association, and was elected an Ashoka Fellow in 2016.

Salman Hussain, PhD, is the Coordinator of TEEB (The Economics of Ecosystems and Biodiversity') which is a UNEP-hosted Initiative that aims to recognize, demonstrate and capture the values of nature. Salman initiated the development and launch of TEEBAgriFood in 2014. Salman also is the Head *a.i.* of the Ecosystem Services Economics Unit in Ecosystems Division in UNEP, as well as being the UNEP focal point for Action Track 3 in the UN Food Systems Summit. Prior to joining UNEP, Salman was with Scotland's Rural College, where he directed the University of Edinburgh's Master's programme in Ecological Economics.

Alastair Iles, PhD, is Associate Professor in the Department of Environmental Science, Policy, and Management at the University of California, Berkeley, in the Division of Society & Environment. He is an environmental policy and social science scholar whose research focuses on the intersections of science, technology, and environment, predominantly around industrial chemicals and sustainable food systems. In the food and agriculture domain, Iles has researched and published in the areas of sustainable seafood, aquaculture sustainability standards, public policies for diversified farming systems, food retailer sustainability strategies, and consumer knowledge about food. He now emphasizes public policies for making transitions to a sustainable food system in particular, together with food sovereignty, farmer knowledge systems, and the connections between bio-based chemicals and agricultural production.

Nick Jacobs has served as Director of the International Panel of Experts on Sustainable Food Systems (IPES-Food) since 2015. He previously worked in the support team to the UN Special Rapporteur on the right to food, and as an agri-food journalist. Nick holds an MA in Modern Languages (Cambridge) and an MSc in Globalization and Development (Antwerp). Nick has led editorial and research work on the IPES-Food reports *Towards a Common Food Policy for the EU* (2019), *Unravelling the Food-Health Nexus* (2017) and *From Uniformity to Diversity* (2016). He was a member of the Food 2030 Independent Expert Group convened by the European Commission (2018), and served as expert to the European Economic and Social Committee rapporteur on the EU Farm to Fork Strategy (2020).

Saru Jayaraman is the President of One Fair Wage and Director of the Food Labor Research Center at University of California, Berkeley. After the 9/11 attacks, together with displaced World Trade Center workers, she co-founded the Restaurant Opportunities Centers (ROC) United, organizing restaurant workers, employers and consumers nationwide for improved wages and working conditions and then went on to found One Fair Wage, a national campaign and organization dedicated to improving wages for all service sector workers and ending all subminimum wages nationwide. The story of Saru and her co-founder's work founding ROC has been chronicled in the book *The Accidental American*. Saru is a graduate of Yale Law School and the Harvard Kennedy School of Government. She was profiled in the *The New York Times*' "Public Lives" section in 2005, named one of Crain's "40 Under 40" in 2008, was 1010 Wins' "Newsmaker of the Year" and *New York Magazine*'s "Influentials" of New York City. She was listed in CNN's "Top10 Visionary Women" and recognized as a Champion of Change by the White House in 2014, and a James Beard Foundation Leadership Award in 2015. Saru authored Behind the Kitchen Door (Cornell University Press, 2013), a national bestseller, and has appeared on CNN with Soledad O'Brien, Bill Moyers Journal on PBS, Melissa Harris Perry and UP with Chris Hayes on MSNBC, Real Time with Bill Maher on HBO, the Today Show, and NBC Nightly News with Brian Williams. In January 2018 she attended the Golden Globes as the guest of actor Amy Poehler, where she brought widespread attention to the issue of sexual harassment in the restaurant industry. Saru was one of eight women activists who attended the Globes as part of the #TimesUp initiative. Besides *Behind the Kitchen Door* (Cornell University Press, 2013), and *Forked: A New Standard for American Dining* (Oxford University Press, 2016), Saru's most recent book is *Bite Back: People Taking on Corporate Food and Winning* (UC Press, 2020).

Amanda Jekums is a program coordinator with the Global Alliance for the Future of Food where she focuses on the economics of sustainable food systems. Over the past decade she has worked across diverse sectors and scales to find creative solutions to complex issues through meaningful collaboration and innovation, including: facilitating local and international nutrition and food security working groups at Ryerson University's Centre for Studies in Food Security; using behavioural insights, design thinking, and collaborative engagement to develop better stakeholder relationships at the Ontario Ministry of Health; and increasing civic engagement and political action through the development of interactive food system asset maps with the Toronto Food Policy Council and FoodShare Toronto. She is guided by her research on Brazil's cross-sectoral systems approach to nutrition and food security program and policy development.

Adele Jones is Deputy CEO at the Sustainable Food Trust (SFT). She has been with the SFT since 2013, primarily focusing on projects including true cost accounting in food and farming and the harmonization of farm-level sustainability assessment. In 2020 she undertook a part time secondment with the Welsh Government working on an annual sustainability assessment for Welsh farmers called the "Farm Sustainability Review". Between September 2018 and September 2019 she completed a part-time secondment with The Department for the Environment, Food and Rural Affairs in London working on a project called the "Gold Standard Metric," which aimed to harmonize government-led farm and supply chain sustainability metrics. Previous to these roles, Adele had a background in geography and soil science.

Zoltán Kálmán has been serving as Permanent Representative of Hungary to the FAO, IFAD, and WFP since 2014. He serves as a member of the Advisory Committee of the UN Food Systems Summit. Between 2009 and 2014 Zoltán worked as Head of Department in the Ministry of Agriculture in Budapest, supervising all international (bilateral and multilateral) activities of the Hungarian Ministry of Agriculture. From 2003 to 2008 he had his first assignment as Permanent Representative of Hungary to the Rome-based UN Food and Agriculture Organization. Prior to 2008, Zoltán worked in the private sector, including as managing director of small-medium sized food processing companies. He holds a Master's degree from the University of Agriculture in Gödöllő (Hungary) and received the award of Knight of Cross from the Order of Merit of the Hungarian Republic in 2019. Some of his related publications are available at www.ipsnews.net/author/zoltan-kalman.

Jan Köpper is Head of the Impact Transparency & Sustainability department at GLS Bank in Bochum. In this function, he is responsible for the conception, coordination, and implementation of social impact measurement, sustainability assessment in corporate customer business, and the integration of sustainability processes into internal sustainability management and overall bank management. After working for the CSR Europe corporate network in Brussels and the sustainability rating agency imug in Hanover, Jan has been in his current position since April 2018.

Louise Luttikholt is the Executive Director of IFOAM–Organics International. Before that, she founded HELVETAS Germany, of which she was the Director for several years. In parallel, she functioned as a Senior Advisor on Sustainable Agriculture to HELVETAS Swiss Intercooperation, where she specialized in Nutrition Sensitive Agriculture; she particularly supported her colleagues from five different mountainous countries in their advocacy work on a conducive environment for a diversified and sustainable agricultural policy framework.

Alicia Mastretta-Yanes, PhD, studied Biology at UNAM (2004–09) and completed her PhD in Evolutionary Biology at University of East Anglia (2010–15). The broad aspect of her research is how Mexican biodiversity has evolved from a genetic perspective. This includes changes on species distributions owing to historical climate fluctuations (e.g. the Pleistocene glacial ages) as well as the effect of human management and domestication. She examines these topics as an evolutionary biologist using molecular ecology and genomic tools. Her research is related to conservation issues, and she aims to get local communities involved or to transform basic research into information and tools that can be applied to better use and manage genetic diversity.

Kathleen A. Merrigan, PhD, is an expert in food and agriculture, celebrated by *Time Magazine* as one of the 100 Most Influential People in the World in 2010. Currently, she serves as Professor in the School of Sustainability and executive director of the Swette Center for Sustainable Food Systems at Arizona State University. From 2009 to 2013 Kathleen was deputy secretary and COO of the United States Department of Agriculture, where she led efforts to support local food systems. She is known for authoring the law establishing national standards for organic food and the federal definition of sustainable agriculture. Merrigan holds a PhD in Public Policy and Environmental Planning from the Massachusetts Institute of Technology, Master's degree in Public Affairs from University of Texas at Austin, and BA from Williams College.

Laura Mervelskemper, PhD is Co-Head of the Impact Transparency & Sustainability Department at GLS Bank. As a "person of conviction," she is passionately committed to all topics related to sustainability, sustainability risks, social transformation and the political demands of GLS Bank. Previously, she spent several years studying the link between sustainability and finance from an academic perspective and also completed her doctorate on the subject. In addition, she has been active in both sustainability consulting and sustainability management. Since joining GLS Bank in 2018, she has found the perfect combination of topics that are close to her heart.

Adrian Muller is a senior researcher at the Research Institute for Organic Agriculture and the Swiss Federal Institutes of Technology ETHZ in Switzerland. His current work focuses on sustainable food systems modelling, on sustainability assessments of agricultural systems, on climate change adaptation and mitigation in organic agriculture, and on climate policy in agriculture in general.

Alexander Müller is the Managing Director of TMG - Thinktank for Sustainability in Berlin Germany, he has served as a member of the municipal council of the city of Marburg, as State Secretary in the Ministry of Youth, Family Affairs and Health in the State of Hessen, and was elected as a member of the Parliament of Hessen. From 2001 to 2005 he was State Secretary in the Ministry for Consumer Protection, Food and Agriculture in the Federal Republic of Germany. From 2006 until 2013 he served as Assistant-Director General of the FAO. In 2008−09 he was nominated as Secretary General of both the High-Level Conference on World Food Security: The Challenges of Climate Change and Bioenergy and the World Summit on Food Security. As a member of the Advisory Group on Energy and Climate Change, he advised the Secretary-General of the UN. Additionally, he chaired the United Nations System Standing Committee on Nutrition. From 2014 until 2018 he led the UNEP project The Economics of Ecosystems and Biodiversity Agriculture and Food. He is member of the Board of ICRAF, The World Agroforestry Centre, Kenya, the Center for International Forestry Research, Bogor, Indonesia.

Rainer Nerger, PhD, joined Soil & More in 2014. He is mainly responsible for R&D, knowledge management and project implementation in agriculture and agroforestry, especially in Latin America and Sub-Saharan Africa. This includes on-site and remote assessments of soil fertility and raw material security. It also comprises designing the scientific basis of the company's projects, carbon sequestration projects and submodel development of a risk assessment tool. His previous work included roles at Perspectives Climate Change GmbH, where he contributed to international carbon projects. After graduating at Technical University of Dresden (Germany) in Physical Geography (2005), he developed a science background at universities in Germany and Spain in soil science, soil carbon, and sampling campaigns. In 2020 he successfully finished his PhD on the detection of changes in agricultural soils of a German soil monitoring network.

Carl Obst is a Director at the Institute for Development of Environmental-Economic Accounting—IDEEA Group. Carl spent many years at the Australian Bureau of Statistics, including time at the OECD before becoming editor of the United Nation's System of Environmental-Economic Accounting. Carl was the co-author, with Kavita Sharma, of the evaluation framework described in the *TEEB AgriFood Foundations Report* and is an Advisory Board member of the Capitals Coalition. He is a leading player in closing the gap between government and corporate approaches to accounting for natural capital and its integration with other capitals.

Joanna Ory, PhD, is a postdoctoral scholar at UC Berkeley researching the different factors that affect the adoption of agricultural practices that promote soil health in California. She completed her PhD in Environmental Studies at University of California, Santa Cruz. Her dissertation compared the impacts of different policies in the USA and the EU related to the use of herbicides in agriculture, as well as the water quality implications. As part of this comparative study, she studied environmental policy and farmer decision-making as a Fulbright Scholar in Italy. More recently, she has worked on issues related to organic farming and how to encourage greater adoption of organic and diversified farming systems.

Vatsal Patel has completed a Master's degree of Environmental Science program at the University of South Australia. He has worked as a research assistant at the this University contributing to the project related to livestock, soil carbon, and climate change. He is a mechanical engineer by training.

Michael Quinn Patton, PhD, is a senior evaluation consultant for the Global Alliance for the Future of Food is the Founder and CEO of Utilization-Focused Evaluation, an independent organizational development and program evaluation organization. After receiving his doctorate in sociology from the University of Wisconsin–Madison, he spent 18 years on the faculty of the University of Minnesota (1973–91), including five years as Director of the Minnesota Center for Social Research and ten years with the Minnesota Extension Service. Patton has written many books on the art and science of program evaluation, including *Utilization-Focused Evaluation* (4th ed., 2008), *Developmental Evaluation: Applying Complexity Concepts to Enhance Innovation and Use* (2010), *Principles-Focused Evaluation: The GUIDE* (2017); and *Blue Marble Evaluation: Premises and Principles* (2019). Michael is a former president of the American Evaluation Association.

Saiqa Perveen is currently working as a Research Associate at the University of South Australia, Adelaide. She has completed a Master's degree in Environmental Science at this University. Her research interests center around natural resource management and inclusive development. Her current research focus is on greenhouse gas accounting in the livestock sector in South Australia. Saiqa is also part of the Environment for Development laboratory that focuses on the co-creation of knowledge required for equitable and sustainable development.

Rex Raimond has more than 20 years of experience developing strategy and building effective partnerships for sustainable and equitable food systems. His current focus is on building a movement among concessional, impact, and commercial investors, companies, civil society organizations, and other stakeholders to increase investing in principles-aligned food system businesses that are economically viable and deliver sustainable impact. Rex has worked on many aspects of food and agricultural systems transformation. Rex is a multi-lingual, international environmental lawyer.

Courtney Regan, PhD, is Research Fellow in agricultural, forestry and resource economics. Working as an agricultural extension officer during Australia's Millennium Drought, Courtney saw first-hand the challenges posed to traditional farming systems by climate and volatile markets. These experiences formed his interest in how we create resilient and sustainable productive landscapes while building a financially viable future for farmers. Specifically, his research investigates the incorporation of alternative land uses such as carbon forestry/agroforestry into agricultural landscapes, how farmers make investment decisions in uncertain conditions, and sustainable forestry operations in an increasingly resource constrained world.

Sarah Reinhardt, MPH, RD, is a Senior Analyst for the Food and Environment Program at the Union of Concerned Scientists, where she evaluates the public health implications of food, nutrition, and health policies and programs, including the *Dietary Guidelines for Americans*, the Supplemental Nutrition Assistance Program (SNAP), and child nutrition programs. Prior to joining UCS, Sarah served the United Way for Southeastern Michigan as a nutrition consultant. She received a BA in Women's Studies and MPH in Human Nutrition from the University of Michigan and is a Registered Dietitian.

Cecilia Rocha, PhD, is the Director of the Centre for Studies in Food Security and a Professor in the School of Nutrition, Ryerson University, Canada. A member of the International Panel of Experts on Sustainable Food Systems (IPES-Food), she was the lead author of the report *Unravelling the Food-Health Nexus: Addressing practices, political economy, and power relations to build healthier food systems* (2017). Cecilia has participated in the development of the Toronto Food Strategy (2008–10), and was a member of the Toronto Food Policy Council (2005–11). She has done research on the role of food in the lives of recent immigrant women and on food insecurity among Latin American recent immigrants in Toronto. She was the director of the project Building Capacity in Food Security in Brazil and Angola (2004–10), and co-director of the project Scaling up small-scale food processing for therapeutic and complementary food for children in Vietnam (2015–18). Considered an expert on the innovative approach to food and nutrition security in the city of Belo Horizonte, Cecilia is the author of a number of scholarly papers and reports on food policy and programs in Brazil.

Ricardo Salvador, PhD, is Director and Senior Scientist for the Food and Environment Program at the Union of Concerned Scientists. He is an agronomist with prior positions at the W.K. Kellogg Foundation, Iowa State University, and Texas A&M University. He is a member of the National Academy of Sciences Board on Agriculture and Natural Resources and of the International Panel of Experts on Sustainable Food.

Harpinder Sandhu, PhD, is a transdisciplinary scientist with research interests in studying the interactions between society and the environment. His current research focuses on measuring social, human, and natural capital for the transformation of agriculture and food systems towards sustainability. He is a global expert in developing models for food and agriculture sustainability.

Nadia El-Hage Scialabba has been a civil servant of the UN Food and Agriculture Organization (FAO) for 33 years, working at the interface between environmental, social, and economic challenges of food systems. As the first trained ecologist at the FAO, she introduced new concerns and new approaches to the agriculture, forestry, and fishery industries, such as: integrated approaches to conflicting use of coastal resources; responsible fisheries with regards entanglement of living resources in fishing gear and plastic marine debris; ecosystem approach to agrobiodiversity and landscape management; participatory approaches to development; integrated policy and planning in Small Island Developing States (triggering the creation of SIDS as a political entity in the UN); and Sustainable Agriculture and Rural Development (SARD was introduced in Agenda 21). As Senior Officer from 2000 to 2018, Nadia led FAO's inter-disciplinary programme on organic agriculture, including policy advice and technical support to member countries, as well as normative activities, such as conceptualization and leadership of the FAO/IFOAM/UNCTAD Task Force on Harmonization and Equivalency of Organic Regulations. Also, Nadia led the FAO's work on sustainability metrics, including Environmental Impact Assessment in the 90s, Sustainability Assessment of Food and Agriculture systems around Rio+20 and at last, full-cost accounting methodologies for food wastage, with special emphasis on assessing food system impacts on social wellbeing and health. Currently retired, Nadia volunteers to the work of international and Italian non-governmental organizations and charity foundations involved in transforming the food system and advancing biodynamic agriculture.

Christian Schader heads the sustainability assessment activities at the Research Institute of Organic Agriculture in Switzerland. Christian's work encompasses evaluations of environmental, economic and social aspects of food production and consumption.

Arno Scheepens, PhD, is leading the impact measurement and valuation team in the Dutch Climate Change and Sustainability Services department at Ernst & Young (EY). Within impact measurement and valuation, Arno specializes in applying Life-Cycle Assessment and True Cost Accounting approaches to help businesses, organizations and governments with the practical implementation of truly sustainable innovation. Arno is also a PhD researcher at the Delft University of Technology, where he also obtained his BSc and MSc in Sustainable Industrial Design Engineering. The topic of his PhD research is eco-efficient value creation: sustainable innovation from the design perspective, titled "More with Less".

Thoraya Seada graduated in Aeronautical Engineering at Egyptian Aviation Academy, Egypt. In 2009 she joined SEKEM's sustainable development team and became a specialist in ecological studies and carbon footprint assessment. In 2014 she joined The Carbon Footprint Center (CFC) as project Manager and analyst. CFC is an environmental research center in Heliopolis University for Sustainable Development, which provides environmental performance assessment, carbon and water footprint assessment, carbon neutrality, carbon sequestration, true cost accounting, and working to raise the environmental awareness in Egypt. In 2016 she participated in the future of agriculture study for calculating the true cost accounting of the strategic crops in Egypt version 1 and updated version. Thoraya is currently concluding a Master's program for education for sustainable development in Heliopolis University and Frederick University.

Julia Sebastian is the Research Director at One Fair Wage, a national coalition, campaign and organization seeking to lift millions of tipped and subminimum wage workers out of poverty, catalyze civic engagement, and reduce workplace sexual harassment by requiring all employers to pay the full minimum wage plus tips. With over a decade of experience in research and advocacy, Julia's work focuses on the intersections of racial and economic justice, specifically in the realms of food policy, workplace discrimination, labor, and immigration. Prior to joining One Fair Wage, Julia was the Manager of Research, Impact Planning, and Evaluation at Race Forward, a national movement and research and media organization for racial justice. While at Race Forward, Julia produced numerous reports and advocacy tools and led trainings for private, non-profit, and government actors around the country. Julia holds a Master's of Public Policy from the Goldman School of Public Policy at UC Berkeley (2021). She graduated with a Bachelor's degree in Anthropology from Stanford University (2011).

Margaret Stern is a tropical forest biologist and long-term resident of Ecuador, where she has a farm on an Andean hillside outside of Quito. She received her PhD (1992) in Evolution and Ecology at the University of California, Davis. Margaret has 35 years' experience with scientific research on forest dynamics, forest resource management, and watershed conservation throughout Latin America, with a focus on Andean-Amazon countries. She maintains a foothold in academia, teaching intensive field courses to diverse audiences, and lately, virtual courses. She is a consultant with EcoDecision.

Esmeralda G. Urquiza-Haas is an independent consultant with a broad experience in topics related to the evaluation of policy instruments for the conservation of biodiversity and agrobiodiversity, and ecosystem services valuation. She studied physical anthropology at the National School of Anthropology and History of Mexico (2002) and holds an MSc in human ecology (2005) at the Research and Advanced Studies Center of the

National Polytechnic Institute. She is currently a PhD candidate in biology at the University of Vienna. She has participated in multiple internationally-funded projects implemented by the National Commission for the Use and Knowledge of Biodiversity (CONABIO).

Federica Varini has an MSc in agroecology and organic agriculture with a multi-disciplinary background and a specialization in public policies and global advocacy. Federica has been working at the intersection of sustainability, research, and outreach for over five years as part of the team of IFOAM−Organics International, a leading organization of the organic sector worldwide. She has been involved in collecting evidences and analyzing public policies that support organic agriculture and sustainable food systems worldwide. In 2018, on behalf of IFOAM−Organics International and in collaboration with the World Future Council and the UN Food and Agriculture Organization, she has been co-organizing the Future Policy Award on Scaling Up Agroecology, which has identified and celebrated policies promoting agroecology committed to bring structural changes in our food systems.

Pablo Vidueira, PhD, is an evaluation consultant, researcher, and professor. With the Global Alliance, Pablo uses evaluation and systems concepts and tools to help the Alliance to make sense of, learn from, and adapt to the complexities of global and local food systems. As a researcher, Dr. Vidueira is affiliated with the Technical University of Madrid where he co-leads two research projects funded by the European Commission on using system concepts and tools to improve food systems in Africa and to optimize innovation processes in rural areas of Europe. Pablo teaches and supervises doctoral students in areas related to evaluation, systems thinking, food systems, international development, and research methods. Pablo served as program co-chair of the Systems in Evaluation Topical Interest Group of the American Evaluation Association and has been elected as president of a regional professional evaluation association in Spain and Portugal, APROEVAL.

Stephanie White, PhD, was an Assistant Professor of Community Sustainability at Michigan State University (MSU) and a researcher with the Global Center for Food Systems Innovation before transitioning to her current position with the State of Michigan as an Environmental Health Educator. Her research focuses on the intersection of cities, climate change, and food systems. Specifically, she has focused on community-based research in the Malawian capital of Lilongwe's highly decentralized urban food provisioning and exchange systems. Her dissertation research focused on urban agriculture and the ways urban residents in M'Bour, Senegal use it to improve individual, family, and community well-being. She earned an MSc degree in Sustainable Systems/Agroecology from Slippery Rock University in Pennsylvania and a BA in Political Science from Colorado State University. She was a Peace Corps Volunteer in Senegal, West Africa from 1992–96.

Gyde Wollesen joined Soil & More Impacts in 2019. She is engaged in creating the link between the work in the field and the company's books through True Cost Accounting. Her work ranges from connecting a True Cost Accounting methodology to existing accounting standards and to customized implementation for the clients. In addition, she supports the team in sales and project management. After graduating in Business Administration with a focus on International Trade, she acquired a Master's degree in Global Sustainability Science. Prior to her position at Soil & More Impacts, Gyde worked in the apparel industry for three years. She is motivated by the fact that everyone is talking about sustainability nowadays, especially when it comes to food and what impact it has on our earth. For her, providing valuable information, advice, and tools to players in the industry is key for driving change.

Foreword

Why True Cost Accounting?

Guillermo Castilleja

In the 1970s the emergent field of ecosystem ecology gave us a framework to understand interdependence in natural (biotic and abiotic) systems, allowing us to map energy and nutrient cycles at different scales in natural and managed systems alike. Ecosystem modeling laid out the basis for sustainability by developing and quantifying nutrient "budgets" across systems where deficits and surpluses indicate departures from balanced baselines with foreseeable systemic implications. Mapping the carbon cycle helped us to understand and become aware of the consequences of releasing more carbon through fossil fuels burning than what can be absorbed by the biosphere globally. The nitrogen cycle helped us to understand the key role of bacterial nitrogen fixation in building plant and animal nutrients and the consequences of excessive runoff of industrially fixed nitrogen into freshwater and coastal ecosystems. Impacts and dependencies across sectors became evident through ecosystem ecology in ways that had not been clear before.

But as compelling and foundational as ecosystem ecology has been for natural resource management, it is safe to say that, by itself, it has not resulted in significant improvements in the two most environmentally impactful human activities on nature: energy generation, and food production and consumption. Power relationships derived from economic dominance supersede the necessity to balance ecosystem budgets and avoid the most nefarious consequences for the planet. Climate change is challenging not because we do not understand why and how carbon emissions accumulate in the atmosphere, but because acting accordingly would imply subverting power relationships that benefit from the system's own malfunction. Eutrophication and coastal dead zones are challenging not because we do not understand the impact of fixed nitrogen runoff from farms, but because a whole sector that depends on industrially produced chemical fertilizers benefits from and promotes a regulatory system that considers eutrophication an externality, not a core function of the operations that industrial agriculture shareholders control and benefit from. Understanding our planetary boundaries is necessary to guide change but not sufficient to effect action.

The ecosystems approach to natural resource management is hampered by the dissonance between ecosystem imperatives and the parochial interests that sectorial institutions serve. The dissonance is particularly concerning when it comes to global ambitions of a sustainable future such as the 2030 Sustainable Development Goals (SDGs). It is safe to say that the majority of the 17 goals are dependent to varying degrees on the future of the food system, as food is a key driver of habitat loss, carbon emissions, disease, and migration. Solutions to these challenges have been designed and applied, but typically in a fragmented fashion: the prevailing system of industrialized agriculture focuses on increasing crop yield through chemical inputs, but externalizes environmental impact; land reform in much of Latin America has been instrumental in giving the landless access to land, but the policies facilitating this social gain have not been supported by the investment needed to build social capital; dietary guidelines respond to the latest advances in food science, but are mute regarding the food deserts that marginalized urban communities live in or the sustainability of the production of food; subsidies to support corn production in the USA result in lower costs of high-level fructose used in beverages and cheap packaged goods, but do not account for the health costs for treating the resulting obesity.

Nations organize the governance and administration of their productive activities, encompassing inherently complex socioeconomic systems, into distinct "sectors." The majority of nations exercise this public function through codes, regulations, and ministries of agriculture, industry, trade, health, social services, and, depending on the economic importance given to other productive activities, fisheries, forestry, mining, and tourism. Food systems, like most socioeconomic and ecological systems, cannot be constrained to a single sector, which means that the policies that influence their behavior end up being formed by a patchwork of sectoral interests (agriculture, environment, fisheries, land reform, trade, industrial processing, health, etc.). Sectoral policies allow decision makers to enhance the effectiveness with which individual components along the food supply chain (production, processing, distribution, retail, and consumption) deliver value to the shareholders relevant to each sector. However, there are implications from the resulting compartmentalization that are worth considering.

First, this sectoral approach means actors lose sight of, and remain ignorant about, the interdependence of the different components of what in reality is an integrated whole. Ignoring underlying interdependencies when fixing or enhancing a given part of the system often results in "unintended," and unattended, consequences elsewhere in the system. Typically, if the fix or enhancement applied is deemed to be successful within a sector's values, any ensuing unintended consequence is considered as external or collateral and not accounted for, or rather discounted from, the value generated by the fix or enhancement. The use of agrochemicals is the poster child of this. Added value from increased crop productivity is not discounted by the loss of biodiversity or human health exposure that the use of agrochemicals might result in.

Second, this sectoral approach helps to perpetuate political control structures through institutions dominated by vested interests to which policies and investment end up serving. The "sectoralization" of a system tends to lock in policies and practices whose cost-benefit is assessed primarily within the logic and values of the sector as currently structured, with varying degrees of integrity and transparency. The expansion of shrimp farming in South East Asia relies on a highly structured subsector, based on the low rent of small farms, employment of migrant workers under slavery conditions, and a financial system supporting a handful of powerful traders ready to tip the scale when regulations are proposed. Furthermore, if such logic and values are solely determined in terms of financial (produced) capital, concomitant losses in other forms of capital (natural, social, or human), even within the same sector, become collateral to the main objective as long as those losses do not impact the financial bottom-line of the sector in question.

And third, it prevents public agencies and other actors from purposely seeking positive outcomes at a truly systemic level, beyond the specific concerns of each sector. Striking the balance between costs and benefits at a systemic scale could be hard to achieve, given the complexity involved, but the opportunities for reaching more widespread benefits multiply as the scope of potential solutions is broadened. Public and international institutions seeking sustainable, socially just, and regenerative solutions would be better served if they were to consider all the values generated by food systems, including sectors, levers, and actors all along the supply chain.

For example, deforestation in the Amazon (with its concomitant loss of biodiversity and increased carbon emissions) is largely driven by soy and beef exports to satisfy the increasing appetite for meat of the growing middle classes in Asia and elsewhere. In turn, the doubling of meat consumption per capita in China in the last ten years is causing an unprecedented epidemic of cardiovascular disease in that country. Without policy coordination and commonly sought outcomes, it is unlikely that the sectoral agencies and actors currently involved will, by themselves, forge the multiple outcome solutions needed to confront these linked problems.

To address these implications of narrow sectoral organization of a given system and to devise systemic solutions, we need to start by defining the system's structure and function, the main interdependencies and energy/capital flows within that system, and the boundary conditions of that system. This characterization will help to identify those pressure points beyond which the system becomes unsustainable, unstable, or unraveled. Such a framework will help to reveal critical points where interventions can result in system-level outcomes deliberately sought, as opposed to the arbitrary addition of outcomes established sector by sector and each limited to a given form of capital.

In summary, the integration of multiple and mutually supportive outcomes across the food system faces two key challenges. The first is the current lack of an operational framework and standards that capture the main impacts and dependencies across the food system structurally along the supply chain and functionally across the four capitals upon which the system runs (natural, produced, human, and social). The second is the lack of intersectoral transparent mechanisms (such as integrated cost-benefit analysis and reporting) to align governance, investment, and administration to support the operationalization of such a systemic approach. Without these transformations, it is unlikely that the current sectoral policies and power dynamics will ever deliver the outcomes needed to meet the SDGs.

True Cost Accounting (TCA) is emerging to address these two challenges and orient public policies and investment towards an integrated approach for delivering food system-level value. TCA has been proposed as a tool to make visible and quantifiable food system dependencies, flows, and externalities, both negative and positive, across four capitals: natural, social, human, and produced. It aims to provide frameworks for holistic food system evaluation at different scales. Using relevant metrics, TCA could establish a baseline upon which strategies to improve system-level value can be developed, implemented, and serve to monitor impacts and changes in capital stocks over time to inform adaptive management.

Although it is still early days to assess the effectiveness of TCA to bring transformation at the scale needed, versions of the tool have already been used in different and increasingly visible ways. Advocacy organizations use it to make consumers aware of the environmental and social externalities embedded in the food that they buy. Food companies use it to stimulate change in business behaviors through a structured assessment of where best to intervene in their operations to minimize negative and enhance positive impacts. Investment advisers use it to guide socially conscientious investors to create and share benefits equitably and sustainably. Individual farmers are increasingly using TCA as a means to account for the full set of benefits that their agronomic methods bring. The trend in the marketplace of ideas is clearly pointing towards TCA as a holistic food system assessment and design tool.

TCA, like ecosystem ecology, is an essential tool to operationalize sustainability writ large in that it takes a quantitative approach to assess capital accrual and depletion resulting from the rates at which finite stocks are being used. The difference, however, between ecosystem ecology and TCA, is that TCA's capital stocks include, beyond natural capital, financial (produced), human and social capital. This difference is significant because the inclusion of all forms of capital across the full supply chain implies seeking consensus among shareholders around a common vision of sustainability, making TCA not just a technical exercise in optimization, but ultimately a political process to balance multiple outcomes and trade-offs within the system's boundaries.

The chapters in this volume explore frameworks such as TEEBAgriFood to account for positive and negative externalities in food systems and true price to calculate and internalize the costs of impacts along the supply chain. Collectively, the chapters provide a useful snapshot of the state of the art of TCA. This assessment highlights some of the gaps that we need to work on, particularly on standards and metrics; the practical challenges of keeping a system's perspective when valuing impacts and dependencies on the basis of four capitals; the controversial implications of "dollarizing" vs. valuing natural and human capital; and the role of the public and private sectors in forging a common vision to which commitments need to be made.

Two main areas for further TCA development that emerge from these chapters are the need for consensus on what to measure in assessing stocks and flows in the four capitals and the processes to bring key shareholders to the table in a collective effort to align desired outcomes across the food system. These challenges are neither trivial nor new; they are in many ways the core of the resistance that TCA is likely to encounter as proponents try to scale its adoption. The two challenges stem from what was highlighted above as the constraints of the "sectoralization" of the food system. For instance, the more tenaciously agriculture sector shareholders remain attached to their belief that the most useful metric to establish the success of investment and policy inter-ventions is yield and productivity, the harder it will be to move the goal post towards sustainability. And it will require that new governance and adminis-tration arrangements be established to give the food system organizational coherence based on verifiable sustainability standards across the system. This kind of public sector leadership will be needed to bring down the silos estab-lished by sectoral policies and open up the options with a systemic perspective.

The chapters in this book highlight progress made on these two challenges and others, giving us cause for optimism. Changes envisaged in the United Kingdom's post-Brexit agricultural policy are excellent examples of what we hope will rapidly become a trend followed more broadly, together with the transparency this would bring to the political process required by the shift toward sustainability. Three other developments showing potential and momentum to TCA adoption are: the interest that the investment community is showing in better agricultural practices and healthier food; the progress that some countries have started to make aligning policies across current silos to show the larger gains that can be obtained; and the availability of data gener-ated by new technologies to better integrate information across the supply chains. We need to seize these opportunities to strengthen and mainstream TCA and accelerate the pace of this unprecedented but urgent transformation of our food system.

Acknowledgments

This book is a project of the True Cost Accounting Community of Practice and Accelerator, which are supported by the Global Alliance for the Future of Food. The idea for the book was born at the Community of Practice's first meeting in Minneapolis at the McKnight Foundation in 2017. It is exciting to see the project come to fruition, and the editors (each of whom are members of the Community of Practice) would like to thank the Global Alliance for the Future of Food and Community of Practice members for their support and contributions.

The editors would also like to gratefully acknowledge the many people involved in the preparation of this publication. The Global Alliance's True Cost Accounting Impact Area Committee must be acknowledged for their support of this work, including colleagues from Clarence E. Heller Foundation, McKnight Foundation, Grace Communications, Thread fund, Barilla Center for Food and Nutrition, Christensen Fund, IKEA Foundation, and Crown Family Philanthropies. Their support is deeply appreciated. The McKnight Foundation provided funding for the Community of Practice convenings and to make this publication open access. This book would not have been possible without the outstanding support of Sue Woodard from the Center for Good Food Purchasing and Amanda Jekums from the Global Alliance for the Future of Food. Their careful attention to detail and project coordination proved invaluable. Thank you!

Disclaimer

This book was supported by the Global Alliance for the Future of Food, a strategic alliance of philanthropic foundations working together and with others to transform global food systems now and for future generations. It reflects the work and viewpoints of independent authors. Any views expressed in the book do not necessarily represent the views of the Global Alliance or of any of its members.

Introduction

The Urgency of Now

Barbara Gemmill-Herren, Lauren E. Baker and
Paula A. Daniels

There has been no dearth of global reports published over the past decade on the failures of our current food systems. These reports clearly document how food systems are generating widespread degradation of land, water, and eco-systems, significant greenhouse gas emissions, major contributions to biodiversity losses, chronic overnutrition, malnutrition, and diet-related diseases, and livelihood stresses for farmers around the world.

A cohesive focus and consensus on the solutions and policies that will genuinely transform our existing food systems is still underdeveloped, much less taken up in food systems policy and practice. In this volume, a diverse group of contributing authors articulate the many cogent efforts to introduce a new economic paradigm that offers new opportunities and pathways: True Cost Accounting (TCA) in agriculture and food systems has the power and potential to catalyze the transformations needed to address our multiple and interconnected crises.

Why TCA, Why Now?

Behind all the food that we eat is a vast realm of unaccounted for interactions: the diversion of water from rivers; the extraction of nutrients from soil; the discharge of pollutants to air and water; the exaction of labor to grow, manage, pick, and package; the release of carbon dioxide to transport and deliver; and so on. When we shine a light on these interactions it becomes clear that a 99¢ hamburger costs all of us a lot more than the dollar placed by a consumer into the hands of a cashier. The singular focus of the business model that made cheap food possible overlooks the multiple costs to society related to suffering with or cleaning up pollution, the cost of social assistance or food charity for large segments of the population who are not paid enough to buy the food they grow, manage, pick, or package, and the public health costs from the diet-related disorders that are a direct consequence of industrially created highly processed food, to name just a few. In the end, this "cheap" hamburger is extremely expensive, but most of the cumulative cost is borne by all of us as a global community. A central challenge is that the uninformed choices that we as consumers, as policymakers, and businesses make perpetuate the problem.

TCA (sometimes referred to as "full" cost accounting; here, we use "true") provides a framework for systemic shifts across food systems. It allows for aggregation of information across affected economies and aspects of the food supply chain (production, processing, distribution, and retail). It intends to create transparency for regulatory decision-making that can realign subsidies in a more balanced direction. It facilitates broad engagement from farmers to consumers, bridging practice and policy. What sounds, on the surface, like a complicated tool relegated to accountants is ultimately a clarion call for a new economics of food and a new relationship with the land and the food that we eat, starting with a holistic view of a system out of balance and ending with a new approach to business and integrated reporting. As TCA enters our vocabulary, readers might find it helpful to turn to the glossary as developed by the Global Alliance for the Future of Food (Eigenraam *et al.*, 2020).

Holistic Framing

True Cost Accounting: Balancing the Scale includes a review of the theoretical and ontological roots and tensions within prior systems of accounting for the "externalities" of agriculture and food systems: those impacts that are a direct result of system activities, but whose consequences and corrections are not borne by the original parties. The more recent evolution of TCA, extended beyond environmental economics and accounting to include social, human, and health aspects, is traced. What emerges strongly throughout many chapters of this book is the need for integrated systems-based framing, taking a holistic view of all interactions comprising a food system.

Measurement and Metrics

As we struggle to find ways to manage our food system for public and planetary health, we need different ways of measuring, which in turn lead to more inclusive and holistic metrics. Currently, our ability to trace economic flows that create negative or positive consequences across the food system is hampered by opacity, as inscrutable as a compressed line on a corporate ledger. It remains more profitable to damage the environment and negatively impact human health, than to protect either of these. Yet the resultant costs are not evident to citizens or consumers, who pay at least twice or three times: at the checkout; for their poor health; and in loss of biodiversity (as one example). The organic food sector has long sought to translate principles and standards into the labeling of food, giving price incentives for food produced in ways that add value to the public good ("Incentives to Change: The Experience of the Organic Sector").

Measuring costs—and benefits—is far from an exact science, with inherent uncertainties and approximations ("Incentives to Change: The Experience of the Organic Sector"). Developing values for what has not previously been measured involves the engagement of communities throughout food systems and across value chains, as clearly illustrated in Chapter 3 by the process of bringing people together to account for water use and costs in the Andes ("Upstream, Downstream: Accounting for the Environmental and Social Value of Water in the Andes"). It demands that we respect the diverse ways of assigning value, which have often been characterized through monetary units, but do not need to be. TCA is subject to the critique that it is another extension of neoliberal policies that unduly quantify nature, thus constraining a more comprehensive respect for its life-giving pricelessness. Monetary units reflect a current societal norm, but this book's authors argue that this is often a poor choice that can and should be reconsidered ("The Economics of Ecosystems and Biodiversity"). More holistic and inclusive measurements of value can consistently help to identify which pathway we are on when considering all food system impacts, whether in a positive or negative direction. In the search for universal application and local contextualization, a number of authors in this volume stress that the value of this realigned framework of measurement and metrics is to arrive at estimates that point us in the right direction, rather than (perhaps impossibly) striving for the exact or perfect information in finite degree ("From Practice to Policy: New Metrics for the 21st Century," "Harmonizing the Measurement of On-Farm Impacts").

The specific challenges of identifying "true" values are explored in many chapters of this book, in manifold ways, and there is wide agreement that TCA in agriculture and food systems should not be mistaken as a ploy to "put nature up for sale" ("Methods and Frameworks: The Tools to Assess Externalities," "The Economics of Ecosystems and Biodiversity"). In many places within this volume, authors reflect on how TCA's application can avoid the strictures of "financialization" of nature and other public goods. Salman Hussain from the United Nations Environment Programme emphasizes this point in his text box on the TEEBAgriFood Evaluation Framework. Incorporating tangible and intangible values is key to learning to manage the complexity of food systems, in transparent ways so that social and environmental considerations are made evident ("Upstream, Downstream: Accounting for the Environmental and Social Value of Water in the Andes"). Marta Echavarria and her co-author describe how traditional Indigenous water management practices represent intangible values and contextually important socially and culturally appropriate approaches. This demands that we—and our decision-makers—consider the implications of our choices across all four capitals (understood as ways of framing the various stocks that embody the streams of benefits contributing to human well-being).

The application of this approach is illustrated by Kathleen Merrigan in the questions that she asks about how we evaluate meat compared with plant-based or cellular alternatives ("Trade-Offs: Comparing Meat and the Alternatives"), showing the utility of TCA to illuminate the degree to which commonly held perceptions might not match with reality across all sectors of concern, thus revealing unanticipated impacts. For example, as noted in Chapter 16 ("Trade-Offs: Comparing Meat and the Alternatives") somewhat ironically, when viewed through a holistic TCA framework, a wholesale shift to faux meat production might actually be viewed as a shift from farm to factory in the face of social push-back against factory farming ("Trade-Offs: Comparing Meat and the Alternatives").

Nadia El-Hage Scialabba and Carl Obst in Chapter 1 on metrics ("From Practice to Policy: New Metrics for the 21st Century") provide an eloquent argument for capturing such system-based perspectives in broad standardized assessment metrics that can facilitate progress toward local and global sustainability goals. The criteria that such metrics must meet are well detailed in this chapter. A key point is that far from being an accounting scheme, or an assessment exercise, TCA is inseparable from a call for collective action, to address the costs so revealed and build benefits for the public good.

Measurement needs to be sensitive to the goals of food systems transformation, and appropriate to the relevant communities and component parts, including the farmers, growers, abattoirs, packhouses, processors, financial institutions, distributors, and retailers all contributing to its overall health. As we are reminded in the Chapter 6, "Harmonizing the Measurement of On-Farm Impacts," in this immensely complex yet interconnected system, each component needs to be healthy to restore the overall system to full vitality, beginning with the earth's farms and pastures. In Chapter 8, "Transforming the Maize Treadmill: Understanding Social, Economic, and Ecological Impacts," featuring four different approaches to TCA of maize farming, it is striking to reflect on the broad range of biodiversity benefits and ecosystem services that one crop can generate, as well as how government policies that do not sufficiently support smallholder farmers can serve as an impediment to the flow of such services.

Yet in order for the application of TCA at each level to be successful, we must first agree on a common language, framework, and metrics ("Harmonizing the Measurement of On-Farm Impacts"). While recognizing different contexts and complexities, essentially everyone working on TCA within this volume notes the compelling need for finding common frameworks and approaches ("Transforming the Maize Treadmill: Understanding Social, Economic, and Ecological Impacts").

The challenges and intricacies of price and valuation come to the fore in a number of chapters. The ability, for example, of municipal institutions to increase their purchases to reflect key values (local economies, environmental sustainability, fair labor, animal welfare, and nutritional health) is hampered by the higher cost of food products that have been certified as having been produced through certified organic standards, fair trade, or human and equitable practices. Such socially and environmentally certified food products are priced closer to "true cost" than industrially produced food. However, the price premiums are difficult for a school district or public institution to bear within current budgetary constraints ("True Cost Principles in Public Policy: How Schools and Local Government Bring Value to Procurement"). Similarly, as explored in Chapter 13 on international policy venues for TCA ("International

Policy Opportunities for True Cost Accounting in Food and Agriculture"), it is well documented that while most of the poor people around the world can afford an energy-sufficient (in terms of calories) diet, they cannot afford either a nutritionally adequate or a healthy diet. The cost of a nutritionally adequate diet is estimated to be about 60% higher than the cost of a diet that is only energy-sufficient. The cost of healthy diets—one that is both nutritionally adequate but also includes a more diverse intake of foods from several different food groups—costs five times that of a diet that is only energy-sufficient, evidence that the true cost of food has much to do with structural inequalities and income equity.

The development of such broad TCA assessment metrics, or a composite index, requires the participation of a range of actors, from innovative practitioners interacting with governmental representatives, health, finance, and economy experts, and farmers and food producers along the food value chain ("From Practice to Policy: New Metrics for the 21st Century," "Harmonizing the Measurement of On-Farm Impacts"). While a diversity of viewpoints and purposes will inform the development of TCA metrics, the authors of this volume argue that TCA's universal application, with local contextualization, could be facilitated through the development of internationally accepted benchmarks.

Engagement Along the Food Value Chain

This is undoubtedly an exciting and critical moment for TCA—a field on the cusp of greater recognition and harmonization across approaches, poised for wider uptake. Recognizing that our food system is economically entrenched and resistant to change in many respects, we have sought in this volume to highlight a number of levers that may be essential for engaging all actors along food value chains. True "costing" cannot be successful if it remains a niche idea. The question remains how to amplify and accelerate changes along the food value chain. The chapters in this book provide us with many intriguing sources of inspiration that we hope might serve as a motivation for actors to engage. A number of areas where the approach can impact (health, power and equity, risk, and investment, engagement of farmers, environmental health and governance) are highlighted below.

Human Health and Consumer Concerns

As TCA in agriculture and food has evolved, an important element has been the incorporation of health and consumer concerns. A good example of how TCA can be relevant to consumers is the presence of sugar in sodas. Sugar-sweetened soda marketed aggressively to children is causally linked by public health officials to the alarming increase in obesity and diabetes among the youth of the world. Yet the purveyors of soda do not bear the medical costs of addressing the health problems that their products create. Who does? Most often, the public does, through the subsidized health care system. The medical costs are external to the price of the soda paid by the consumer and the profits received by the soda company. Yet they are a significant consequence of the transaction.

The centrality of health impacts to TCA is reflected in two chapters in this book. One provides a framework to understand the multiple channels of health impacts: occupational hazards, environmental contamination, contaminated/unsafe/altered foods, unhealthy dietary patterns, and food insecurity ("Health Impacts: The Hidden Costs of Industrial Food Systems"). This chapter helps us to think systemically about health impacts and links them to the political economy of food systems. The other, through a deep dive into food and health implications in the USA, provides striking documentation of the costs of current diets and proposes several solutions and policy mechanisms. This chapter illuminates the fact that even in the face of strong evidence, there remain substantial individual and societal barriers to food system change, whose root causes (income constraints, systemic racism, political influence of the food and beverage industry, to name a few) need to be addressed ("The Real Cost of Unhealthy Diets").

Power and Structural Inequities

TCA, if it is to be transformative, must address issues of power and existing structural inequities in food systems that impose the greatest costs on the most vulnerable members of society. Historic injustices—colonization, slavery, and racism—have long led to extractive relationships between the Global North and Global South. Current economic structures are built in part on exploitative relationships with women and people of color. This demands that we scrupulously include issues of power and equity in our true cost frameworks. The case of maize in Malawi in Chapter 8 illustrates how historical injustice and power imbalances are at the root of much food policy and practice; through a singular focus on yields, without consideration of impacts on other social, human, and natural capital, power has been consolidated, and smallholders disenfranchised without achieving the stated aims of food and nutrition security ("Transforming the Maize Treadmill: Understanding Social, Economic, and Ecological Impacts").

Equally compelling is to understand how labor—in restaurants and along food value chains in general—has been deeply undervalued throughout history. One of the greatest blind spots in true costing is labor. Throughout food value chains, labor is poorly compensated, and people are often obliged to work under unsafe conditions—an issue that has only been amplified in fields and meatpacking plants during the coronavirus (COVID-19) pandemic. The occupational hazards faced by farmworkers, farmers, and smallholders the world over affect populations that are already vulnerable, owing to modest income levels. The service sector in the USA is one of the few industries in which the majority of labor costs are not actually reflected in the cost of the meal, but are expected to be paid in tips. The cultural changes needed to reverse these biases and cultural norms will take concerted education campaigns and policy measures, but at the same time, the pandemic and the global reckoning with race is "both the gravest crisis in the service sector's history in the United States and also the greatest moment for transformation—for building power among workers and change among employers toward a sustainable future of equity and collective prosperity" (see Chapter 17, "Dining Out: The True Cost of Poor Wages").

Several chapters make a strong case for building new compensation practices, creating new business models, and building the power of farmers, workers, and local communities through the application of TCA and resulting action.

De-Risking the Future

Risk is another theme underscored by authors in this volume—a compelling motivation for the engagement of the private sector. For many companies, it is the risk of supply chain disruptions owing to dependencies on natural, social, and human capital that brings them to focus on how they can build, rather than draw down these capitals ("From Practice to Policy: New Metrics for the 21st Century"). It is a challenge to introduce wholesale systems change through TCA to companies and institutional investors who logically seek enterprise or fund-level success. Addressing risk and risk exposure could have positive impacts at the enterprise/investment level and for food systems more broadly if the right metrics are considered ("Investing in the True Value of Sustainable Food Systems"). As noted in Chapter 14, "The Business of TCA: Assessing Risks and Dependencies Along the Supply Chain," although the importance of the state of nature, ecosystems, and employees for the success of companies is undeniable to the corporate sector, these impact drivers have not been sufficiently considered in quantitative risk management. Redefining risk, reward, efficiency, and the issue of scale to align with the systems approach of TCA is both a challenge and a potential game-changer ("Investing in the True Value of Sustainable Food Systems").

Chapter 14 ("The Business of TCA: Assessing Risks and Dependencies Along the Supply Chain") reminds us of the importance of looking beyond conventional food and agriculture stakeholders, as important as they are (farmers, food processors, markets, etc.) to those in the banking, finance, and insurance industries that can also drive change. If at present it is the general public—communities and taxpayers—who are paying for the negative externalities in our food system, the transformations required for our food systems to meet the synergistic needs of humans and nature will require those in investment and finance to understand and address such costs. The chapters on investment and supply chains note that a TCA assessment for those in the private sector must demonstrate benefits of better practices not only using sustainability language but in tangible financial terms and incorporate these in credit ratings, insurance policies, annual accounts, and company valuations ("The Business of TCA: Assessing Risks and Dependencies Along the Supply Chain" and "Investing in the True Value of Sustainable Food Systems"). In Chapter 15, a strong need is articulated for a common approach and metrics to measure impact and to have those metrics align with accounting standards ("Investing in the True Value of Sustainable Food Systems).

Governance for the Public Good

Societal support for TCA and the potential public good arising from this approach is critical at all levels of governance. This is illustrated in the chapters in this book that explore local ("True Cost Principles in Public Policy: How Schools and Local Government Bring Value to Procurement"), subnational ("Fostering Healthy Soils in California: Farmer Motivations and Barriers"), national ("Cotton in Egypt: Assisting Decision Makers to Understand Costs and Benefits," "The Economics of Ecosystems and Biodiversity"), and international levels ("International Policy Opportunities for True Cost Accounting in Food and Agriculture") of action. In Indonesia, for example, a TCA study carried out under the Economics of Ecosystems and Biodiversity (TEEB) framework convinced the government to include cacao agroforestry in their 2020 Five-Year Development Plan for Indonesia for the first time ("The Economics of Ecosystems and Biodiversity"). Exactly how governments can most effectively design and implement policy and programs to support TCA is still being explored. Although public policy responses to current health (and diet-related) crises are evolving with relative urgency, governments are often reluctant to regulate the private sector, even when the influence of both is paramount to dietary choices. As pointed out in Chapter 10, the real cost of unhealthy diets is often directed at consumers who can hardly be expected to counter the full weight of the food industry. A number of tools are available to the public sector, explored in this volume, including encouraging and rewarding various incentives ("Fostering Healthy Soils in California: Farmer Motivations and Barriers"), taxes and subsidies ("The Real Cost of Unhealthy Diets"), public procurement policies ("True Cost Principles in Public Policy: How Schools and Local Government Bring Value to Procurement"), and greater support for the organic sector ("Cotton in Egypt: Assisting Decision-Makers to Understand Costs and Benefits," "Incentives to Change: The Experience of the Organic Sector"). The multilateral system provides a number of entry points for governments to discuss and explore the implications of TCA. Given the complex and interconnected nature of the global food system, progress will require significant commitment on an international level as well ("International Policy Opportunities for True Cost Accounting in Food and Agriculture").

One concrete measure proposed here is for policymakers to embed TCA in decision-making as an administrative process, as seen in Chapter 12 ("Embedding TCA Within U.S. Regulatory Decision-Making"), as a way of assessing a country's stock flows through international trade, and as a means of monitoring international commitments, as seen in Chapter 13 ("International Policy Opportunities for True Cost Accounting in Food and Agriculture").

Illuminating the cost to society of negative externalities through TCA can be a way to rework this unintentionally reinforcing system, to re-order policy priorities and to bring the system back into balance by promoting and incentivizing the positive benefits of food systems when they are managed for health and sustainability.

The Externalities of Farming, Positive and Negative

TCA does not end with farmers, but it begins with them. As described in the "Harmonizing the Measurement of On-Farm Impacts," farms are the basis of all food systems, containing the key to their vitality. While industrial agriculture might be responsible for excessive use of toxic chemicals, pollution of waterways, and sterile soils, ecologically based farming systems are capable of minimizing or eliminating such inputs, restoring soil fertility, fostering diversity, and creating building resilience against the shocks anticipated with climate change, while producing nutritious food and providing a decent quality of life for farmers and farm workers. But none of this can be done in isolation; working with farmers to recognize these positive externalities is central to TCA in agriculture and food.

True Cost Accounting as a Transparent Process

Negative externalities are not new; they are a form of market failure that has existed as long as markets have existed. But current global economic structures, with vastly increased technological growth, international trade and the institutionalization of the price-based market model has led to a massive externalization of costs in food and agriculture, through the drive for higher shared public costs and lower food prices throughout the world ("True Price Store: Guiding Consumers"). The author of Chapter 18 ("True Price Store: Guiding Consumers") proposes that if true costs are the problem, true prices (the market price, together with added external costs) and transparency regarding both of these are the solution. Establishing what is "true" and how to measure it naturally has challenges but provides important opportunities to link to consumers, to highlight environmental impacts and human rights ("True Price Store: Educating Consumers"). It also provides scope for dialogue across value chains and for companies to communicate differently about their impacts, positive and negative. Chapter 9, "Fostering Healthy Soils in California: Farmer Motivations and Barriers," on building incentives for sound agricultural practices points to the disconnect between consumers and the producers and processors that, without ways to communicate more transparently, hamper the ability of consumers to make choices that internalize true costs. We are invited to see "true pricing" is as an activity rather than an analysis, as a decision-making process in which everyone can participate as a consumer, citizen, business owner, employee, investor, and policymaker ("True Price Store: Guiding Consumers").

Conclusion

As a whole, the chapters in this book point toward the potential for our food system to be more human centered than profit centered; and toward a food system that has a more respectful relationship with the planet. The authors outline a path forward based on TCA for food. This path seeks to broaden, expand, and fix our current food metrics, in policy and in practice, by applying a holistic lens that evaluates the actual costs and benefits of different food systems, as well as the impacts and dependencies between natural systems, human systems, and agriculture and food systems. Most importantly, this path acts upon this integrated understanding to create an economic system that respects true costs and results in a more balanced relationship with respect to its role in the world.

References

Eigenraam, M., McLeod, R., Sharma, K., Obst, C., & Jekums, A. (2020). *Applying the TEEB AgriFood Evaluation Framework*. United Nations Environment Programme and IDEEA Group.

Section 1

The Power and Potential of True Cost Accounting

The window for change in our farming and food systems can be opened through declarations, meetings, and summits, but actual change will occur only when we operationalize those ideas and aspirations. This initial section of this volume stakes out the very real potential and game-changing power in applying True Cost Accounting (TCA).

Despite all that we might wish for, the value system underpinning current food systems continues to place a greater weight on the high production of commodity crops throughout the value chain. In order to broaden the viable range of choices for decision-makers, whether in government, in local communities, or in the private sector, there is a clear need for standardized assessment metrics that recognize environmental and social consequences. Chapter 1 in this section ("From Practice to Policy: New Metrics for the 21st Century") lays out this compelling need, who must be involved, and what practical steps are required to arrive at an agreed universal conceptual framework and associated metrics.

Essential to the adoption of TCA will be proof of its wide applicability. As we speak of the price of food being undervalued, there is a risk that the concept sounds as though it is only relevant where people are able to pay more, rather than less, for food. This construct will be examined closely in many subsequent chapters and sections, including those that disaggregate the many health, social, and environmental costs of current food production systems. In two subsequent chapters of this section, a close look is taken at how TCA concepts have been applied to cotton and other crops in Egypt (Chapter Two, "Cotton in Egypt: Assisting Decision-Makers to Understand Costs and Benefits) and on community water management in the Andes (Chapter 3, "Upstream, Downstream: Accounting for Water Use and Costs on Agricultural Landscapes in the Andes"). Both address how benefits for society as a whole, over the long term, are revealed through TCA.

1 From Practice to Policy

New Metrics for the 21st Century

Nadia El-Hage Scialabba and Carl Obst

Context

The authors of the 2016 Nature article, *Fix Food Metrics* (Sukhdev et al., 2016), start by describing the significance that food systems have for humanity, especially small-scale agriculture, in terms of nutrition and employment but also by recognizing the massive environmental costs of commercial food systems with respect to the effects of biodiversity loss, greenhouse gas (GHG) emissions, soil degradation and the decline in fish stocks. They conclude their introduction by saying:

> Current metrics for agricultural performance do not recognize or account for any of these costs or benefits. The emphasis on yields or profits per hectare is as reductive and distorting as is gross domestic product, with its disregard for social and natural capital. Food metrics must be urgently overhauled or the United Nations Sustainable Development Goals will never be achieved.
>
> (Sukhdev et al., 2016)

The general appreciation of the need to go beyond measures of agricultural yield is widely recognized and indeed is the motivation for this book. The need for systems-based framings of agriculture, the need to recognize multiple capitals, the need to recognize the true or full costs of agricultural activity, and the need to progress towards actions, which are consistent with the objectives of sustainable development, have collectively spawned a mass of metrics and indicators. These metrics, indicators, and associated frameworks exist at global, national, and community level and focus on different aspects of the multiple dimensions of agricultural systems. Often, they are also related to alternative management practices and policy solutions. However, the sheer breadth and richness of well-intentioned alternatives has driven home the singular advantage of the incumbent performance assessment measure or metric—"yield per hectare"—simplicity. The *status quo* of the agricultural yield metric—and policies that use it as their key performance indicator (KPI)—can be maintained because there is no simple alternative.

Thus, somewhat paradoxically, the variety of solutions that are available to take into account the diversity of food systems, has paralyzed the potential for change and has ensured that the substantive knowledge that exists "beyond yield per hectare" is not widely translated into on-the-ground change in agricultural practice and associated economic, environmental, and social outcomes. This chapter discusses what might be done differently.

Role of Metrics and Policy

The primary motivation for agricultural systems is the provision of food and fiber to satisfy the human needs of current and future generations. The ongoing changes in these systems date back thousands of years, as farmers have sought to make the most of varying environmental, economic, and social contexts. Over the past 100 years, as populations have grown, and life expectancy and purchasing power have increased, the demands on the system have risen and been accompanied by significant changes in technology and management practices that have increased production.

In this framing of the role of agriculture, agriculture policy and decision-making have a clear end goal of ensuring sufficient production to meet current demand. Moreover, in this framing, a metric of "yield per hectare by type of agricultural output" seems sufficient, especially because related environmental and social costs often have effects over a medium-to-longer-term period, and by their incremental nature are often considered small. Thus, where alternative farm practices are being considered, the common measuring stick of whether the same volume of food is being produced seems appropriate (i.e., the incremental effects can be ignored or assumed away).

Of course, the reality is that there are environmental and social constraints on the ongoing or future supply of food in particular places (e.g., owing to the loss of soil quality or loss of local farming knowledge, together with considerable negative consequences of certain production practices). This has led to the development of alternative, more sustainable, farm management practices over many years. Examples include organic and regenerative agriculture, fair trade, holistic management, and community-based natural farming, among many others. Each of these has sought to go beyond yield per hectare as the common measuring stick and instead highlighted the need for a more balanced assessment of outcomes.

However, as each alternative management practice has adopted different measures for performance assessment, the policymaker is left with a choice (between the *status quo* and the alternative) without a standardized information set on which to make the comparison, other than yield per hectare. The lack of a standardized assessment metric that recognizes the need to consider environmental and social consequences thus takes on great significance in changing the *status quo*.

The barrier is not that the environmental and social consequences cannot be assessed (indeed there are many ways in which this has been done) but rather, there is no standardized approach by which a decision-maker can readily understand, compare, and evaluate the merits of different approaches. These decision-makers include politicians and government officials but also individual farmers making choices about the management of their land and other agri-food sector businesses making decisions about the sustainability of their supply chains.

Furthermore, as the focus is commonly on the adoption (or not) of specific land management practices, success might be measured in terms of the extent of adoption of the practice (e.g., hectares under certified organic practices) rather than on agreed, independently determined assessment metrics—for example, concerning environmental impacts, decent work, and community health. As a single farming practice will not be ideally suited to all contexts or for all agricultural output types, the lack of a standardized assessment tool makes it relatively straightforward to claim that a specific practice is unsuited and hence justifies maintaining the *status quo.*

Standard, independent assessment metrics allow each decision-maker to make their own appraisal of the merits of different practice solutions. This situation is taken for granted in the economic and financial space, where financial and accounting data are organized following standard principles and definitions, which allow arms-length assessments of the performance of companies and government policies, notwithstanding the great variety of economic activities and practices that take place. Of course, these metrics suffer from a lack of integration of social and environmental consequences, and that is the issue to be tackled.

The development of broad, standard, independent assessment metrics encompassing economic, environmental, and social dimensions would thus drive:

- Local assessment of how alternative practices can be best adapted to the local context
- Clear messaging to consumers on the relative performance of practices for their chosen agricultural output with respect to economic, environmental, and social impacts
- Access to finance, as investors are able to compare alternatives on a common basis
- Development of a common language to exchange experiences, reduce capacity-building costs, and provide a platform for innovation and adaptation
- Standard and trusted sources of data and methods to support decision-making

Fundamentally, a focus on alternative solutions without allowing individual stakeholders to reach their own conclusions cannot drive change. A focus on alternative solutions together with standard, independent metrics might change the balance.

State of the Art: Instruments for True Cost Accounting

The Sustainable Development Goals (SDG) indicators for measuring progress towards the SDG targets, such as SDG2.4.1 regarding sustainable agriculture areas and SDG12 on sustainable consumption and production, struggle to define metrics that are straightforward to translate in a consistent manner into policy analysis and that support effective monitoring of the food and agriculture system and its impact on the overall ecological and social environments. Metrics based on True Cost Accounting (TCA) could resolve the sectoral divide of current indicators, thus offering support for policy and decision-making, as well as the evaluation of progress towards the SDG goals and targets.

Given the potential for TCA metrics to be applied at farm and business level, as well as at regional and national scales, their use to evaluate progress towards the SDGs can be relevant for the reporting of governments and the agri-food sector alike. In particular, the use of a common framing supports clear articulation of the agri-food sector's contribution to achieving the global goals and targets.

Beyond SDG reporting, the agri-food sector faces a range of challenges to their operation, including the risk of supply chain disruption and the need to respond to consumer concerns about the sustainability of food systems. These challenges are described in more detail in other chapters of this book. As part of the response, companies are developing a range of related but different full-cost protocols for managing risk and dependency on natural, social, and human capital. In the past few years, a number of TCA tools and methods have emerged, including impact frameworks, footprint calculation, capital changes, databases of valuation factors, and ecosystem models. The main methods specific to food systems are impact frameworks, but these do not provide specific, comparable metrics, nor monetary valuation. More importantly, there is no standard scheme of food system footprints. Decision-makers are thus confronted with a confusing range of well-intentioned approaches.

Key Criteria for a True Cost Accounting Tool

Ideally, a True Cost Accounting tool should meet the following criteria:

- Provide a systems' perspective, including upstream and downstream components of the value chain (i.e., input sourcing, production, processing, storage, transportation, consumption, disposal, or recycling) and related societal elements and their area of influence, such as institutions/policies, socio-cultural norms and public health.
- Allow for universal application for local contextualization, with internationally accepted benchmarks: must be applicable across the whole diversity of socio-economic and environmental circumstances that exists in the food and agriculture sector worldwide.
- Support integration across all sustainability dimensions or multiple capitals: economic, environmental, and social, the latter including also human and

intellectual capitals and good governance within the sphere of influence of the entity, or national boundaries.

- Apply to multiple scopes and purposes, from enterprise performance assessment, through product standards, to organizational strategies for policy, planning and reporting.
- Reflect various value use perspectives, from government, through B2B (business to business) and B2C (business to consumer), to small and medium-sized enterprises (SMEs) and local communities.
- Allow for indicator selection, including quantitative, qualitative, and monetary valuation: indicators vary from being outcome-oriented (indicating trend and status) to process indicators (assuming that management systems are in place for better management, but without precise causal link), with different data quality levels (related to the timeframe, type, and methodology), and rating, weighting, and aggregation levels; and
- Work within a common standardized reporting framework to drive comparability, translating the complex true costing process into a simple dashboard or index.

While these are ideal criteria, there is a substantial amount of work that has been completed to suggest that satisfying these criteria is within reach. To see this through, however, it is necessary to pull apart the components of an ideal TCA tool, each of which can be standardized in progressive fashion and designed to connect over time.

Leading Pathways Towards Standardization

Standardization is required in five areas: conceptual frameworks, metrics, data, valuation, and reporting. This section considers each area in turn, highlighting the advances that have been made and describing the "best in class" approaches that are currently evident. There are undoubtedly overlaps, as the approaches do not neatly belong in any single area but in fact, that itself has been part of the challenge. Where a single approach attempts to standardize across multiple areas, they generally fall short, as each area raises different standardization challenges and involves different stakeholders and expertise.

Conceptual Frameworks

Frameworks are the conceptual structure that provide the basis for the selection and combination of variables under a fitness-for-purpose principle. Frameworks including non-financial values drive improved organizational behaviors, but results depend on context-specific assumptions and on what is considered most relevant for different stakeholders.

Several frameworks have been developed by international accountants and industry coalitions in the past seven years, including: PwC's Total Impact Measurement and Management framework (n.d.); KPMG's True Value framework; EY's Total Value framework; the Natural Capital Coalition's Natural Capital Protocol; WBCSD's Social and Human Capital Protocol; the Roundtable for Product Social Metrics' Product Social Impact Assessment; and the Impact Institute's Framework for Impact Statements (de Groot Ruiz, 2019).

The only framework that was developed in cooperation with inter-governmental entities and that contemplates policy usage is the TEEBAgriFood framework (The Economics of Ecosystems and Biodiversity, 2018). The TEEBAgriFood framework is comprehensive and inclusive, offering a common language to describe diverse and complex food and agriculture systems coherently and comparably across many spatial scales (national, regional, farm), and to account for hidden costs and benefits of these systems. The strength of the TEEBAgriFood framework is the systematic categorization of the "what" of evaluation, so that assessments of food systems can be seen within a broader context. However, it is not prescriptive on metrics and units, and, in order to retain universality across a range of evaluation approaches, it does not formalize impact pathways.

Metrics

Metrics, often expressed as key performance indicators (KPI) are revealed by identifying the relevant themes and impacts to be the focus of assessment in any given context. TCA is better understood by distinguishing two steps: classical impact assessment (the well-known "environmental impact assessment" extended into sustainability assessment, complete with environmental, social, and economic assessments) and the more recent expression of sustainability assessments in monetary terms. It is the first step that is the focus at this point.

TCA-based KPIs should be identified based on issues of risk and importance to stakeholders, thus selecting indicators that address the highest-priority issues. A meaningful way forward in terms of standardization is for impact assessment to follow the Food and Agriculture Organization of the United Nations (FAO) Sustainability Assessment of Food and Agriculture systems (SAFA) Guidelines (El-Hage Scialabba *et al.*, 2014) that suggest universal impact themes. The SAFA Guidelines developed targets and indicators specific to enterprises' performance, and hence the use of SAFA for TCA-related policy will require the identification of appropriate policy targets. A study comparing SAFA and the 2015 SDGs showed a high level of convergence between the SAFA themes and the SDG targets (El-Hage Scialabba et al., 2016). Thus, using the TEEBAgriFood Evaluation Framework, the SAFA themes, and the SDG targets offers a sound basis for the identification of TCA KPIs, reflecting the relevant themes and impacts.

While existing literature is not short on methods and processes, common KPIs have yet to be defined for comparative TCA assessments. Table 1 presents a synthesis of KPIs structured according to the four capitals of the TEEBAgriFood framework. It is based on a short analysis by the authors of selected agri-food sector company reports and various valuation and impact frameworks and protocols. It is not intended to present a definitive list but rather give an indication of the KPIs found in most agri-food business sustainability evaluations that might form the basis for standardized core KPIs for TCA assessment and reporting. While Table 1 gives a potential core set of KPI, there must always remain room for customization for different stakeholders and situations.

Also required for inclusion in policy discussions are agreed thresholds. In this space, progress by the Science Based Targets Network (https://sciencebasedta rgets.org) and relevant SDG targets provide a starting point for standardization concerning environmental themes and could point to the potential for other themes. It is important to note that TCA-based indicators will run across all SDGs, thus providing an opportunity to integrate economic, natural, human, and social accounting within a unified framework.

Data

The next challenge is data. The choice of data and evaluation methods used depends on the purpose of evaluation, spatial scale, and scope of the value chain, which depends on the expected application. Furthermore, data quality, valuation factors, scoring system, time, and budget constraints affect output accuracy, data correctness, compatibility, user-friendliness, and transparency. All these factors make it difficult to interpret and compare TCA assessments.

There is simply no space in this article to give credit to the tremendous amount of data and modelling that exists for capital changes associated with food system activities. Thus, spatial and contextual boundaries for measurement do not need to be considered from scratch, nor the understanding how the footprints align to capital changes, and capital changes to subsequent capital changes, along the chain of outputs and outcomes. For instance, the integrated assessments models that are used to determine the social cost of carbon connect economic modelling to climate modelling. A similar process of attaching

Table 1.1 Potential Core TCA Key Performance Indicators

Economic	Natural	Human	Social
Wages	Air (GHG, air pollutants)	Health and safety	Nutrition and
Taxes	Water (use and pollution)	Education, skills,	food security
Employment	Soil (occupation and	and knowledge	Overall mission
Profit	pollution)	Fair treatment of	Corruption
Investment	Biodiversity (land use	workers	Provision of
Intangibles	change, eco-toxicity, eco-		infrastructure
(brands,	system complexity, habitat		and technology
transparency)	encroachment, regulation)		
Livelihoods	Raw materials (food and		
	fiber) and energy		
	(provisioning)		
	Waste		
	Recycling		

Source: Authors' synthesis.

economic modelling (or collating agreed economic valuations from literature) to food system modelling is a feasible start for food impact costing (Lord, 2020). The data challenge is rather to establish the agreed definitions for measurement to work within the conceptual framework and facilitate the derivation of KPIs.

Agreed data definitions are the focus of work on the United Nations System of Environmental-Economic Accounting (SEEA), which has a component for Agriculture, Forestry and Fisheries (SEEA−AFF) (Food and Agriculture Organization of the United Nations and United Nations Statistical Division, 2020). It provides definitions underpinning environmentally focused national aggregates that provides a baseline for TCA. It also supports accounting of carbon footprints, water footprints, flow accounts for nitrogen and phosphorous, pesticide use, and food loss. Generally, however, this footprint accounting is too coarse for impact valuation, as its contextual scope is limited for land areas used for agriculture, forestry, aquaculture, and the maintenance and restoration of environmental functions.

Thus, to complement the SEEA−AFF, the SEEA Ecosystem Accounting (SEEA−EA) (United Nations Statistics Division, 2021) can be used. It is spatially specific to ecosystems, while recognizing that capital changes and impacts do not accord with national boundaries. This is essential for accounting for food system impacts, as the spatial resolution at which to measure footprint for food system impact valuation is important. The SEEA−EA includes both accounting of quantities and qualities of capital and valuation in its scope. The SEEA−EA offers a conceptual discussion about non-financial capital accounting that could underpin a version for food systems.

Valuation

Here, the challenge of monetary valuation is distinguished from data definition and management recognizing the additional considerations that are involved in establishing appropriate monetary valuations, especially concerning environmental and social dimensions. Nonetheless, as with the data area, there are widely practiced approaches in the valuation space, and many different monetary evaluation techniques are available[1], for instance from the Natural Capital Protocol Annex. Furthermore, with regard to standardization, there are a range of valuation databases that can be further developed and applied (such as TEEB ecosystem services valuation database (www.es-partnership.org/esvd/) that draw upon a combination of pre-loaded, or user-defined, data sets to model the distribution of ecosystem services across areas of interest) and also progress in discussions via the SEEA, projects on wealth-accounting, such as being led by the World Bank through their Changing Wealth of Nations (Lange *et al.*, 2018) work and the corporate-level Value Balancing Alliance (www.value-balancing.com) that is aiming to harmonize impact valuation factors.

Initially, general valuation factors could be determined, with appropriate caveats. This might build into a database of shadow prices at a pragmatic level of resolution—that is, having sufficient spatial, temporal, and contextual detail to avoid gross errors but coarse enough to make compiling the database feasible. In broad terms, it is more important to get estimates that point in the right direction, with enough resolution to distinguish sustainable production methods and gather a collective weight willing to promote and use scientifically based food impact costings, rather than wait for synthesized and standardized modelling efforts to emerge from a myriad of scientific projects.[2]

Reporting

The general disclosure of content, according to the scope of reporting, is an important step towards credibility, let alone comparability of TCA outcomes. Currently, reporting on impact valuation is as diverse as there are entities engaging into impact valuations, making comparisons and "fair play" impossible.

The Global Reporting Initiative (GRI) offers Sustainability Reporting Standards (www.globalreporting.org/standards/), which consists of different modules: Principles, disclosures, and management approach (GRI 101–103); Economic (GRI 201–206); Environmental (GRI 301–308); and Social (GRI 401–419). SDG reporting for businesses leverages the GRI Standards, which are the world's most widely used sustainability reporting standards, and the Ten Principles of the United Nations Global Compact.

However, countries' evaluation of SDG progress remains challenged by a lack of benchmark data on the SDGs and how to report on their achievements. The SDG National Reporting Initiative (www.sdgreporting.org) provides information about key policy and technical considerations for SDG reporting.

For policy analysis, decision-making and monitoring of effective food system transformation, a dashboard would boil down the complexity of non-financial reports into clear messages, as hotspots, trade-offs and synergies are easily visualized. Furthermore, the aggregation of quantitative, qualitative, and monetary valuations into an index offers the kind of simple metric that policymakers need, in the same fashion as gross domestic product.

The Way Forward for Scaling-Up TCA Adoption by Policymakers

Towards a TCA Index

Traditionally, TCA translates non-financial environmental and social costs and benefits into monetary terms. It is intended that individual monetary metrics "speak" to decision-makers, especially when resource allocation needs to be made. For instance, quantifying excess fertilizers' effect on future drinking water quality and the attendant mitigation costs decreases interest in subsidizing fertilizers to boost yields, as costs are shown to exceed benefits.

For businesses, integrated profit and loss statements (IP&L) give an overview of all material impact that results from the organization's activities. The impact addressed in the IP&L is usually organized by type of capital, thus stating the triple bottom line, while addressing different stakeholders: investors value-creation statement; stakeholders value-creation statement (license to operate); external cost statement (do no harm of operations); and SDG contribution statement (de Groot Ruiz, 2019).

Given the variety of challenges in valuation for a range of social and environmental impacts, the integrated evaluation and comparison of TCA assessments by governments will require the development of a suitable composite index, built by scoring, weighting, and aggregating indicators. Robust indexes are useful to policymakers, owing to their ease of understanding of complex issues.

Widely known indexes include the Human Development Index, the Quality of Life Index, the Gender Empowerment Measure, the Environmental Sustainability Index, and the Global Ecological Footprint. Indexes are constructed for easy interpretation of trends covering a number of KPIs.

If reliable, scientifically sound, and transparently developed KPIs are available, composite indexes can be developed through Multi-Attribute Utility Theory (MAUT) techniques (Talukder et al., 2016). MAUT is a branch of multi-criteria decision analysis—a structured approach that quantitatively evaluates alternatives in decision-making by considering indicators and their weighting alike, thus allowing incorporating multiple indicators into an evaluation process. The MAUT technique can generate indexes on a 0 to 1 scale, where a score near 0 indicates bad performance and near 1 indicates good performance in terms of targets and indicators.

Developing a TCA Composite Index requires the following steps: (i) main thematic targets are defined within an underpinning conceptual TCA framework; (ii) default KPIs (metrics) for the defined TCA targets are selected; (iii) data for the selected KPIs is collected; (iv) application of the MAUT technique to generate an index score for each target; and finally, (v) transformation of the index results into a dashboard.

This proposed approach to developing indexes and dashboards is capable of handling the typical incommensurability (or lack of common measures) of TCA KPIs, as this method can aggregate indicators by considering only indicators' scores—that is, the score of the index depends solely on the relative performance of the indicators and it is transparent and replicable.

Food system assessments have historically been challenged by aggregation, double counting, and bias in under or overestimating costs of impacts (depending on temporal, spatial and contextual details). These challenges are unlikely to be completely resolved by TCA. Thus, what TCA should aim for is robust and reliable descriptions of trends, in order to understand systemic improvements. This argument favors an index approach over monetary valuations, exact shadow pricing, and value transfer.

Food impact costing does not need to get the perfect answer, and food systems in particular include myriad connections that are impossible to precisely determine. Agreement, credibility, and the opportunity to intervene in market failure in the direction of food system transformation are the guiding principles for assessment and reporting.

Who Needs to be Involved?

To date, each stakeholder group in each area has defined their own approach, according to specific yet varying scopes. While this diversity is natural and has been welcomed, wider adoption requires increased harmonization and coordination. Luckily, all existing frameworks are tending to converge on similar processes (such as described in the natural, social, and human capital protocols) to be followed in terms of setting the scope and acquisition of data.

However, the selection of appropriate indicators, measurability, and valuation remains a stumbling point. An eventual agreement on what KPIs to select and how to define them will require the cooperation of all public and private parties. Furthermore, what should be included in valuation and measurement choices depends on stakeholders' agreement, with science guiding, rather than establishing, pathways.

Thus, despite the methodological feasibility of a TCA Composite Index, its development process will require the participation of a range of actors, from innovators such as the current TCA practitioners gathered under the umbrella of the Global Alliance for Agriculture and Food Community of Practice, through government and intergovernmental institutions concerned with food and agriculture sustainability, as well as finance and economy experts, as comparable food system impact valuation is the main innovation brought by TCA.

Overall, the participation and leadership of governments in developing TCA standards is crucial, especially in terms of ensuring public ownership, accountability, and enforcement. As highlighted by Organisation for Economic Cooperation and Development Development Cooperation Director Jorge Moreira da Silva:

> The challenge lies in defining and measuring impact … Different countries, public and private organizations are using different yardsticks to measure different elements. To counter the risk of 'impact washing', public authorities have a responsibility to set standards and ensure they are adhered to.
> (Organisation for Economic Cooperation and Development, 2019)

Where to Start?

Ultimately, an agri-food system focused TCA standard would guide practitioners on what to measure and disclose in terms of footprints, while providing a set of quantities on which to base shadow prices. The FoodSIVI report (Lord, 2020) argues that the United Nations SEEA EEA and the FAO SAFA Guidelines offer a blueprint for such a standard for agreeing on the spatially and contextually explicit footprints needed to incentivize impact reduction and track progress toward food system transformation targets.

In the short term, a pragmatic approach could be that stakeholders develop standards for each of the constituent elements of TCA, including agreeing on: a universal conceptual framework, core metrics and KPIs, data definitions, valuation impact factors, and reporting standards. Current stakeholders' activities are pointing at solutions in each of these tasks, but coordination is needed to align efforts. In addition, further work is required to develop a TCA Composite Index, with the committed participation of government officials occupying SDG-related roles.

Among the array of efforts, a pathway towards improved environmental and social outcomes for agri-food systems is emerging. The coordination and standardization of agri-food system assessment using TCA as a focal point can be a core part of the solution.

Notes

1 Monetary valuation techniques include: the market-based approach (market prices, hedonic pricing), revealed preference technique (travel cost method, hedonic price method), stated preference approaches (contingent valuation, choice experiments), subjective wellbeing valuation, cost-based approaches (compensation costs, defensive expenditure, damage/repair costs), and value transfer.
2 Although with appropriate focus the synthesis of results from multiple projects might indeed take less time than imagined.

References

de Groot Ruiz, A. (2019). Framework for Impacts Statements Beta Version (FIS Beta). Impact Institute. Available at: www.impactinstitute.com/framework-for-impact-statements.

El-Hage Scialabba, N., Rocchi, S., & Guttenstein, E. (2016). SAFA for Sustainable Development. Food and Agriculture Organization of the United Nations. Available at: http://www.fao.org/fileadmin/templates/nr/sustainability_pathways/docs/SAFA_for_sustainable_development__01_.pdf.

Food and Agriculture Organization of the United Nations (2014). Sustainability Assessment of Food and Agriculture Systems Guidelines. Food and Agriculture Organization of the United Nations. Available at: www.fao.org/3/a-i3957e.pdf.

Food and Agriculture Organization of the United Nations and United Nations Statistical Division. (2020). System of Environmental-Economic Accounting for Agriculture, Forestry and Fisheries (SEEA AFF). Food and Agriculture Organization of the United Nations. Available at: https://doi.org/10.4060/ca7735en.

Lange, G., Wodon, Q., & Carey, K. (2018). The Changing Wealth of Nations, 2018: Building a Sustainable Future. World Bank. Available at: http://hdl.handle.net/10986/29001.

Lord, Steven. (2020). Valuing the impact of food: towards practical and comparable monetary valuation of food system impacts. A report of the Food System Impact Valuation Initiative (FoodSIVI). University of Oxford, Food System Transformation Group, Environmental Change Institute. Available at: https://foodsivi.org/wp-content/uploads/2020/06/Valuing-the-impact-of-food-Report_Foodsivi.pdf.

Organization for Economic Co-operation and Development. (2019). Impact Investment needs global standards and better measurement. Available at: www.oecd.org/newsroom/impact-investment-needs-global-standards-and-better-measurement.htm.

PwC. (n.d.). Total Impact Measurement & Management. Available at: www.pwc.com/gx/en/services/sustainability/total-impact-measurement-management.html.

Sukhdev, P., May, P., & Müller, A. (2016). Fix Food Metrics. *Nature*, 540(7631), 33–34. doi:10.1038/540033a.

Talukder, B., Hipel, K.W., & van Loon, G.W. (2018). Using multi-criteria decision analysis for assessing sustainability of agricultural systems. *Sustainable Development*, 26(6), 781–799. https://doi.org/10.1002/sd.1848.

The Economics of Ecosystems and Biodiversity (TEEB). (2018). Measuring what matters in agriculture and food systems: a synthesis of the results and recommendations of TEEB for Agriculture and Food's Scientific and Economic Foundations report. Available at: http://teebweb.org/wp-content/uploads/2018/10/Layout_synthesis_sept.pdf.

United Nations Statistics Division. (2021). System of Environmental-Economic Accounting – Ecosystem Accounting. Available at: https://unstats.un.org/unsd/statcom/52nd-session/documents/BG-3f-SEEA-EA_Final_draft-E.pdf.

2 Cotton in Egypt

Assisting Decision-Makers to Understand Costs and Benefits

Helmy Abouleish, Thoraya Seada and Nadine Greiss

Introduction

Agriculture and forestry have been the largest contributors to climate damage over the past 200 years. Yet evidence is pointing to the reality that sustainable forms of agriculture can offer solutions. Research on SEKEM fields in Egypt has conclusively shown that biological, organically cultivated crops can be competitive to the conventionally cultivated ones in expenses and benefits since organic produce provides less negative externalities. Here we report on an update of "The Future of Agriculture in Egypt" study, conducted on five strategic crops, one of which is cotton.

Challenges Facing Egypt

Egypt is facing many challenges related to demographics, economy, and public health, which have the potential to become exacerbated unintentionally by new problems related to the environment. Three that will particularly impact the agriculture sector are water scarcity, rural development, and soil health.

Water Scarcity

Water scarcity continues to be a major issue for Egypt, which depends almost entirely on the Nile for the country's water resources. Egypt is facing a current annual water deficit of about 7 billion cubic meters. According to some analysts, the country is on track to reach a threshold of "absolute water scarcity" by 2030. Uneven water distribution, misuse of water resources, and inefficient irrigation techniques are some of the major factors playing havoc with water security in the country. Large amounts of water are also lost through evaporation every year—something that climate change will worsen. Egypt has only about 20 cubic meters per person of internal renewable freshwater resources, and as a result, the country relies heavily on the Nile for its main source of water. The river Nile is the backbone of Egypt's industrial and agricultural sectors and is the primary source of drinking water for the population. Rising populations and rapid economic development in the countries of the Nile Basin, pollution, and environmental degradation are reducing water availability in the country.

Rural Development

Changes to the climate, particularly higher temperatures, are expected to shorten growing seasons and reduce agricultural yields in Egypt. This creates a further challenge for agriculture and rural development in Egypt. As noted by Shalaby *et al.* (2011), Egypt's economy depends primarily on agriculture and rural resources. Agriculture contributes approximately 14% of gross domestic product and absorbs about 31% of the workforce. More than half of the population lives in rural areas where, directly or indirectly, their livelihood depends upon the agricultural sector. The agriculture sector contributes significantly to Egypt's economy and food security, yet faces many challenges which, in turn, impact rural development initiatives. Key among these are land and water issues, a lack of information on alternatives, including sustainable approaches to the use of natural resources and marketing that can support this. In addressing overall challenges of poverty, degradation of natural resources, other environmental issues, and population growth, Egypt faces inadequate support services, framework and institutional constraints, and a lack of agricultural and rural development policies.

Soil Health

Egypt is located in the severely dry region extended from North Africa to West Asia. In this region, soil erosion is regarded as one of the most serious environmental problems associated with land use. In many cases, erosion causes an almost irreversible decline in soil productivity and other soil functions and leads to environmental damage. Wind erosion is considered to be one of the main desertification processes affecting areas exceeding 90% of the state in the western desert, eastern desert and particularly Sinai area. These areas are characterized by a fragile ecosystem, scarcity of vegetation cover, and severe drought.

True Cost Accounting

True Cost Accounting (TCA) is a method to calculate the external effects of agriculture on the environment and society. "External effects" are described as all unintended effects on the life of one person occurring during an action done by another person, which can be any action in daily life, as well as any economic activity. Examples for human actions could include even one person spewing smoke into the air or dumping litter on the highway, although usually they are actions by groups of people.

The methodology of TCA as used here seeks to highlight further hidden costs besides the direct costs of raw material, labor, etc., and to place a value or cost on these. True Costs are described as the sum of internal and external costs, which can be understood for this study as "Direct Costs" and "Damage Costs."

Throughout this study, the most important examples for external costs are soil erosion, atmosphere damage through greenhouse gas (GHG) emissions, and water damage. In this study, the term "Damage Costs" (or "Environmental Damage

Costs") is used as an equivalent for the more commonly used term of "externalities." Right now these damage costs are being paid by the society and future generations. Internalization by, for example, an environmental tax would represent a cost shift from the common responsibility to the responsibility of the polluter.

"The Future of Agriculture in Egypt" Study

"The Future of Agriculture in Egypt" study was first released in April 2016, calculating the "true cost" of cotton, rice, wheat, potatoes, and maize, in the year 2015. The study concluded that at least for the five examined strategic crops it would be economically more expensive to produce crops based on a conventional farming system than to apply organic practices. For Egypt's economy, the concept of true costs is highly relevant in the context of the shortage of natural resources such as land, water and fertile soil. Meanwhile, in conventional farming systems the inputs need to be increased over time to maintain the same output, which consequently increases the cost of production and use of natural resources.

In June 2020 the study was repeated for the same five crops to provide an update after five years and to re-evaluate the costs after the economic changes in Egypt owing to the currency devaluation. Once more, the organic agricultural methods were put to the economic test in comparison to conventional agriculture, using TCA.

Methodology: Calculation and Evaluation

The comparison structure and the calculation for the direct cost parameters is based on the methodology of the Food and Agriculture Organization of the United Nations (FAO) Study "Economic & Financial Comparison of Organic and Conventional Citrus-growing systems" prepared by the University of Valencia in 2000 (Food and Agriculture Organization of the United Nations, 2000), except for the financial investment calculation. This is because the presented study aims to focus on the explanation of the specific damage costs, which would be distorted by integrating financial multipliers. The calculation methodology for the damage cost parameter for water quality, atmosphere damage, GHG emissions, and soil erosion is based on the FAO report on Food Wastage Footprint: Impacts on Natural Resources (Food and Agriculture Organization of the United Nations, 2013). Other calculations are described in detail below.

Carbon Footprint Calculation: The Carbon Footprint assessment is conducted by the Cool Farm Tool (CFT) (https://coolfarmtool.org/news-resources), which is an online greenhouse gas, water, and biodiversity calculator for farmers. The CFT was originally developed by Unilever and researchers at the University of Aberdeen and the Sustainable Food Lab to help growers measure and understand on-farm GHG emissions. The Cool Farm Tool requires general information about farms, such as crop area, yield, soil type, fertilizer, and inputs, as well as some detailed information on electricity and fuel use (for field operations and primary processing). Carbon sequestration is a key feature of agriculture that has both mitigation and adaptation benefits. The CFT includes calculations of soil carbon sequestration, defined as a long-term storage for carbon dioxide or other forms of carbon.

In organic farming, the calculation for the carbon footprint assessment includes the carbon sequestration through the use of compost. The sequestration amount from compost might offset carbon dioxide emitted by other farm operations, such as diesel consumption. Through calculations using the CFT, the results for total GHG emission in organic farming in Egypt are calculated to be negative or zero. Subsequently, in conventional farming, the calculation for the carbon footprint assessment was done by the previously described methodology of the Cool Farm Tool.

Water Footprint Calculation: The concept of water footprint emerged in 2002, and it has been created in an analogy to the ecological footprint. While an ecological footprint measures how much land a human population requires to produce the resources that it consumes, and to absorb its waste, a water footprint measures human demand on freshwater. An updated manual of the *Water Footprint Assessment Manual* was published by the Water Footprint Network (www.waterfootprint.org) (Hoekstra *et al.*, 2011).

The Water Footprint methodology distinguishes three types of water usage:

- Consumptive use of rainwater (green water).
- Consumptive use of water withdrawn from groundwater or surface water (blue water).
- Pollution of water (graywater; the gray water footprint is an indicator of the degree of freshwater pollution that can be associated with the consumptive use of water).

In organic farming, the water calculation was conducted with the previously described methodology "Water Footprint Assessment," to determine the amount of water required per acre (of both green and blue water). The water quality costs (gray water) for organic farming equates to zero, as these costs are related to the usage of pesticides and to the number of nitrates in sources of drinking water.

In conventional farming, the calculation was conducted by using the Water Footprint Assessment to determine the amount of water required per acre (green and blue water). These costs are dependent on the usage of pesticides and the number of nitrates in sources of drinking water, therefore integrating gray water data as well.

Soil Erosion: In organic farming, the soil loss from erosion is 15% less for organic agriculture than for conventional agriculture according to Auerswald *et al.* (2003).

The wind erosion ratio in Egypt is about 5.5 ton/hectare (2.33 ton/acre) a year in the oasis areas in the western desert and 71–100 tons/hectare a year in areas of rain fed agriculture on the northwest coast, with wind erosion risks in these areas wavering between moderate and severe. This information was used to calculate the amount of soil erosion from wind for conventional farming and the cost is calculated according to the FAO (2014).

Parameters

Direct Costs

The direct costs are all variable factors of production, which have been broken down into different subcategories.

Raw Materials Inputs: This category represents the costs generated by inputs—that is the value of all inputs used during the production process.

Irrigation Water: The irrigation cost includes the energy cost such as diesel and electricity used for the irrigation and calculated per acre. As water is freely available to Egyptian farmers, the cost of irrigation is only related to the energy cost. The cost of irrigation water regarding electricity and diesel cost is calculated according to the Ministry of Electricity and Renewable Energy prices in 2019 as follows:

- Irrigation using electricity costs 0.75 Egyptian pound per kilowatt hour.
- Irrigation using Diesel costs 6.75 Egyptian pound per liter.

Fertilizers: This includes the cost of compost for organic farming and the cost of fertilizer for conventional farming. The price is calculated using data from the Ministry of Agriculture and Land Reclamation (MALR) for conventional farming and data from the Egyptian Bio-Dynamic Association (EBDA) for organic farming. The amount of fertilizer usage varies according to the type of crop.

Insecticides, Fungicides, and Herbicides: Conventional systems rely on pesticides (herbicides, insecticides, and fungicides), many of which are toxic to humans and animals. The data for the cost of pesticides for organic farming is assumed to be zero, as no harmful synthetic pesticides are used. The emphasis in organic agriculture is on using the inputs (including knowledge) in a way that encourages the biological processes of available nutrients and defense against pests. Most pesticides are prohibited in organic farming as they can hinder these processes. In organic agriculture, management is directed towards preventing problems, while stimulating processes that assist in nutrition and pest management. Organic agriculture uses biocontrol methods instead of synthetic pesticides.

Seed Costs: The cost of seeds is similar in conventional and organic farming. Prices were taken from the MALR and EBDA.

Other Costs: Costs that are not directly related to the manufacturing of a product or delivery of a service such as maintenance or emergency.

Labor and Machinery: Includes the total cost of labor required during the production cycle to perform farming tasks. It also includes the cost of renting machinery, as this is common in Egypt.

Certification: Cost incurred by the farmer to have his or her land certified as organic by the Organic Farming Board, which is the agency responsible for inspecting land and verifying the nature of the used growing method.

Damage Costs

The damage cost determines the amount of damage to the environment and society caused by agriculture through the unsustainable use of water, atmosphere, and soil. The environmental impacts of food wastage have been monetized. These costs are estimated via the wastage quantities and unit costs of the related environmental and social impacts. This also applies to the categories that are assessed on the basis of per-area cost data, as the area numbers related to food wastage are at the end linked to the food wastage quantities.

Water Quality: Describes the effect on water resources, occurring through the use of pesticides and fertilizer in agriculture.

Pesticides in Drinking Water Sources: These estimates are based on the removal costs of pesticides from drinking water for the United Kingdom and Thailand, as more specific information for Egypt is not available (see FAO, 2014).

Table 2.1 Cotton Cost per Acre: Detailed Breakdown of Costs and Damages

Impact category	Evaluation method	Unit value used (US$, 2012)
Water quality (Nitrate and pesticide contamination of drinking water, N/P eutrophication)	Remediation expenditures (costs of pesticide, nitrate, phosphate removal from drinking water), damage costs, WTP (willingness to pay) to avoid	N Eutrophication (based on $0.286/kg N leached in UK, correction for N input and output levels and agricultural areas in each country, and benefit transfer). P eutrophication (based on $12.32/kg P leached, correction for P input and output levels and agricultural areas in each country and benefits transfer). $78/ha (Thailand) for pesticide contamination (total 264 million in the UK, 14.6 million Thailand, corrected for toxicity levels, area, and benefits transfer[1]).

Nitrate and Phosphate in Sources of Drinking Water: These estimates are based on the removal costs of nitrate from drinking water for the UK (and Thailand), as no other data were available (FAO, 2014).

GHG Emissions: Damage cost of GHG emissions (including deforestation and managed organic soils), based on a range of approaches, damage costs, and remediation expenditure.

Table 2.2 Cost Benefit Analysis - Cotton (Old Land)

Impact category	Evaluation method	Unit value used (US$, 2012)
GHG emissions (including deforestation and managed organic soils)	The social cost of carbon (based on a range of approaches, damage costs, and remediation expenditure)	$113/tCO$_2$e (globally, no benefit transfer needed)

Soil Erosion: The cost of soil loss through wind erosion caused by food production.

Table 2.3 Conventional Total Costs 2015

Impact category	Evaluation method	Unit value used (US & 2012)
Soil erosion (owing to wind)	Damage costs (on-site and off-site)	$27.38/t for wind erosion (US values plus benefit transfer, plus per ha soil erosion levels from 48 countries and regional averages; corrected for soil erosion potential of different cultures)

Total

Total Income: The total income is calculated regarding the crop's revenue per acre and it depends on the market price per each crop. The average premium price between organic and conventional crops was 14.49% in 2019.

Table 2.4 Organic Total Costs 2015

Crops/price[2]	Rice	Maize	Potatoes	Wheat	Cotton
Conventional	6,000 LE	2,714 LE	6,000 LE	4,567 LE	16,680 LE
Organic	7,200 LE	3,250 LE	6,700 LE	5,200 LE	20,016 E

LE: Egyptian pound

Total Expenses: The total expenses are the sum of the total direct cost, which represents all variable factors of production and the total damage cost which determines the amount of damage on the environment and society caused by agriculture through the unsustainable use of water, atmosphere, and soil.

Net Benefit: The net benefit is the result of deducting the total expenses which included the direct cost and the damage cost from the total income.

Data Analysis

Here we present the calculated production costs of the five strategic crops covered by this study. This compares the cost trends of producing these crops under conventional farming and organic farming systems in old land (the Delta region) as well as in new land (reclaimed land from the desert) in Egypt during the past four years. Some crops can only be cultivated in old land; such as cotton and rice, owing to their high water consumption.

The results are presented using the previously described parameters. They include two main components of the production cost: "Direct Costs", which are costs commonly paid by the farmer during production, and "Damage Costs", which are not included in the individual cost calculation.

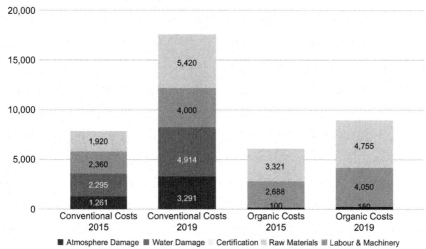

Figure 2.1 Cotton Cost per Acre: Detailed Breakdown of Costs and Damages

Results and Conclusion

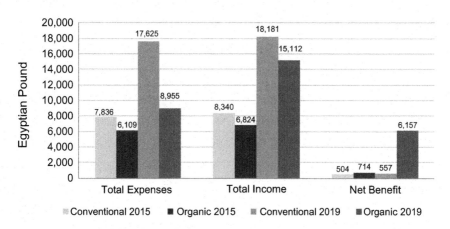

Figure 2.2 Cost Benefit Analysis – Cotton (Old Land)

The four graphs below give an overview and show the main outcome of the previously described results in 2015, while updating calculation for the results of 2019. These figures provide a comparison of the total production costs of organic and conventional farming considering all five evaluated crops.

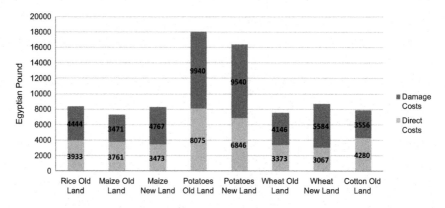

Figure 2.3 Conventional Total Costs 2015

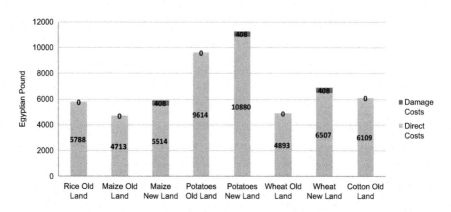

Figure 2.4 Organic Total Costs 2015

The graph outlines the higher costs for the environment and society occurring through the use of conventional farming methods, as they include higher damage costs. Organic farming enables a cost reduction for society per acre for nearly every crop evaluated in this study, because of the lower damage costs.

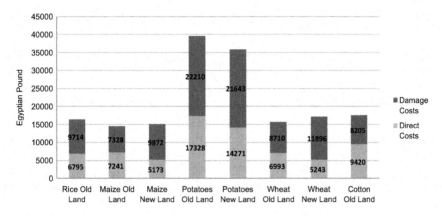

Figure 2.5 Conventional Total Costs 2019

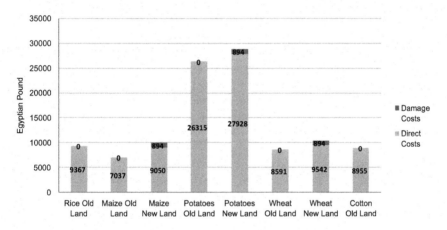

Figure 2.6 Organic Total Costs 2019

To summarize the results of the study, the figures presented give an overview of the five evaluated crops in terms of total income and total expenses, calculating the net benefit and comparing it with organic farming and conventional farming methodologies. The result of this comparison shows that the net benefit for society and the environment using conventional farming methodologies is negative, while organic farming produces a positive net benefit for the most part.

Conclusion

Organic farming has proven to be remarkably effective in reversing the negative impact of agriculture on the environment. The study concludes that although organic agriculture has a slightly higher direct input cost of production, it enables a reduction of the environmental damage costs and therefore results in better cost-effectiveness and profitability in the long term for society as a whole.

The study shows that, with regard to prices, organic food is in fact already cheaper to produce than conventional products, if the externalized costs for pollution, CO_2 emissions, energy, and water consumption are considered. These are currently transferred to society or future generations, but if they would appear on supermarket bills, this would be evident to everyone. Organic agriculture has recently gained importance as an alternative farming system, soil organic matter plays a key role in sustainable agriculture in terms of ecology and farm economics. The agricultural inputs in organic farming systems are not subsidized, but they improve the soil structure, maintain water quality, increase soil organic matter, increase biodiversity and yields while decreasing the total cost to produce one ton of any crop.

As a matter of fact, organic farming enables a cost reduction for the society for every crop evaluated in this study, because of the low damage costs included in the calculation. Even if the selling price of organic products was equal to conventional products, the organic products would still be more profitable for the farmer and cheaper for society, when including the true cost.

The Future of Agriculture in Egypt study, prepared by the Carbon Footprint Center and the Faculty of Agriculture of Heliopolis University for Sustainable Development, is endorsed by the World Future Council, IFOAM–Organics International, and the Biodynamic Federation–Demeter International e.V., which call on policymakers to advance these recommendation based on the outcomes of the study:

- 100% organic agriculture in Egypt.
- Future studies to include a more comprehensive set of indicators to measure the true cost of products (for instance results of TEEB or DALY studies).
- The Egyptian government to implement a polluter tax to reveal the true price of pollution.
- The government, researchers, and farmers to conduct such a study of all crops in Egypt.
- Governments, researchers, and farmers in all countries of the world to conduct such a study to determine the future of agriculture in their countries.
- All members of the organic and biodynamic agroecology and sustainable agriculture movement worldwide to promote these kind of studies.
- Researchers worldwide to study the economic impact of organic agriculture regarding water consumption, health and social impacts, and climate mitigation and adaptation.
- Entrepreneurs and companies worldwide to include the True Cost Accounting approach in their business models.
- Governments and education institutions worldwide to increase the education and training opportunities in organic agriculture to enable more farmers

worldwide to benefit from the economic, ecological, cultural, and social benefits of organic agriculture.

● Governments and all media representatives to increase awareness of all the benefits of organic agriculture among their citizens; and
● All consumers of the world to contribute to a better future for forthcoming generations by their educated and responsible purchase decisions today.

Notes

1 Benefit transfers are carried out as region-wide as possible. Where values for the UK and Thailand are given, UK numbers are used for developed country benefit transfer and Thailand numbers are used for developing country benefit transfer (FAO, 2014).
2 This table of prices refers to the Egyptian market prices in 2019 for organic and conventional mentioned crops per ton.

References

Auerswald, K., Kainz, M., & Fiener, P. (2003). Soil erosion potential of organic versus conventional farming evaluated by USLE modelling of cropping statistics for agricultural districts in Bavaria. *Soil use and Management*, 19(4), 305–311.

Food and Agriculture Organization of the United Nations . (2000). Economic And Financial Comparison Of Organic And Conventional Citrus-Growing Systems In Spain. http://www.fao.org/3/ac117e/ac117e00.htm.

Food and Agriculture Organization of the United Nations. (2013). Food wastage footprint: Impacts on natural resources. Available at: www.fao.org/3/i3347e/i3347e.pdf.

Food and Agriculture Organization of the United Nations. (2014). Food wastage footprint: Full-cost accounting. Available at: www.fao.org/3/a-i3991e.pdf.

Heliopolis University for Sustainable Development. (2019). Future of Agriculture in Egypt Carbon Footprint Center Egypt. Available at: www.hu.edu.eg.

Hoekstra, A.Y., Chapagain, A.K., Aldaya, M.M., & Mekonnen, M.M. (2011). *The Water Footprint Assessment Manual: Setting the Global Standard*. London: Earthscan.

Shalaby, M.Y., Al-Zahrani, K.H., Baig, M.B., Straquadine, G.S., & Aldosari, F. (2011). Threats and challenges to sustainable agriculture and rural development in Egypt: implications for agricultural extension. *The Journal of Animal & Plant Sciences*, 21(3), 581–588.

3 Upstream, Downstream

Accounting for the Environmental and Social Value of Water in the Andes

Marta Echavarria and Margaret Stern

Water and irrigation are fundamental to food production globally. Institutional and regulatory frameworks for water in Latin America prioritize agricultural use, creating perverse incentives to maximize water use without control. In general, water quantity and quality are not effectively measured nor paid for, and management decisions are neither transparent nor accountable. Therefore, True Cost Accounting (TCA) can provide a framework to better measure water stocks and flows, understand environmental limits, and inform financial decisions over time.

Three case studies from Colombia, Peru, and Ecuador and additionally Mexico, Rider B provide examples of social and environmental values of water recognized by voluntary advocacy. Payments have been accompanied by grassroots decisions to develop regional and national policies for reliable access to water. They describe how agriculture water users in the Andes, working with governmental water regulators and non-governmental organizations, have internalized the costs of source-water protection into their practices, rates, and policies. Sugarcane farmers and processing plants in Colombia's Cauca Valley pay a voluntary tariff through a centralized water fund to support upstream conservation practices that protect the source of their irrigation water. Agricultural communities in Ecuador and Peru have been empowered to work with water management districts to approach water use and protection in a more holistic fashion. Their ancestral practices and recognition of water's multiple values—a TCA approach—can improve water availability and crop production over the landscape.

Introduction

Farmers around the world are experiencing diminishing water flow, quality, and timeliness. As their lands are interconnected to the ecosystems that sustain them, cumulative environmental impacts generate unforeseen results, affecting farming viability itself. A great example of this feedback loop is water use in farming, which is the principal use of water throughout the world. The water cycle, if not well understood and managed, can lead to extreme impacts and a "tragedy of the commons" that threatens the viability of the individual. Therefore, there is a fundamental need to better understand, quantify, and price nature—in this case, water—to guarantee safeguards and warning signals so that individual and collective decisions are sustainable (Daily and Ellison, 2002).

The Andes of Colombia, Ecuador, and Peru are part of this reality. As a vital agricultural region where water is traditionally abundant and managed, people assume that water should be available for everyone, all the time. Traditional highland communities have long-term visions and values that allow them to recognize the signs of deterioration of their water sources. Over the past four decades, agriculture and urbanization have intensified across the region, forest loss has increased, and water pollution has gone unchecked. As a result, water quality and year-round flows are greatly threatened. For example, Quito, the capital of Ecuador, has seen water flow reduced by 12–34% in its more than 200 intakes (R. Osorio; personal communication, January 2020). Consequently, water has to be brought to the city from more distant sources: 24 km in 1957, 73 km in 2011, and now an estimated 110 km for drinking water projects that are being planned for the future. Command-and-control policies have been established to address these problems, but enforcement has been limited, and cultural norms tend towards informality. Therefore, water management continues to be overregulated for agricultural users who respect the law. In other cases, regulation is not respected, creating free-for-all situations. These parallel worlds generate greater disparities and promote social conflict, creating unjust and unsustainable conditions for the very poor (Boelens, 2008). Transparency is needed so that all water users and regulators work from the same information, take collective action, and enforce the same set of rules for everyone.

Water Must be Priced to Include Source Protection

Agriculture fails to pay for the true cost of water, both in terms of quantity and quality, and the protection of the ecosystems that source it. The agricultural sector, being the leading user of water, pays the least per unit of water. Typically, some components of the total cost of agricultural water, for example, source protection measures and wastewater discharge impacts, are not included nor estimated completely and effectively. Water rates tend to be flat, and there is little innovation in costing.

In the northern Andes, if agricultural water is charged, enforcement might be lax. Furthermore, the irrigation rate does not cover all its costs; the capital costs of gray infrastructure such as dikes, canals, and uptake structures required to distribute the water, are not always included. If groundwater is being pumped, depletion rates are not respected. Usually subsidized by central governments, irrigation costs do not include operation and management, and even less often are the environmental costs of irrigation systems considered. Generally, agricultural water rates do not give a price signal to promote efficiency, nor do they account for the costs of protecting water sources, meaning the range of activities needed to conserve the ecosystems sustaining them. In addition, water rates do not include the environmental impact of the discharge, and agriculture can be a substantial source of water pollution, both as point and nonpoint sources.

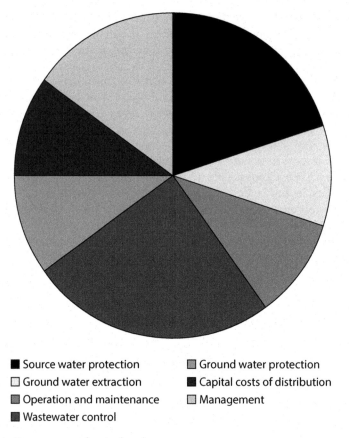

Source water protection Ground water protection
Ground water extraction Capital costs of distribution
Operation and maintenance Management
Wastewater control

Figure 3.1 Components of agricultural water cost.

Prices Must Include Ecological Values

As will be discussed in Chapter 4, there has been an evolution of terms and methodologies of TCA. As global studies have highlighted the value of nature to the global economy, greater interest has been directed towards quantifying these benefits—also known as ecosystem services—and finding ways to monetize them. Several Andean countries began promoting payments to landowners that apply sustainable management practices. Many of these beneficiaries are agricultural producers committed to protecting these landscapes (Daily and Ellison, 2002). All the ecological considerations imply values which the formal economy does not internalize. As the TEEBA-griFood framework discussed in Chapter 4 illustrates, different methods of TCA include different values in order to be more comprehensive. All methods indicate that prices of agricultural water might need to increase, which can be controversial and politically sensitive. Unlike other product prices, water rates for agriculture are not updated regularly, and the legislation does not have mechanisms in place for evaluation and renewal. For example, in Colombia rates have been litigated against for decades. The only region in the country where agricultural water rates were established—the Cauca River Valley—has also been the region where more innovative management measures have been incorporated. Strong user participation, particularly by farmers, is fundamental to price water effectively and consider additional benefits and costs. Some of the first investments to protect water supplies were established in Latin America and were promoted by farmers aware of the impending water limitations, such as the Water User Associations in Colombia or the Water Producers Program in Brazil.

Colombia: Source Water Protection by Agricultural Users

Colombian sugar cane producers in the Cauca River Valley created an institutional arrangement to value source-water protection. Inspired by French watershed organizations, a local sorghum and soy farmer and former Minister of Agriculture, Jaime Uribe, was worried about the reduced water flow of the Guabas River, owing to upstream deforestation. He invited his neighbors and friends to form a water users association, a legal construct under Colombian law, and ASOGUA-BAS was born in early 1988. This association brought together farmers with established water rights. Each farmer agreed to pay an additional voluntary fee based on the volume of water allocated in their water rights, which would go to a common fund to support upstream conservation actions. The collected money was managed by the farmers themselves, through an elected board that worked closely with the local water and environmental agency to co-finance and collaborate on activities as part of an integrated watershed management plan.

What began as a focus on purchasing land so that poor upstream farmers would not expand the agricultural frontier, evolved into a more holistic approach towards community organization and development to improve local livelihoods. Colombia has faced social conflict for more than four decades, during which time much of its rural population has migrated to urban areas, furthering the economic and social demise of rural communities in the Andean highlands. The case of ASOGUABAS inspired the creation of eight more associations on surrounding watersheds that involved hundreds of sugar cane and grain farmers and sugar processors in the area.

Currently, there are more than ten associations that work throughout the Cauca river valley. Over time, the association of sugar cane producers (ASOCANA) supported numerous initiatives to improve watershed management and socioeconomic studies and projects to improve local livelihoods. For example, the Association of water users of the Desbaratado River (ASODES) developed a new watershed management plan that involved the active participation of local communities. Rather than focusing solely on maintaining forest cover, the integrated approach taken by ASODES also empowered local stakeholders to be active participants in the planning process. As a result, the association has evolved into a comprehensive watershed management entity that has received various national and international recognitions.

In 2010 The Nature Conservancy proposed to ASOCANA to bring the disparate associations together into a common water fund to create the "Water Fund for Life and Sustainability." This fund provides financial resources to water user associations that in turn leverage more money from other entities, including multinational companies, foundations, and other donors. CENICANA, the industry's research center, coordinates hydrological monitoring throughout the Cauca river valley. Therefore, the price of water for sugar cane producers in Colombia includes the protection of its regeneration.

Water Stocks Depend on Ecology

Agriculture uses about 70–80% of water worldwide, and it is now commonly accepted that increased water scarcity is negatively affecting food production. The causes for water shortages are varied, but as human populations grow, there is greater demand for water that is often supplied with little concern for environmental safeguards, thus depleting aquifers, and leading to rivers, lakes, and wetlands drying up. This phenomenon goes largely unperceived and is happening at an alarming rate (Rodell *et al.*, 2018). As global water consumption continues to increase, more aquifers are being depleted, and more water is diverted to urban areas, contributing to a growing irrigation water deficit that impacts arid regions and countries particularly hard. We use water every day but, astoundingly, most of us do not know where our water comes from, although farmers might know better and recognize the rivers, forests, or wetland that sustain their livelihood. The ingrained assumption is that water is renewable and available at all times, especially for food production. Food provision is vital, but its production can put local water sources at risk. This is the case in the northern Andes, a water superpower.

Northern Andean Ecosystems are True Water Factories

Humid Andean grasslands, known locally as *paramos*, run from Venezuela to northern Peru and are also present in some of the mountainous regions of Costa Rica. *Paramos* are veritable water factories owing to their deep soils with high organic content that store and filter water. There is a direct relationship between healthy natural ecosystems and good water flow and quality, particularly evident in these *paramos*. The amount of rainfall and soil moisture that infiltrates to become stream flow is measured as the runoff coefficient, which tends to be at 20–40% in most soils. In *paramo* soils that act like a sponge, this value can go up to 50–70% (Buytaert, 2018). Rainfall is retained by the vegetation and soils and is released slowly over time; therefore, water management decisions in the highlands must ensure the protection of *paramos*. Another important characteristic of this region is an abundance of lakes and shallow wetlands, which also store large amounts of water and release it slowly. These upland wetland areas are vital sources of water for migrating birds and other animals and plants, food production, as well as drinking water and industrial and recreational uses downstream.

Threats to Livelihoods

Agricultural development, urban sprawl, and mining are causing the loss of vegetative cover on upper watersheds, which has a negative impact on rural communities and their landscapes and puts water flow and quality at risk. Traditional rural livelihoods that have been marginalized and threatened by modern development are losing standing, creating a rural underclass of predominantly older women and children who are left behind to eke out a living from a deteriorated environment. Communal decision-making structures common to the Andes come in conflict with institutional and legal frameworks. Ancestral water management technologies, such as water harvesting, agricultural terracing, or recharge ditches, are being left unused, leading to further water and soil loss. To increase yields, new fields are cleared for cattle and crop production, often draining wetlands, which in turn affects the viability of soils and water for future use.

Current global and regional uncertainties like rising temperatures and erratic and more intense climatic events further affect the sources of water for local communities and food production, competing with downstream demands of urban users and other agricultural producers. McDonald and Shemie (2014) found that the largest 25 cities in Latin America—a total of more than 100 million people—relied on drinking water from landscapes that are more than 40% forests, 30% cropland, and 20% native grasslands and pastures. So, it is clear that food production is a key component of the water protection discussion.

Peru: Water Has Cultural and Spiritual Values

Water is considered a living being as well as a necessity for Quechua communities in the Peruvian Andes, so its management and protection are a top priority. Farming families in Ayacucho grow native crops and raise domestic animals, principally alpacas, in arid areas above 3,500 m elevation. They were severely impacted by regional violence in the 1980s and 1990s, which seriously affected their cultural identity. These families are undergoing a process to recover their identity that has led them to reassess their traditional knowledge and apply it to improve their living conditions (Romero, 2012). Farming families in the Quispillacta region, for example, receive assistance from local professionals of the Aripaylla Bartolomé Association (ABA) who understand and identify with the daily problems and needs of families in the area. ABA's primary focus over the past 20 years has been to ensure access to clean water for domestic use, crop irrigation, and the conservation of communal pastures. Ancestral practices that have been passed on through generations and are still being used have helped these communities to maintain a water supply throughout the year (Ochoa-Tocachi and Buytaert, 2020).

In the local world view, water, soil, and other components of nature are considered to have life. Through ceremonial songs, it is necessary to "call", "store" and "carry" to receive water (Romero, 2012). The planting and harvesting of rainwater consist of the following ancestral activities:

- Rainwater storage in small lakes or water holes is surrounded by embankments built of stone and clay. This activity is ceremonial and is carried out with deep respect for the deities of the place where the water holes will be built. The objective is to plant water so that it infiltrates through the soil and feeds the aquifers that give rise to springs and the wetlands of the Andean *puna* (high-elevation dry grassland).
- Source water protection and conservation of riparian zones and other strategic sites are on the upper watershed.
- Sowing specific plants near water holes and wetlands that "call water" or are "mothers of water." Water holes and wetlands are "bred" with plants that in turn "raise water" as their presence promotes the appearance of water in new places and increases the volume of water in springs. They are protected from domestic animals with stone barriers.
- Wetland restoration and protection through activities that promote the formation and expansion of wetlands in Andean *puna* and the maintenance of water holes and the underground connections among them.
- Festive maintenance of water holes. The use and care of water is linked to ceremonial rituals celebrated between May and September. It is also linked to a high level of community organization, led by young people and children, who are responsible for the festivities and cleaning of the water holes.

The recovery of these water conservation activities has been carried out with the support of ABA and other non-governmental organizations and local government authorities, specifically decentralized water and sanitation agencies. It was necessary to revive such activities, as farmers had begun to forget them owing to their reliance on piped water and the abandonment of traditional practices.

Most of Peru's coastal and mountainous regions are extremely dry and water is critical during many months of the year. A national law (2015) designed to improve water security featuring the Mecanismo para la Retribución de Servicios Ecosistémicos (MERESE)—a mechanism to provide compensation for ecosystem services—was established to recognize the value and promote watershed conservation and management. MERESE works through voluntary agreements that support actions for the conservation, recovery, and sustainable use of natural infrastructure (or "nature-based solutions") that improves year-round water security for agriculture, industry, and domestic use. In the case of Ayacucho, MERESE activities (e.g., source-water protection, restoration of small lakes for water storage) are underway on the Cachi watershed—the result of agreements reached between the local water company and upstream communities. This renewed interest in traditional practices for water security and other ecosystem services such as agrobiodiversity conservation has re-energized local participation and the belief in ancestral knowledge transmitted through generations.

Rivers Run Dry

We have emphasized that upstream natural landscapes protect and regenerate source water for a region's water budget, which national water authorities should monitor and regulate. Water authorities should also be gauging the condition of a country's water supply, both in terms of quantity and quality, and comparing this to the needs of the population and the ecosystems that it harbors. Understanding the ecological basis of water supply and demand is an integral part of TCA, and this is beginning to happen in Latin America where instream flow regulations have been instituted in some countries but have been difficult to enforce, owing to poor capacities to manage and evaluate information on water quantity and demand. Notably, Mexico has recognized the value of environmental flows and has innovated to create water reserves that set aside an amount of water that cannot be allocated to any human use (see Water Reserves Program Mexico box below). This is an amount of water that offers protection to a river to ensure its ecological viability and its continuous provision of environmental and social benefits downstream. In contrast, most governments of Andean countries have narrowed their water management focus to investing in gray infrastructure projects to benefit, more often than not, entrenched interests. Rivers tend to be overallocated, and the water rights that are handed out exceed the river's water supply.

Box 3.1 Water Reserves Program Mexico

Water needs go beyond humans and must be considered for sustaining all life on the planet. To that end, the national water agency CONAGUA with World Wildlife Fund Mexico has been working over the past decade to redefine water management in the country by promoting the use of an environmental allocation, or now named "water reserve." This entails setting aside a volume of water in a watershed that cannot be allocated for human use, but is instead dedicated to protecting or restoring river health and ecosystems functions. These functions are now recognized as part of the Sustainable Development Goal (SDG) 6.

Currently, the Mexican Government's National Water Reserves Program (NWRP) protects environmental flows (also referred to in other legal contexts as eFlows) in 300 rivers of the country's 765 water catchments, which represent over 50% of the total runoff in the country. The program began in river basins with little conflict over freshwater resources, making it easier to establish a water reserve to maintain healthy ecological characteristics and build environmental resilience. The NWRP's success has been due to the effective collaboration among government, non-governmental organizations, and academic institutions working towards a common goal: to set aside water for nature now and to mitigate problems in the future. This innovative approach to water allocation uses science to inform public policy which in turn provides the legal basis for a water management strategy that incorporates ecosystem and biodiversity values within the value of water (SDG 15), which has great potential for replication in Latin America.

Water Management Decisions Should Respect the Watershed

Existing water management frameworks tend to be rigid and prescriptive without the necessary mechanisms for periodic evaluation and accountability, and water allocation might not be equitable or transparent. With limited monitoring capacity, water agencies can allocate water without understanding actual supply or might be lacking the necessary data. Rarely do they consider the hydrological processes supported by natural ecosystems, and sometimes watershed boundaries that demarcate the planning unit fail to be soundly incorporated into the decision-making process. Meanwhile, demand is driven by water user groups that lobby to ensure their own allocation. This sectoral approach limits an understanding of the whole and the impacts that users are likely to cause to each other.

Water management is inherently a collective process. In every watershed there are many stakeholders with distinct water needs; therefore, conflicts must be resolved, and agreements made. New digital and information technologies allow access to large amounts of data needed for water budgeting at a reduced cost. A holistic approach to water budgeting implies understanding supply and demand and their ecological limits and can improve decisions. However, this will succeed only if there is the active participation of users who commit to the application of sustainable water practices. And farmers have to be protagonists in this story, as illustrated by the cases discussed here. In the words of Brian Richter (2014), a global water expert, "To be durable and effective, water plans must be informed by the culture, economics, and varied needs of affected community members, through open, democratic dialogue and local collective action."

Ecuador: Agricultural Users are Key Players in Water Governance

An illustrative example of agricultural water governance is the El Angel watershed in northern Ecuador, where the concept of a "social watershed" was used (Jaramillo *et al.*, 2020). Water availability for agricultural use is a constant concern for thousands of rural farmers who require water for food security and family income on this watershed that covers about 75,000 hectares in the Andes. Understanding the farmers' perspective, their continuous interactions with water, and their daily exercise of governance, highlighted its complexity. The water board's commitment to daily governance stood out, although it was clearly dominated by men, based on an unspoken assumption that women should be excluded from decision-making related to water and other natural resources. Specifically, the water board was expected to be filled by long-standing male farmers who were knowledgeable about irrigation and water resource regulations and had the expertise to manage a group of agricultural water users. The work of the water board requires a considerable amount of time in meetings, procedures, tours of the premises, and visits to the water authorities, and their time and efforts should be valued in TCA. Other factors that affected water provision and governance, such as irrigation infrastructure, pollution, water theft, and delinquent payments, generated conflict on the social watershed. These issues and the actions taken towards their resolution need to be better understood and included in the cost equation:

- *Poor irrigation infrastructure*: Most field systems, irrigating through gravity or flooding, are highly inefficient and have structural flaws, having been built many decades ago and not updated or improved, having reached the end of their usefulness.
- *Inadequate waste management*: The rural communities and dispersed dwellings on the El Angel watershed have limited coverage of basic services including sewers and solid waste collection, so waste is dumped into canals and ditches, as well as natural waterways, polluting them and creating social conflicts. Irrigation water was polluted by sewage, industrial waste, agricultural waste, and sediments, which had a negative impact on crops, as well as the animals that drink the water.
- *Water theft*: The improper use of irrigation water or water theft was a generalized conflict, with greater occurrence in the dry season. Although laws and regulations established the responsibility to state entities and irrigation organizations, it was difficult to punish offenders, so frequently legal procedures fizzled out.
- *Delinquent payments*: Despite the committed work of leaders, there were delays in farmers' payments for the irrigation service, which limited investment possibilities in vital infrastructure to improve the operation of irrigation systems. Administrative and financial management experience is required to plan and assign costs for emerging actions.

Conclusion

The experiences discussed here are just samples of a long list of cases that illustrate how communities are internalizing environmental and social values within agricultural water management. ASOGUABAS was a pilot case that has evolved to become a trend, and now there is a critical mass of experiences as tracked by the environmental markets information service Ecosystem Marketplace. As reported in the State of Watershed Payments, investments surpassed $24 billion worldwide in 2016, including public sector and privately led programs. In the Latin America and Caribbean region, investments were $65.9 million, which is a relatively small amount used to finance a large number of projects (Bennett and Ruef, 2016).

As the debate evolves away from just water scarcity toward the application of the Sustainable Development Goals, there is greater impetus to apply more integrated approaches. In particular, the region has been the testing ground for the evolution and proliferation of the water fund model as a vehicle for collective action, and this is scaling globally (Ziegler *et al*, 2014).

This chapter has illustrated the watershed services that nature provides to agriculture. The incorporation of tangible and intangible values in a TCA framework is a step towards learning to manage the complexity and uncertainty of water management. Incorporating social and environmental considerations to quantify water budgets and pricing is imperative for governments to ensure water security. The process is complex because of the multiplicity of systems that are interconnected, but there are emerging lessons and models for effective collective action, with strong civil society and private sector participation that can drive innovation to scale TCA for the agricultural sector. In the spirit of the development expert Paul Polak (2008), the practical solutions presented here are based on real experiences that respect the local context.

References

Bennett, G. and Ruef, F. (2016). *Alliances for Green Infrastructure: State of Watershed Investment 2016*. Washington, DC: Ecosystem Marketplace.

Boelens, R.A. (2008). *The Rules of the Game and Game of the Rules: Normalization and Resistance in Andean Water Control*. Wageningen: Wageningen University .

Buytaert, W. (2018). Hidrología de sistemas alto andinos. Congreso Ecosistemas Acuáticos Tropicales en el Antropoceno- AQUATROP, Quito, Ecuador, June 29–30, 2018.

Daily, G.C. & Ellison, K. (2002). *The New Economy of Nature: The Quest to Make Conservation Profitable*. Washington, DC: Island Press.

Jaramillo P., Poats, S.V., & Valdospinos, C. (2020). *La gobernanza del agua de riego en la subcuenca del Río El Ángel, Ecuador*. Quito: Corporación Grupo Randi Randi.

McDonald, R.I. & Shemie, D. (2014). *Blueprint de Agua Urbana: Mapeo de Soluciones de Conservación para el Desafío Mundial del Agua*. Arlington, VA: The Nature Conservancy.

Ochoa-Tocachi, B.F. & Buytaert, W. (2020). Ancient water harvesting practices can help solve modern problems. *The Science Breaker: Science Meets Society*, 1–2.

Polak, P. (2008). *Out of Poverty: what works when traditional approaches fail*. San Francisco, CA: Berrett-Koehler Publishers.

Richter, B. (2014). *Chasing Water: A Guide for Moving from Scarcity to Sustainability*. Washington, DC: Island Press.

Rodell, M., Famiglietti, J.S., Wiese, D.N., Reager, J.T., Beaudoing, H.K., Landerer, F.W., & Lo, M.-H. (2018). Emerging trends in global freshwater availability. *Nature*, 557, 651–659. https://doi.org/10.1038/s41586-018-0123-1.

Romero, D. (2012). *Perú: Siembra y cosecha de agua en Ayacucho*. Servindi–Servicios de Comunicación Intercultural (www.servindi.or).

Ziegler, R., Partzsch, L., Gebauer, J., Henkel, M., & Lodemann, J. (2014). *Social Entrepreneurship in the Water Sector: Getting Things Done Sustainably*. Cheltenham: Edward Elgar Publishing.

Section 2

Thinking Systemically

The SARS-Coronavirus-2 disease (COVID-19) pandemic has mainstreamed our collective global need for thinking systemically. The two chapters included in this section, complemented by two text boxes, challenge us to think systemically and recognize the interrelated and intersectional dynamics across food systems. This includes not only understanding and accounting for impacts (positive and negative) across food systems and supply chains, but also acknowledging and assessing the relationships and power dynamics across the system.

What would a more systemic, relational approach to food systems look like? The contributions in this section of the book provide important insights on how and why systems thinking is a capacity that needs to be nurtured and developed over time. In Chapter 4, Harpinder Sandhu and colleagues take readers on a journey through time to look at how ecological economics and ecological accounting are being adapted and extended through True Cost Accounting (TCA) to include social, human, and health factors, together with environmental and ecological impacts. For a systems perspective, inter- and trans-disciplinary, integrative approaches and methods are required.

Salman Hussain, Michael Quinn Patton, and Pablo Vidueira in their text boxes on the United Nations Environment Programme's TEEBAgriFood initiative and Blue Marble Evaluation describe how systems thinking is not just a tool but a process. The TEEBAgriFood Food Systems Evaluation Framework and "Blue Marble Evaluation" set out guidelines and criteria to help us to think holistically and systemically about the full costs, benefits, and impacts of transformational initiatives. To illustrate this, Hussain describes how narrow assessments and singular indicators like yield and profit have led to destructive agricultural practices at the expense of more holistic solutions that provide greater value overall.

In Chapter 5, Cecilia Rocha and colleagues model systemic thinking in their analysis of the health impacts of food systems. Instead of a narrow view of health, they take an integrated, holistic view of health impacts across the food system linking human, animal, ecosystem, and planetary health. Their analysis identifies a set of practices across industrial food systems that result in negative health impacts and explores five leverage points to transform food systems. Yes, we need to correct for externalities, as the authors emphasize, but if we do this without addressing issues of equity and power, we will fail to understand and communicate the connections between impacts and across food systems.

A key theme in Chapter 5—and indeed throughout this book—is how negative impacts and externalities across the system disproportionately accrue and affect the most vulnerable and precarious workers, communities, and countries. These impacts and dynamics across the system, as the authors of Chapter 5 illustrate, are reinforced and entrenched by actors and practices across the system that benefit from the *status quo*. Systems thinking requires us to dismantle structural barriers that limit our ability for integrated analysis and action. The chapters in this section point to promising approaches, frameworks, criteria, methods, and tools that will facilitate the adoption of TCA more widely.

4   Methods and Frameworks

The Tools to Assess Externalities

Harpinder Sandhu, Courtney Regan, Saiqa Perveen and Vatsal Patel

Introduction

Demand for nutritious and sustainably grown food is increasing worldwide and is likely to grow in the future as 2 billion people are added to the existing human population by 2050 (Food and Agriculture Organization of the United Nations *et al.*, 2019). Meeting this demand for food without impacting the environment and human health is a common goal for humanity and a priority for most governments (Sandhu *et al.*, 2019, 2020; Sukhdev, 2018); The Economics of Ecosystems and Biodiversity, 2018). Yet to many observers, increasing production alone is not the solution; in fact, overproduction in the developed world is responsible for many of the negative impacts already felt. Industrial farming practices have reduced farm produce to commodities that are traded around the world. Global trade and mass movement of food as a commodity often results in highly distorted markets, volatility in food prices, and has caused massive changes in diets. Global trade often puts pressure on land use in many parts of the world and has been linked to deforestation and loss of biodiversity. All value chain stages of global food systems and many small scale agroecological systems are affected by global distortions and result in negative social and economic impacts in addition to the growing environmental impacts (The Economics of Ecosystems and Biodiversity, 2018).

In addition to the economic impacts, climate change poses one of the biggest risks to current food systems (Sandhu *et al.*, 2012). Agricultural activities are not only affected by climate change but are also responsible for one-quarter of the total greenhouse gas emissions that contributes to global warming (Intergovernmental Panel on Climate Change, 2019). Farming and food systems can play an important role in reducing impacts on the natural environment, as agriculture occupies about 38% of Earth's terrestrial surface, consumes more than 70% of global freshwater, provides employment to more than 1 billion people, and produces food for all (Food and Agriculture Organisation of the United Nations *et al.*, 2019). Farming and food system systems play a dominant role in shaping the landscapes, social, and economic aspects of communities and people around the world. However, the impacts of farming, both positive and negative, on people lives, health, social networks, and natural resources are not captured comprehensively (The Economics of

Ecosystems and Biodiversity, 2018). Such an understanding is required in order to fix deficiencies in global and local food systems and farming practices.

Unaccounted impacts of agriculture and food systems, positive and negative, need to be captured comprehensively in order to respond to the global goals of operating within the planetary boundaries (Rockstrom *et al.*, 2009). Operating within the nine planetary boundaries is essential to achieve the Sustainable Development Goals (SDGs) that include an end to poverty, zero hunger, good health and well-being, gender equality, clean water and sanitation, climate action, responsible consumption, life on land, etc. among others that are linked to agriculture (United Nations, 2015). In order to achieve the SDGs that are associated with agriculture, it is important to first examine the impacts of agricultural activities, then develop incentives to reduce them so that social, human, and natural capital in agriculture and food systems can be maintained and enhanced.

There are several approaches, methodologies, and tools that can be applied at the farm, landscape, and regional level and across food and agriculture value chains to understand the comprehensive costs and benefits of farming and food systems. One such approach that captures significant impacts and dependencies of agricultural and food systems on natural, social, and human capital is known as True Cost Accounting (TCA) (Aspenson, 2020; Lord, 2020).

This chapter describes the genesis of the TCA approach and its theoretical foundations that are established in the disciplines of Ecological Economics and Environmental Accounting. We then review and update recent scientific and economic literature in order to identify gaps in our current knowledge regarding TCA as applied to agriculture and food systems. We develop conceptual foundations of TCA in agriculture and food systems by defining key concepts, terms and methods, utility, and challenges in its application. We conclude by summarising current and ongoing work and initiatives that promote TCA.

Genesis of TCA

The discipline of economics clearly established the role of the market in economic development that also formed the basis of capitalism in the 18th century (Smith, 1776). However, by middle of 20th century, a greater realization of scarce natural resources and environmental pollution from industrial activities prompted expansion of neo-classical economics to include the impact of economic activities on the natural environment. The discipline of Environmental Economics included environmental pollution, whereas Natural Resource Economics started examining the supply and demand of natural resources (Daly and Farley, 2010). Both these disciplines encouraged Environmental Accounting, which is a sub-discipline of Accounting to account for any costs associated with the impacts of economic activities on natural resources. However, Environmental Accounting is limited in its scope, as it includes direct and indirect costs associated with the environment and does not fully capture impacts on environmental and societal health (Jasinski *et al.*, 2015).

At national and global scale, one widely used measure of wealth—gross domestic product (GDP)—uses principles of accounting to capture all goods and services produced in a given country annually. GDP is often criticized for not being inclusive in its reporting, as the significant impacts on nature and society are not part of national accounts (Costanza *et al.*, 2009). Moreover, GDP does not adequately measure well-being.

By pushing the narrowly defined boundaries and by addressing the limitations of neo-classical Economics, including Environmental Economics and Natural Resource Economics, the discipline of Ecological Economics has established the broader foundations for capturing social, environmental, and economic sustainability by focusing on the global environmental limits and societal well-being (Daly and Farley, 2010). The principles of Ecological Economics provide the scientific and economic foundations for the initiatives and approaches such as TCA, which are inclusive and orientated towards societal well-being.

Based on these foundations, some initiatives have been undertaken over the past two decades to recognize and value the benefits that biodiversity provides to people, through including natural capital into national accounting process. One such global initiative led by the United Nations Environment Program known as the Millennium Ecosystem Assessment (Millennium Ecosystem Assessment, 2005) described the importance of natural resources by highlighting the role of ecosystem services as the life support system of Planet Earth. Another process led by the United Nations (UN), widely known as the System of Environmental-Economic Accounting, is developing tools to measure the contribution of the natural resources to the economy and the impact of the economy on these resources (United Nations, 2014). Inclusive Wealth Index is another such initiative by the UNEP to estimate the comprehensive wealth of countries by expanding the scope to include natural and human capital in addition to the produced capital.

The UN Environment Program has examined these issues through the Economics of Ecosystem Services and Biodiversity (TEEB) initiative. From 2014 to the present this initiative has delved into the development of scientific and economic foundations that capture the positive and negative impacts of agriculture and food systems through a project known as The Economics of Ecosystem Services and Biodiversity in Agriculture and Food systems (see Box 4.1, TEEBAgriFood; The Economics of Ecosystems and Biodiversity, 2018). It provides a comprehensive framework to analyze costs and benefits of global farming and food systems for appropriate policy responses. The TEEBAgriFood framework provides the direction to apply TCA in agriculture and food systems to evaluate food products, agricultural systems, diets, national accounts, and policy options.

Box 4.1 The Economics of Ecosystems and Biodiversity

Salman Hussain

The Economics of Ecosystems and Biodiversity (TEEB) and TEEBAgriFood feature prominently in the current chapter and indeed the current volume.

This box aims to set out the origins of TEEB, the rationale for the TEEB-AgriFood Evaluation Framework, and—perhaps most importantly—to provide clarity on two key misconceptions. First, TEEB *is not a technical methodology*; TEEB is first and foremost a *stakeholder-led approach* to mainstreaming the values of nature. It *applies* methods from, *inter alia*, environmental science, ecological economics, and social anthropology. Second, TEEB does not commoditize nature. Nature is not "priced," and it is not "for sale." Even though TEEBAgriFood is a TCA approach, and we seek *where possible* to monetize changes in capital stocks, in some cases it is neither appropriate nor possible to do so. And yet these changes that cannot be monetized remain relevant (and are included in TEEB) as they affect human welfare.

The Origins of TEEB and TEEBAgriFood

Inspired by the Stern Review on the Economics of Climate Change (Stern, 2007), which revealed the economic inconsistency of inaction with regard to climate change, Environment Ministers from the governments of the G8+5 countries agreed at a meeting in Potsdam, Germany in 2007 to "initiate the process of analysing the global economic benefit of biological diversity, the costs of the loss of biodiversity and the failure to take protective measures versus the costs of effective conservation". Aiming to address the economic invisibility of nature, TEEB emerged from that decision.

Although its genesis is linked to climate change, and indeed "carbon sequestration and storage" is part of the TEEB typology of ecosystem services, advocacy for better outcomes is very different for biodiversity compared with climate change. Biodiversity is the living fabric of our planet, including all its ecosystems, species, and genes, in all their quantity and diversity. There is no apex indicator—no equivalent to ppm CO_2-equivalent to rally around. From the start, TEEB had to consider how to deal with *trade-offs*, and this remains the case for TCA via TEEBAgriFood.

The remit of TEEB was to "correct the economic compass"—that is, the entire economy with its many industrial sectors. The agri-food sector is an apt choice for TEEB to focus on, given its impacts and dependencies on nature. The agri-food sector encompasses areas of economic activity beyond farm operations to include farm-related activities, such as processing, manufacturing, and transport. This sector is underpinned by complex ecological and climatic systems at local, regional, and global levels, and overlaying these natural systems are social systems. These systems (economic, ecological and climatic, and social) interface and interact with each other, and that is why TCA via TEEBAgriFood assesses the "eco-agri-food systems complex."

TEEBAgriFood as a Process

The TEEBAgriFood Evaluation Framework is set out in this chapter. It can and is being applied by different stakeholders—governments, businesses, communities, and farmers—although we focus below on implementation by governments. Before the initiation of TEEBAgriFood, TEEB had developed and implemented a more generic Six-Step Approach for producing tailored economic assessments of ecosystems and biodiversity and supporting the mainstreaming of this information in policymaking on a country level included in a Guidance Manual (www.teebweb.org/media/2013/10/TEEB_GuidanceManual_2013_1.0.pdf). The TEEB six-step approach is as valid for TEEB country studies unrelated to agri-food (such as the application to land reclamation options in Manilla Bay in the Philippines) as it is to TCA/TEEBAgriFood application in-country. When presenting TCA on a national level, it is a fair *a priori* assumption that senior-level policymakers in Ministries have any number of approaches that are "pitched" to them as novel solutions to issues at the environment/social/economic interface. Among the important challenges in terms of deciding which TCA approach to adopt and which methods and data to input, it is important that TCA practitioners see the wood from the trees and are guided by its *purpose* and *audience*. The unique approaches of TCA/TEEBAgriFood (as set out in this chapter) address these challenges and are part of the "pitch" to decision-makers.

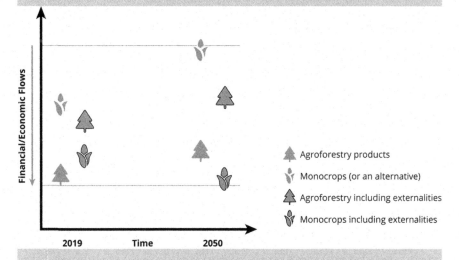

Figure 4.1 'Pitching' TEEBAgriFood to ministries: agroforestry vs monocrops.

The schematic in Figure 4.1 is the culmination of a series of slides used by the TEEB Office to highlight the TEEBAgriFood approach. The narrative that it summarizes is as follows: (i) if only financial flows that are marketed—that is, not including any externalities—are included in policy assessment then a government acting in the best interests of its citizens would pick monocrops over agroforestry today as the unoutlined maize icon (monocrops financial flows) is higher than the unoutlined tree icon (agroforestry financial flows) for 2019. However, if all positive and negative externalities and impacts are included (so the "economic" as opposed to "financial" flows), then the situation now changes—that is, in 2019 "agroforestry including externalities" is a better option than "monocrops including externalities."

If we look to 2050 projections, some modelling (or mere supposition) would show financial flows from monocropping even further outperforming agroforestry compared with 2019, but this is flawed. By 2050 monocropping will likely have depleted soil health and pushed ecosystems beyond planetary boundaries, causing declines in yield, and as a result, even these projected financial flows in 2050 likely overstate the superiority of monocropping. If we once again for 2050 (as we did for 2019) also include externalities, then agroforestry even further out-competes monocropping. In such TEEB "pitches," there is a need to validate this schematic representation above with evidence, and, depending on context, results from one of the TEEB-AgriFood are presented, for example, the World Agroforestry Centre-led study for TEEBAgriFood (http://teebweb.org/agrifood/home/agroforestry).

This is just the start–to convince Ministries that TEEBAgriFood can be a useful approach compared with the myriad alternatives. From this, the TEEB Six-Step Approach is applied to determine via a stakeholder-driven participatory process what policy options TEEBAgriFood should be applied to. This process is critical. TCA/TEEBAgriFood should not be a technical analysis looking for a question; rather, it should be a policy question as formulated by an end user to which the TCA/TEEBAgriFood framework is applied, adapted to the specific economic, political, social, and ecological context. If (and only if) stakeholders have been involved in and thus take ownership of this process from project inception will the results of the TCA/TEEBAgriFood application have any chance of creating material change.

A part of the TEEB Six-Step Approach is to determine the constituency of "winners" and "losers" were a policy intervention to be adopted. This is important for two reasons: first, losers will tend to resist and/or block change, and in pragmatic terms it is important to be aware of this, as all decisions have a political dimension: just because the TCA reveals (say) that agroforestry improves net natural, social, human, and social capital compared with oil palm does not mean that it will be promoted; second, if the constituency of losers includes those in society that are poor with few or no alternative livelihood options then this is important vis-à-vis the changes that we advocate for.

The TEEB process has been successful. The interim TEEBAgriFood study for Indonesia contributed to cacao agroforestry being included in the 2020 Five Year Development Plan for Indonesia—the first time that it has been. TCA and TEEBAgriFood can make an impact, but only if considered as a process rather than just a technical methodology.

References

Stern, N.H. (2007). The Economics of Climate Change: The Stern Review. Cambridge: Cambridge University Press. doi:10.1017/CBO9780511817434.'

Conceptual Foundations of TCA in Agriculture and Food Systems

Agriculture and food systems, being extremely diverse and complex, require an assessment approach that can capture all impacts and dependencies (The Economics of Ecosystems and Biodiversity, 2018). Unlike extractive industries, agriculture, and food systems include physical, human, and social inputs. The outcomes of current agriculture and food systems are increasingly linked to various chronic diseases such as cancers, obesity, pesticide poisoning, etc. Agriculture and food systems are also embedded in social systems (Pretty, 2003). Therefore, a transdisciplinary approach is required to estimate all costs and benefits of agriculture and food systems (Sandhu *et al.*, 2019).

The TEEBAgriFood framework provides a conceptual basis of TCA (Figure 4.2). It extends our current understanding of estimating environmental accounts and includes social and human health impacts. It is based on a systems approach. All economic, biological, and social components of agriculture and food systems are part of the TCA method. Four forms of capitals that are associated with TCA in agriculture and food systems are described below (Table 4.1).

Produced Capital

Produced capital is based on the concept measured in the Inclusive Wealth Report by the UN University's International Human Dimensions Programme on Global Environmental Exchange and the United Nations Environment Programme (United Nations University – International Human Dimensions Programme on Global Environmental Exchange and United Nations Environment Programme, 2014) and defined by the TEEBAgriFood Report (The Economics of Ecosystems and Biodiversity, 2018). The stocks and flows associated with produced capital are measured by concepts and definitions of accounting standards at farm level, landscape level, and corporate level (processing), by using definitions from the System of National Accounts.

Natural Capital

Natural capital includes natural resources such as air, water, soil, and biodiversity associated with agriculture. Natural capital can be measured by using

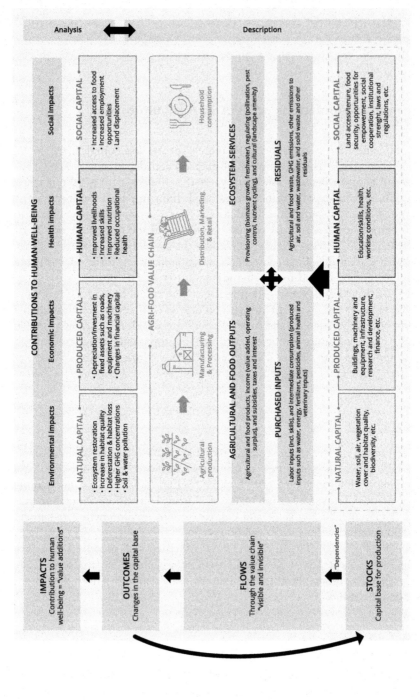

Figure 4.2 Elements of the TEEBAgriFood evaluation framework.

the United Nations System of Environmental-Economic Accounting (United Nations, 2014).

Social Capital

Social capital is defined as the features of social life, networks, norms, and trust that enable participants to act together more effectively to pursue shared objectives (Putnam, 1993; Sandhu *et al.*, 2020; The Economics of Ecosystems and Biodiversity, 2018). Its four key features are relations of trust; reciprocity and exchanges; common rules, norms, and sanctions; and connectedness in networks and groups (Pretty, 2003).

Human Capital

Human capital comprises an individual's health, knowledge, skills, and motivation that are essential for productive work. It is based on the premise that individuals and society derive economic benefits from investments in people (Sandhu *et al.*, 2020; Sweetland, 1996; The Economics of Ecosystems and Biodiversity, 2018).

Utility of Environmental Accounting Approaches

This section summarizes current environmental-accounting approaches that are being used to measure and understand the environmental impacts of different operations, as can and are being used in TCA. There are several approaches, methods, and models that are currently being used to capture environmental impacts of processes and products in various industries (Bebbington *et al.*, 2001; Elkington, 1999; United Nations, 2014). Over the past several decades, Environmental Management Accounting (EMA) systems have been developed that use several approaches such as environmental cost accounting, full cost accounting, and environmental balanced scorecard (Bebbington *et al.*, 2001, 2007; Jasch, 2003; Jasinski *et al.*, 2015). These are summarized in Table 4.2.

Application of TCA Approach

The TCA approach can be used to identify benefits and costs associated with agriculture and food systems that are not captured in general accounting frameworks and tools. TCA includes all environmental, social, and health-related costs and benefits of agriculture and food systems. TCA often uses damage function approach (damage costs) and the cost of control approach (avoidance, restoration, abatement, and maintenance costs) to estimate the true cost of food production through the value chain, as demonstrated by the TEEBAgriFood framework. It can be used to:

- Develop sustainable agriculture and food systems by first understanding all externalities and then by reducing them;
- Develop sound policy responses for just agricultural and food policies;

Table 4.1 Four Types of Capital, Stocks, and Flows Associated with Agriculture and Food Systems

Capital	Stocks	Flows (+/-)
Produced	All manufactured/built capital such as farm buildings, machines and equipment, physical infrastructure (roads, irrigation systems), processing plant, storage, warehouses, retail stores etc; knowledge and intellectual capital embedded in, for example, seed development, fertilisers, agrochemicals, GM/hybrid seed, etc.; and financial capital such as farm loans, investment, insurance, etc.	Rent, all inputs, output
Natural	Soil, water, biodiversity, atmosphere	Water runoff, aquifer recharge, local climate regulation by carbon sequestration, regulation of atmospheric chemical composition, soil erosion control, role vegetative cover plays in soil retention, nitrogen fixation, nutrient cycling, biological control of pests/diseases, greenhouse gas emissions from farm, damage to water resources, soil resources, ecosystems and biodiversity, honey bee and pollination losses and gains, loss of beneficial predators by pesticides application, biological control of pests and diseases, fish kills owing to pesticides, bird kills owing to pesticides
Social	Farming group networks, partnerships with research and development, individual links, market linkages	Loss of labour, small family farms
Human	Farmers knowledge, proficiency in farm practices, use of software, health, employment opportunities	Loss of traditional knowledge, impacts on health of farmers, consumers

Note:
* Stocks of capitals are accumulated over time.
* Flows are processes over a period of time. Flows can be described in the form of ecosystem services, agricultural inputs and output, and any residual flows such as pollution and greenhouse gas emissions.

- Reduce impacts on natural resources and operate within the planetary boundaries;
- Justify better appreciation, valuation, and payments to farmers;
- Enable consumers to support food that is sustainably grown and has lower impacts;

Table 4.2 Types of Accounting Approaches

Approach	Definition	Costs	Reference
Cost accounting	Accounting for direct and indirect costs.	Direct + Indirect costs	United Nations, 1953
Environmental cost accounting	The system of accounting for estimating direct environmental costs of a product or a process. These costs include fossil energy use, materials obtained from nature, wastewater and solid waste that directly affects the environment	Internal (Direct) + External (Indirect) costs	Bebbington *et al.*, 2001; Madu, 2001; United Nations *et al.*, 2014
	Life cycle costing includes all stages of the value chain. It includes production, processing, manufacturing, distribution, consumption and recycling. This approach is focused on the use of resources and reflects both internal and external costs. Internal costs include cost of materials, energy, labour, capital etc., whereas external costs include environmental impact of the processes, cost of pollution, cost of health-problems and social costs.		Howes, 2003
	Activity based costing includes the costs of each activity that is required to produce a product. This approach helps to divide environmental costs (by products), composition of the environmental costs and strengthen the environmental cost management of the operations.		Almeida & Cunha, 2017; Hoozee & Hansen, 2018.
	Material flow costing measures the flows and stocks of materials in manufacturing. It produces accounts in both physical and monetary units. Material cost accounting helps organizations to improve their business efficiency and reduce environmental impacts.		Christ & Burritt, 2015

(Continued)

Table 4.2 (Cont.)

Approach	Definition	Costs	Reference
Full cost accounting	A system that explicitly includes all direct and indirect costs and benefits of a transaction. Most of the tools in EMA measure direct costs of pollution, but full cost accounting includes indirect costs as well. These indirect costs and benefits are incurred by the direct beneficiaries or any third parties involved in a transaction. It includes conventional business costs, environment costs and social costs of products or services	Internal (Direct) + External (Indirect) costs + Social costs	Bebbington et al., 2001; Elkington, 1999; Gale & Stokoe, 2001; Jasinski et al., 2015
Environmental balanced scorecards	An accounting method to assess social, economic, and environmental performance of an organization. It includes both financial and non-financial performance and allows organizations to look at their business from five perspectives; economic, social, internal business, learning, and growth and environmental.	Economic + Social + Learning and growth + Internal process + Environmental	Al-Zwyalif, 2017; Kalender and Vayvay, 2016; Kaplan and Norton, 1992; Krivo-kapić and Jovanović, 2009; Sandhu, 2013; Sandhu et al., 2014

- Incentivise agricultural practices that are less detrimental to the environment and human health and to penalise those that have high impacts;
- Help better accounting of natural capitals in national accounts for further investment in their management;
- Protect traditional food systems;
- Enable integration of biodiversity into agricultural landscapes;
- Help achieve Sustainable Development Goals.

Challenges in Applying TCA

TCA is a more comprehensive and up to date approach that can be used from farm scale to national policy level. However, there are several limitations and challenges in the adoption of the tool that are further discussed here.

Data Source and Collection

One of the key challenges in applying TCA is data source and data collection. There are lots of data that are already available in scientific literature and in farm accounts that are very helpful in TCA analysis (Soil & More Impacts and

TMG Thinktank for Sustainability, 2020). However, a process to streamline data collection is required with some uniform standards at local, regional, and global scale. Working with farmers to share their farm accounts is also a nuanced process with many safeguards required.

Complexities of Value Chains

Agriculture products and food systems have extremely complex value chains, ranging from locally produced to locally consumed fresh food to global chains such as those for cereals, coffee, cocoa, cotton etc. Owing to these complexities, it becomes difficult to trace the entire chain from production to consumption to develop a comprehensive understanding of all positive and negative impacts. Even if the impacts are estimated for one stage of value chain, the response might not be sufficient to reduce overall impacts.

Inclusion of Health and Social Impacts

Impact assessments by business are mostly focused on improving the natural capital base for their businesses as a part of creating positive value for shareholders. Therefore, they focus on tracking and reducing carbon emissions, as this is the global focus as well. However, the impacts on biodiversity, water, oceans, health of their workers, health of consumers, etc. is rarely on the agenda of such impact assessments. A transdisciplinary approach is required to expand the scope from biophysical impact assessment to health and social impacts through the value chain.

Target Audience

TCA is a tool that can be useful for producers, supermarkets, agri-businesses, and governments. However, there are challenges in scoping each study based on its target audience. If farmers want to use it to correct their detrimental farming practices, then the scope is limited to farm scale. In contrast, if a supermarket wants to raise awareness of the food that they sell, they need to have a wider scope for applying TCA. At a governmental level, TCA needs to include policy assessment.

Uniform Standards and Practices

A lack of uniform international standards is one of the key challenges for TCA application in agriculture. There are several iterations and ways in which TCA approaches are being used. Some organizations are conveniently using it to create value for their organization and demonstrate positive values created by their operations, in terms of water as a capital stock, for example. Other organizations use it to start a conversation with prospective sustainability issues. A lack of international standards allow organizations to use and misuse TCA for their own advantage. Such practices can reduce the utility of TCA.

Incentives in Market and Policy

Currently, some markets provide incentives to organic food with premium prices that are paid by consumers. Beside this there are no such incentives at farm, market, or national agricultural policies to apply TCA and to understand impacts.

Consumer Awareness

Consumers are not aware of all impacts of food systems. However, there is an increase in the number of consumers who demand full disclosure in how food is produced. TCA can help them to understand these impacts and then support food products that are less damaging. But there is a need to raise awareness amongst consumers about the utility of TCA as a comprehensive tool.

Legal Framework

A lack of policy at national and global level also means that there is no existing legal framework to advance the use and implementation of TCA through the value chain of each agriculture and food product.

Conclusion

TCA uses a systems approach, building on the existing environmental cost accounting framework. It extends its scope to include social, human, and health impacts in addition to environmental impacts in order to develop more sustainable agriculture and food systems. Development of a TCA approach is a first step in advancing methodology to analyze current systems to better understand and improve them. Further development of international standards followed by policy response through appropriate market incentives and national agriculture policies will help in the adoption of TCA applications more widely. This has the potential to assist the global community with operating agriculture and food systems within planetary boundaries and advancing the well-being of farming communities around the world.

Farming and food systems can be made more resilient to climate change. There are several ongoing initiatives that promote resilience to climate change and consider agricultural landscapes as multifunctional landscapes that provide multiple ecological, social, and community benefits. Transition to agroecological, regenerative, and circular agriculture are ongoing efforts in many parts of the world. This is a positive outcome of applying a TCA approach to identify and minimize the negative impacts of human, social, and natural capital in agriculture and food systems. Recognizing and measuring all positive and negative externalities by using TCA is not an end but a beginning towards more equitable and sustainable agriculture and food systems and the achievement of the well-being of society at large.

References

Almeida, A., & Cunha, J. (2017). The implementation of an Activity-Based Costing (ABC) system in a manufacturing company. *Procedia Manufacturing*, 13, 932–939. doi:10.1016/j.promfg.2017.09.162.

Al-Zwyalif, I. M. (2017). Using a Balanced Scorecard Approach to Measure Environmental Performance: A Proposed Model. *International Journal of Economics and Finance*, 9(8), 118–126. doi:10.5539/ijef.v9n8p118.

Aspenson, A. (2020). "True" Costs for Food System Reform: An Overview of True Cost Accounting Literature and Initiatives. Baltimore, MD: John Hopkins University. Available at: https://clf.jhsph.edu/publications/true-costs-food-system-reform-overview-true-cost-accounting-literature-and-initiatives.

Bebbington. J., Gray, R., Hibbitt, C., & Kirk, E. (2001). *Full Cost Accounting: An Agenda for Action*. London: Certified Accountants Educational Trust.

Bebbington, J., Brown, J., & Frame, B. (2007). Accounting technologies and sustainability assessment models. *Ecological Economics*, 61(2–3),224–236. doi:10.1016/j.ecolecon.2006.10.021.

Christ, K. L., & Burritt, R. L. (2015). Material flow cost accounting: a review and agenda for future research. *Journal of Cleaner Production*, 108(B), 1378–1389. doi:10.1016/j.jclepro.2014.09.005.

Costanza, R., Hart, M., Posner, S., & Talberth, J. (2009). *Beyond GDP: The Need for New Measures of Progress*. The Frederick S. Pardee Center for the Study of the Longer-Range Future. Available at: https://pdxscholar.library.pdx.edu/cgi/viewcontent.cgi?article=1010&context=iss_pub.

Daly, H.E., & Farley, J. (2010). *Ecological Economics, Second Edition: Principles and Applications*. Washington, DC: Island Press.

Elkington, J. (1999). *Cannibals with Forks: The Triple Bottom Line of 21st Century Business*. London: John Wiley & Son.

Food and Agriculture Organization of the United Nations, International Fund for Agriculture Development, UNICEF, World Food Programme, & World Health Organization (2019). *The State of Food Security and Nutrition in the World 2019. Safeguarding Against Economic Slowdowns and Downturns*. Rome: FAO. Available at: www.fao.org/3/ca5162en/ca5162en.pdf.

Hoozee, S., & Hansen, S.C. (2018). A Comparison of Activity-Based Costing and Time-Driven Activity-Based Costing. *Journal of Management Accounting Research*, 30(1), 143–167. doi:10.2308/jmar-51686.

Howes, R. (2003). *Environmental Cost Accounting: An Introduction and Practical Guide*. Wokingham: CIMA Publishing.

Intergovernmental Panel on Climate Change. (2019). Climate Change and Land: an IPCC special report on climate change, desertification, land degradation, sustainable land management, food security, and greenhouse gas fluxes in terrestrial ecosystems [Summary for Policymakers].Available at: www.ipcc.ch/site/assets/uploads/sites/4/2020/02/SPM_Updated-Jan20.pdf.

Jasch, C. (2003). The use of Environmental Management Accounting (EMA) for identifying environmental costs. *Journal of Cleaner Production*, 11(6), 667–676. doi:10.1016/S0959-6526(02)00107-5.

Jasinski, D., Meredith, J.O., & Kirwan, K. (2015). A comprehensive review of full cost accounting methods and their applicability to the automotive industry. *Journal of Cleaner Production*, 108(A), 1123–1139. doi:10.1016/j.jclepro.2015.06.040.

Kalender, Z.T. & Vayvay, Ö. (2016). The Fifth Pillar of the Balanced Scorecard: Sustainability. *Procedia - Social and Behavioral Sciences*, 235, 76–83. doi:10.1016/j. sbspro.2016.11.027.

Kaplan, R.S. & Norton, D.P. (1992). The Balanced Scorecard—Measures that Drive Performance. *Harvard Business Review*, (January–February 1992). Available at: https:// hbr.org/1992/01/the-balanced-scorecard-measures-that-drive-performance-2.

Krivokapić, Z. & Jovanović, J. (2009). Using Balanced Scorecard to Improve Environmental Management System. *Strojniški vestnik, Journal of Mechanical Engineering*, 55(4), 262–271.

Lord, Steven. (2020). *Valuing the impact of food: towards practical and comparable monetary valuation of food system impacts. A report of the Food System Impact Valuation Initiative (FoodSIVI)*. University of Oxford, Food System Transformation Group, Environmental Change Institute. United Kingdom. Available at: https://foodsivi.org/wp-content/uploads/2020/06/Valuing-the-impact-of-food-Report_Foodsivi.pdf.

Madu, C.N. (2001). Environmental Cost Accounting and Business Strategy (pp. 119–136). In *Handbook of Environmentally Conscious Manufacturing.*. New York, NY: Springer. doi:10.1007/978-1-4615-1727-6_6.

Millennium Ecosystem Assessment. (2005). *Millennium Ecosystem Assessment Synthesis Report*. Washington, DC: Island Press.

Pretty, J. (2003). Social Capital and the Collective Management of Resources. *Science, 3029*(5652), 1912–1914. doi:10.1126/science.1090847.

Putnam, R. (1993). The prosperous community: Social Capital and Public Life. *The American Prospect*, 13, 35–42.

Rockstrom, J., W. Steffen, K. Noone, A. Persson, F.S. Chapin, III, E. Lambin, T. M. Lenton, M. Scheffer, C. Folke, H. Schellnhuber, B. Nykvist, C.A. De Wit, T. Hughes, S. van der Leeuw, H. Rodhe, S. Sorlin, P.K. Snyder, R. Costanza, U. Svedin, M. Falkenmark, L. Karlberg, R.W. Corell, V.J. Fabry, J. Hansen, B. Walker, D. Liverman, K. Richardson, P. Crutzen, & J. Foley. (2009). Planetary Boundaries: Exploring the Safe Operating Space for Humanity. *Ecology and Society*, 14(2), 32. Available at: www. ecologyandsociety.org/vol14/ iss2/art32.

Sandhu, H., Nidumolu, U., & Sandhu, S. (2012). Assessing Risks and Opportunities Arising from Ecosystem Change in Primary Industries Using Ecosystem Based Business Risk Analysis Tool. *Human and Ecological Risk Assessment: An International Journal*, 18(1), 47–68. doi:10.1080/10807039.2012.631469.

Sandhu, S. (2013). Towards an Integrated Conceptual Framework for Corporate Social and Environmental Sustainability. In Wells, G. (Ed.), *Sustainable Business: Theory and Practice of Business Under Sustainability Principles* (pp. 19–38). Cheltenham: Edward Elgar.

Sandhu, S., McKenzie, S., & Harris, H. (Eds.) (2014). *Linking Individual and Global Sustainability*. Dordrecht: Springer Netherlands. doi:10.1007/978-994-017-9008-6.

Sandhu, H., Müller, A., Sukhdev, P., Merrigan, K., Tenkouano, A., Kumar, P., Hussain, S., Zhang, W., Pengue, W., Gemmill-Herren, B., Hamm, M.W., Tirado von der Pahlen, M.C., Obst, C., Sharma, K., Gundimeda, H., Markandya, A., May, P., Platais, G., & Weigelt, J. (2019). The future of agriculture and food: Evaluating the holistic costs and benefits. *The Anthropocene Review*, 6(3), 270–278. doi:10.1177/ 2053019619872808.

Sandhu, H., Scialabba, N.E., Warner, C., Behzadnejad, F., Keohane, K., Houston, R., & Fujiwara, D. (2020). Holistic costs and benefits of corn production systems in Minnesota, US. *Scientific Reports*, 10(3922), 1–12. doi:10.1038/s41598-020-60826-5.

Smith, A. (1776). *An Inquiry into the Nature and Causes of the Wealth of Nations.* Chicago, IL: University of Chicago Press.

Soil & More Impacts & TMG Thinktank for Sustainability. (2020). True Cost Accounting: Inventory Report. Global Alliance for the Future of Food. Available at: https://futureoffood.org/wp-content/uploads/2020/07/TCA-Inventory-Report.pdf.

Sukhdev, P. (2018). Smarter metrics will help fix our food system. *Nature*, 558, 7. doi:10.1038/d41586-018-05328-1.

Sweetland, S.R. (1996). Human Capital Theory: Foundations of a Field of Inquiry. *Review of Educational Research*, 66(3), 341–359. https://doi.org/10.3102%2F0034 6543066003341.

The Economics of Ecosystems and Biodiversity. (2018). TEEB for Agriculture & Food: Scientific and Economic Foundations. UN Environment. Available at: http://teebweb.org/agrifood/scientific-and-economic-foundations-report/.

United Nations. (1953). *A System of National Accounts and Supporting Tables, Studies in Methods.* United Nations Department of Economic Affairs, Statistical Office. Series F, No 2. Rev. 1. New York, NY: United Nations Publications.

United Nations. (2015). Transforming our world: the 2030 Agenda for sustainable development. Available at: www.un.org/ga/search/view_doc.asp?symbol=A/RES/70/1&Lang=E.

United Nations, European Commission, Food and Agriculture Organization of the United Nations, Organisation for Economic Co-operation and Development, & The World Bank. (2014). System of Environmental-Economic Accounting 2012–Experimental Ecosystem Accounting. Available at: https://unstats.un.org/unsd/envaccounting/seearev/seea_cf_final_en.pdf.

United Nations University–International Human Dimensions Programme on Global Environmental Exchange and United Nations Environment Programme. (2014). *Inclusive Wealth Report 2014.* Cambridge: Cambridge University Press.

5   Health Impacts

The Hidden Costs of Industrial Food Systems

Cecilia Rocha, Emile Frison and Nick Jacobs

Health Impacts: The Hidden Costs of Industrial Food Systems

Voluntary exchanges in food markets should benefit everyone. However, the price we pay for food often does not adequately reflect the full cost to society of producing or consuming it. Economists use the term "externality" to refer to the costs that are not incorporated into market prices. They are the hidden costs of our food systems. Such externalities are evidence of "market failure," meaning that the market has failed to convey—through prices—the true social value of food. An example would be the health consequences (skin lesions, respiratory problems, cancers, miscarriages, and birth defects) of high levels of toxic agrochemicals used in banana production in many areas of Latin America. The price of these bananas, sold in supermarkets around the world, does not reflect the health costs borne not only by the workers in banana plantations, but also by members of their families, and their communities.

In *Unravelling the Food-Health Nexus* (2017), IPES-Food, in collaboration with the Global Alliance for the Future of Food, undertook a systemic analysis of the negative health impacts of industrial food systems, occurring through five channels: i) occupational hazards; ii) environmental contamination; iii) contaminated, unsafe and altered foods; iv) unhealthy dietary patterns; and v) food insecurity. Below we summarize these findings, underline the challenges with regard to taking action to correct these market failures, and reflect on leverage points for overcoming these challenges and building healthier food systems.

Five Channels of Impact

Food systems impact human health through five broad channels, with most of these impacts linking back to the same underlying *industrial* food system practices (IPES-Food, 2017):

Impact Channel 1: Occupational Hazards

Firstly, a variety of occupational hazards can be identified across food systems. Farmers' and agricultural workers' exposure to harmful substances, dangerous working conditions and general insecurity have been extensively documented and associated with a range of health problems such as increased rates of musculoskeletal disorders, respiratory diseases, skin disorders, certain cancers, poisoning by chemicals and heart-related illnesses (Anderson and Athreya, 2015; Cavalli *et al.*, 2019). Acute accidental and intentional (suicidal) pesticide poisoning is common, especially in developing countries.

The risk of occupational injury and death is much higher in agriculture, fishing, and forestry than in any other work environment. In the USA, fatal work injury rates per 100,000 full-time equivalent workers in 2018 were 23.4 in agriculture and almost 80 in fishing, which is well above the average for all industries at 3.5 deaths (US Department of Labor, 2019). But the food manufacturing sector also presents high rates of injuries and fatalities (Neff, 2014), as a result of high-stress environments that can cause anxiety, depression, mental illness, and even lead to suicide (Lunner Kolstrup *et al.*, 2013).

The plight of migrant workers is of particular concern. The harsher working conditions, language and communication barriers, lack of safety instructions and training, and poorly maintained equipment are factors accounting for the higher occupational injury rates for foreign workers in many countries (Moyce and Schenker, 2018).

Impact Channel 2: Environmental Contamination

Severe health impacts arise via the exposure of whole populations to contaminated environments "downstream" of food production, including the pollution of soil, air, and water resources.

Endocrine disrupting chemicals (EDCs) pose one of the greatest challenges for public health; EDCs are chemicals that interfere with hormonal systems, and there are almost 800 chemicals known or suspected to function as EDCs (World Health Organization and United Nations Environment Programme, 2013). EDCs are found in the pesticides used in conventionally grown crops; hormones used in the production of meat, poultry, and dairy products; chemicals used to coat canned foods and in some plastic containers; compounds used as food preservatives; and even substances in non-stick cookware[1] (Wielogórska *et al.*, 2015). Potentially harmful effects can be generated from very low concentrations and from very short periods of exposure (Khetan, 2014). Growing scientific evidence shows that exposure to these chemicals can lead to adverse reproductive outcomes (infertility, cancers, malformations), as well as impacts on thyroid function, and neuroendocrine and neurodevelopmental functions (Gore *et al.*, 2015; World Health Organization and United Nations Environment Programme, 2013). Studies have also linked EDCs with increased rates of obesity and susceptibility to type-2 diabetes (Gore *et al.*, 2015).

Agriculture also has an important impact on air quality, potentially causing respiratory illness, cerebrovascular disease, ischemic heart disease, chronic obstructive pulmonary disease, and lung cancer (Lelieveld *et al.*, 2015). Agricultural emissions of ammonia through livestock production and fertilizer use are an important factor in the high levels of air pollution in densely populated urban areas (Gu *et al.*, 2014; Paulot and Jacob, 2014). Agriculture has been identified as the largest contributor to air pollution in many regions of the world, including Europe, Russia, Turkey, Korea, Japan, and the Eastern USA (Lelieveld *et al.*, 2015). In several European countries, agricultural sources are responsible for as much as 40% of air pollution and its associated health burden (Lelieveld *et al.*, 2015).

Box 5.1 Global Systems Evaluation and True Cost Accounting

Michael Quinn Patton, Pablo Vidueira

Transformation has become the clarion call. Humans are using the Earth's resources at levels, scales, and speed that are changing Earth's ecological systems and, in so doing, warming, polluting, and degrading the environment at a level that threatens the future survival of humanity. Addressing food systems transformation also means addressing systemic issues like inequality, social justice, climate change, and poverty. Food systems transformation is affected by related global emergency challenges and trends including climate change, virulent infectious diseases, pollution of land, air, and water, millions of displaced people, ever more severe weather, species extinction, and biodiversity loss.

Evaluating these complex and dynamic agriculture and food systems interactions cannot be reduced to simple yardsticks like per hectare productivity of a single crop or the intake of nutrients. The question, then, is what framework and criteria should be used for evaluating food systems transformation. Currently, the most influential and widely used criteria internationally are those adopted and disseminated by the Development Assistance Committee (DAC) of the Organisation for Economic Co-operation and Development. Revised in 2019, the criteria call for interventions to be judged by their relevance, effectiveness, efficiency, impact, sustainability of impacts, and coherence. These mainstreamed criteria carry the message that incremental changes through projects and programs (essentially closed systems) are adequate, as they were decades ago. The DAC criteria are not aligned with the nature of complexity or the magnitude and urgency of transformations needed. Global systems evaluation and True Cost Accounting (TCA), in contrast, offer criteria appropriate for and aligned with transformation. Both advocate comprehensive, multi-faceted, and holistic evaluation through assessing systemic costs and benefits.

Global Systems Evaluation: Introducing Blue Marble Evaluation

Blue Marble Evaluation (BME) (Patton, 2019) has been developed by evaluation leaders as a leading-edge approach supporting the design, implementation, evaluation, and adaptation of systems transformation processes and initiatives. BME encourages users to embrace and act on a whole-Earth perspective by looking beyond nation-state borders and across sectors and issue silos to connect the global and local, the human and ecological, and the macro and the micro. In doing so, evaluative thinking and diverse evaluation methods can be joined with systems thinking to illuminate system dynamics, boundaries, inter-relationships, and diverse perspectives.

BME invites application of four overarching principles to evaluating systems transformations (Patton, 2019).

1   Apply whole-Earth, big picture thinking to all aspects of systems change. Global problems require global interventions and, correspondingly, globally orientated design, implementation, evaluation, learning, and adaptation
2   Know and face the realities of the Anthropocene and act accordingly. Sustainability, resilience, equity, diversity, inclusion, health, and power dynamics need to be addressed to undertake interventions and evaluations knowledgeably and credibly in the context of the Anthropocene
3   Engage consistently with the magnitude, direction, and speed of transformations needed and envisioned. We are no longer in the incrementalism era. Major and rapid systems transformations are urgently needed in response to global, anthropogenic problems
4   Ensure transformational scale and scope. Transformation requires multiple interventions and actions on many fronts undertaken by diverse but interconnected actors

These evaluation principles direct attention to how systems are established and operate, including who benefits and who is disadvantaged.

True Cost Accounting as a Global Systems Evaluation Framework

In evaluating the true costs and benefits of creating goods and offering services, True Cost Accounting (TCA) offers a framework for examining not just input-output efficiency but broader impacts on the environment, economy, polity, society, and culture, especially the interdependencies between human and natural systems. Systems design and evaluation using systems principles and criteria means looking beyond the traditional efficiency criterion to systems transformation criteria. TCA provides a framework for documenting, assessing, and acting on whether, how, and to what extent activities affect economic, social, cultural, and environmental systems. For example, applied to specific farming systems, this means identifying, quantifying (when possible), and making transparent all costs and benefits of producing food in a particular way, beyond the traditional value chain perspective: from the prices farmers receive, to the affordability for consumers, including externalities like impacts on the environment, communities, and human health. This requires evaluation using criteria like equity, ecological sustainability, environmental health, community well-being, resilience, and regenerative/adaptive capacity. That is how global systems transformations can be evaluated comprehensively.

Vignette 1: Removing Regulatory Restrictions on Polluting Rivers, Lakes, and Farmland

In 2020 the Trump Administration asserted that many environmental regulations were too costly to producers, reflecting a closed-system, narrow efficiency judgment. Full-cost accounting from a global systems perspective would take into account the longer-term environmental and human health costs of pollution.

Vignette 2: True Cost Accounting and the COVID-19 Pandemic

Flattening the coronavirus disease (COVID-19) pandemic's curve focused on the health system's capacity to prevent infections and identify and treat those infected. When the magnitude of the pandemic led to economic and social shutdowns, poorer people, especially, were faced with food shortages. A lack of access to health care, increased unemployment, and income losses exacerbated food insecurity around the world. Agricultural and food systems were dramatically affected by food transportation restrictions and broken connections between producers of food and consumers. Hunger and malnutrition increased vulnerability to the coronavirus. TCA and systems perspectives support a broader understanding of the pandemic's impacts and could contribute to holistic solutions that provide short-term relief while contributing to longer-term systemic solutions that address multiple threats and crises.

In summary, global systems evaluation and TCA both advocate comprehensive, multidimensional, and holistic approaches by assessing systemic costs and benefits. TCA is, of course, an ideal—and therefore idealistic. But that is a purpose of criteria—to define standards and ideals. Addressing systems transformation criteria helps us to think holistically and systemically about the full costs, benefits, and impacts of transformational initiatives.

References

Organization for Economic Co-operation and Development Development Assistance Committee Network on Development Evaluation (2019). Better Criteria for Better Evaluation: Revised Evaluation Criteria Definitions and Principles for Use. Available at: www.oecd.org/dac/evaluation/revised-eva luation-criteria-dec-2019.pdf.

Patton, M.Q. (2019). Blue Marble Evaluation: Premises and Principles. New York, NY: Guilford Publishing.

Impact Channel 3: Contaminated, Unsafe, and Altered Foods

Illnesses also systematically arise from the ingestion of foods containing various pathogens, while risks are also linked to compositionally altered and novel foods.

Food-borne disease agents fall into distinct categories—most importantly bacteria (of which many have developed antimicrobial resistance), viruses, chemical agents and toxins (e.g., EDCs), and parasites—and can cause a variety of illnesses upon ingestion, ranging from gastrointestinal and diarrheal illnesses to influenza-like, respiratory, and neurological symptoms, allergies, as well as terminal sicknesses (Newell *et al.*, 2010).

Although all final consumers are exposed to foodborne pathogens, the probability of contracting illnesses with serious consequences is higher for vulnerable populations such as children, pregnant women, the elderly, and immune-compromised individuals (Lund, 2015; Yeni *et al.*, 2016). The World Health Organization also highlights that, in a regional comparison, persons living in low-income sub-regions of the world are disproportionally affected by food-borne diseases (World Health Organization, 2015).

Many bacterial, viral, and parasitic disease agents are zoonotic—that is, they are transmitted through fecal matter or direct contact with animals or their meat (Larsen *et al.*, 2014; World Health Organization, 2015). Therefore, disease outbreaks frequently originate in the consumption of meat, poultry, and animal products such as eggs and unpasteurized (or poorly pasteurized) milk, cheeses, and other dairy products (Doyle *et al.*, 2015; Painter *et al.*, 2013). However, zoonotic pathogens can also spread to non-animal products, for instance through the use of untreated manure on cropland, contaminated irrigation water, the runoff from livestock operations, and wildlife intrusion (Strawn *et al.*, 2013).

Recent changes in the scale and organization of food production and distribution have exacerbated food-borne disease risks. First, the movement of farm animals and foods across a global market makes traceability and the upholding of appropriate microbiological safety procedures increasingly complicated, while increased human mobility also leads to rapid spread of microbes and parasites (Carstens *et al.*, 2019; Manitz *et al.*, 2014; McEntire, 2013; Robertson *et al.*, 2013), as evidenced during the recent COVID-19 pandemic. Second, in many countries, diets are shifting to include greater shares of out-of-home consumption and the use of semi-prepared ingredients, leading to additional disease transmission channels through food preparation in restaurants and food services (Callejon *et al.*, 2015; Doyle *et al.*, 2015; Gould *et al.*, 2015). Finally, the changing climate could bring novel vectors into newly temperate climates or create temperature-related changes in contamination levels (Newell *et al.*, 2010).

In addition to environmental exposure (see above), pesticide residues on foods represent a major health risk, including in regions with stringent controls in place. For example, 83% of EU soils contain one or more pesticides residues, 58% contain mixtures (Silva *et al.*, 2018), and residues are regularly found in samples (European Food Safety Authority, 2020). Furthermore, reliable methods for assessing the health impacts of "cocktails" of different pesticide residues are still lacking (Reffstrup *et al.*, 2010).

Industrial contaminants in foods also represent a major health risk. An estimated total of 98.8% of the food consumed globally is estimated to be grown in unsustainable soils which have been degrading over the years (Kopittke, 2019). The use of wastewater for irrigation, particularly in regions with intensive mining and smelting activities, has led to dangerous levels of heavy metals such as mercury, lead, and cadmium in soils and water sources, while the intensification of livestock agriculture has led to increased concentrations of arsenic, zinc, and copper (Lu *et al.*, 2015).

Impact Channel 4: Unhealthy Dietary Patterns

Health impacts also occur through consumption of specific foods or groups of foods with problematic health profiles. Unhealthy dietary patterns are characterized by high consumption of added sugar, sodium, saturated fat, and transfat, and low consumption of fruit, vegetables, pulses, whole grains, and nuts (Global Panel on Agriculture and Food Systems for Nutrition, 2016; Willett et al., 2019). Unhealthy diets have become increasingly prevalent in recent decades and are linked to the growth of overweight and obesity rates. The growing prevalence of obesity is a global health concern, as it heralds increasing incidence of several debilitating diseases, including type 2 diabetes, hypertension, coronary heart disease, metabolic syndrome, respiratory conditions, cancer, and osteoarthritis, as well as reproductive, gallbladder, and liver diseases (Butland *et al.*, 2007; Grundy, 2016; Swinburn, et al., 2019; Wang *et al.*, 2011). In 2015 dietary factors accounted for 56% of cardiovascular disease deaths in men and 48% in women across Europe (Srour *et al.*, 2019).

Overweight and obesity have reached epidemic levels in many countries. Since 1975 the worldwide prevalence of obesity has nearly tripled, with 39% of adults estimated to be overweight and 13% to be obese in 2016 (World Health Organization, 2017). Some 38 million children under the age of five years (UNICEF *et al.*, 2018) and over 340 million children under the age of 18 are now overweight or obese (World Health Organization, 2017). Non-communicable diseases are now the leading cause of death globally, responsible for 71% of all deaths, 85% of those in low- and middle-income countries (World Health Organization, 2018). The global prevalence of diabetes (closely linked with the rise in obesity) is estimated to be 6.4% among adults aged 20–79 years. The International Diabetes Federation estimates that one in 11 adults have diabetes, with 79% of cases from low- and middle-income countries (International Diabetes Federation, 2019).

Unhealthy diets often contain a significant amount of processed foods that are high in calories but low in nutrients. The increased global availability of ultra-processed "convenience" foods is one of the biggest drivers of dietary change leading to overweight, obesity, and chronic non-communicable diseases (Kaveeshwar and Cornwall, 2014; Monteiro *et al.*, 2013; Rauber *et al.*,

2018). A recent large observational prospective study concluded that higher consumption of ultra-processed foods was associated with higher risks of cardiovascular, coronary heart, and cerebrovascular diseases (Srour *et al.*, 2019); new forms of ultra-processing to produce "fast carbs" are affecting the molecular structure of food in ways that disturb the digestive and hormonal systems (Kessler, 2020). Worldwide, increased globalization of food manufacturers, supermarket chains, and fast-food restaurants has contributed to the increased supply of high-sugar/fat energy-dense foods at reduced prices (An *et al.*, 2019).

Impact Channel 5: Food Insecurity

Finally, health impacts occur through insufficient or precarious access to healthy and culturally acceptable foods and diets.

One of the goals of a sustainable food system is food security—"a situation that exists when all people, at all times, have physical, social, and economic access to sufficient, safe, and nutritious food that meets their dietary needs and food preferences for an active and healthy life" (Food and Agriculture Organization of the United Nations, 1996). People are food insecure when they cannot—or are at risk of not being able to—access a healthy diet. Insufficient diets result from insufficient access to/intake of calories or micronutrients.

In 2010 over 820 million people worldwide were facing hunger, which had been rising in parts of Africa, Latin America, and Asia. Another 2 billion people were experiencing moderate to severe food insecurity (Food and Agriculture Organization of the United Nations, International Fund for Agricultural Development, UNICEF, World Food Programme, and World Health Organization, 2019). These numbers are now expected to increase further as a consequence of the economic crisis in the wake of COVID-19 (Chan, 2020; World Food Programme, 2020). The Zero Hunger target set under the Sustainable Development Goals looks very unlikely to be reached by 2030.

Some of the health effects of undernourishment are well known. Inadequate intake of calories and proteins (protein-energy malnutrition—PEM) is the leading cause of death in children in developing countries (Global Panel on Agriculture and Food Systems for Nutrition, 2016). Low birth weights, stunted growth, and compromised neurodevelopmental capacity are also common consequences of PEM. Long-term health consequences of childhood exposure to hunger include greater risks for conditions such as asthma and depression in adolescence and early adulthood (Kirkpatrick *et al.*, 2010; McIntyre *et al.*, 2013). Micronutrient malnutrition increases the risk of infections and infectious diseases, as it weakens the immune system (Schaible and Kaufman, 2007). Overall, food insecure individuals are more likely to have poorer health (Vozoris and Tarasuk, 2003).

Challenges in Addressing Market Failures in Food Systems

The health impacts of food systems are thus severe, wide-ranging, and costly in human and economic terms, with recent studies such as the FAO's "MARCH" project (Food and Agriculture Organization of the United Nations, 2017) helping to advance the economic valuation of these impacts.

The obvious solution is to correct these market failures through a new set of policy incentives, including taxes on products generating negative externalities (such as sugar-sweetened beverages), and/or regulation or banning of toxic substances (such as proven EDCs and harmful pesticides). Similarly, subsidies for agroecological production or no taxation of "healthy food," such as organically produced fruits and vegetables, should be implemented. While careful consideration of cultural/political environments is required, the precautionary principle in favor of public health must be applied, particularly in cases of toxic substances leading to life-long disabilities and death.

However, comprehensive policy reforms to correct market failures and address health risks in food systems remain the exception around the world. Rather than increasing efficiency and equity, government policies often privilege those generating negative externalities and disadvantage those (businesses, consumers, and civil society organizations), providing extra social benefits through their actions. In other words, market failure is being compounded by policy failure. Often, the point of contention is what is considered "enough evidence." In IPES-Food's 2017 review, we found that the collective strength, consistency, plausibility, and coherence of the scientific evidence has established a solid basis for action. The evidence has since grown in all of these areas. For example, the perilous situation of food- and farmworkers has been broadly recognized in the wake of the COVID-19 pandemic.

This points to a deeper problem rooted in the political economy of food systems. Market failures and policy failures in food systems are underpinned by what could be considered as "evidence failure"—not a *lack* of evidence, but rather challenges with regard to our ability to see the full picture of food system impacts, understand the connections between impacts and across food systems, and communicate them at the science-policy interface (IPES-Food, 2017).

Firstly, there are major blind spots in the evidence base. The biggest health risks tend to accrue to vulnerable groups and precariously employed workers (women, indigenous people, migrant workers, even children), who are less likely to report injuries and illnesses for fear of termination or victimization, or for lack of knowledge of their right to medical services (Cavalli *et al.*, 2019). In turn, employers could have a financial incentive to under-report injuries and illnesses that occur on their premises in order to lower their workers' compensation insurance payments. Risks to farmers and farmworkers in developing countries are generally under-documented. These blind spots make it less likely for problems to be prioritized politically, allowing health risks to continue to afflict marginalized populations. This is compounded by a broader disconnection of the general public from the process of food production.

Secondly, food systems are complex, and health impacts cannot always be traced back to their origin. For example, zoonotic pathogens and antimicrobial resistance can spread through multiple pathways within and around food systems. Chronic exposure to EDCs is hard to trace to specific sources or even to specific chemicals. Many of the health impacts described above are compounded by factors like climate change, poverty, and unsanitary conditions, which are reinforced by industrial food and farming practices.

Thirdly, powerful actors—from multinational agribusinesses to governments and donors—have the power to establish the narratives that frame the problems and the solutions in food systems (IPES-Food, 2016; IPES-Food, 2017). These "solutions"—from biofortification to climate-smart agriculture—are premised on further industrialization and standardization of food systems. They ignore the role of current systems in driving health risks (e.g., by perpetuating poverty and climate change) and thereby obscure the extent and severity of health impacts in food systems.

Leverage Points for Building Healthier Food Systems

What is required in building healthier food systems is nothing short of revisiting the fundamental pillars and underlying assumptions of the industrial food and farming model. Five leverage points were identified in the 2017 IPES-Food report to help to break the current cycles, addressing the deficits of public awareness, scientific evidence, and political will in combination:

- **Leverage Point 1—Promoting Food Systems Thinking:** We must systematically bring to light the multiple connections between different health impacts, between human health and ecosystem health, between food, health, poverty, and climate change, and between social and environmental sustainability.
- **Leverage Point 2—Reasserting Scientific Integrity and Research as a Public Good:** Research priorities, structures, and capacities need to be fundamentally realigned with principles of public interest and public good, and the nature of the challenges we face (i.e., cross-cutting sustainability challenges and systemic risks).
- **Leverage Point 3—Bringing the Alternatives to Light:** We need to know more about the positive health impacts and positive externalities of alternative food and farming systems (e.g., agroecological crop and livestock management approaches that build soil nutrients, sequester carbon in the soil, or restore ecosystem functions such as pollination and water purification).
- **Leverage Point 4—Adopting the Precautionary Principle:** The complexity of food systems is real and challenging, but cannot be an excuse for inaction in the face of threats of serious or irreversible damage to human health.
- **Leverage Point 5—Building Integrated Food Policies under Participatory Governance:** The monumental task of building healthier food systems requires more democratic and more integrated ways of managing risk and governing food systems. A range of actors—policymakers, big and small private sector firms, healthcare providers, environmental groups,

consumers' and health advocates, farmers, agri-food workers, and citizens—must collaborate and take shared ownership of this endeavor.

Collectively, these steps can provide a new basis for action to build healthier food systems, including taxation of externalities. A systemic approach to food and health has been translated into concrete policy proposals in the shape of IPES-Food's advocacy for a Common Food Policy for the EU (see Box 5.2). Integrated food policies, in particular, can provide the framework for ensuring that policies combine in a way that puts health, equity, and sustainability at the heart of our food systems and redistributes power in the process.

Box 5.2 A Common Food Policy for the EU

At the European level, health and nutrition goals have typically been addressed in isolation from broader food and agriculture policies, focused on specific actors and parts of the chain (e.g. retailers, schools, consumers), and limited to guidance and voluntary action. In contrast, the Common Food Policy blueprint—co-developed with over 400 food system actors—establishes *healthy agro-ecosystems* and *healthy sustainable diets for all* as two of the five overarching objectives for EU food systems (IPES-Food, 2019). In doing so, it requires actions on the supply and demand-side to be packaged together, and identifies sector-specific reforms (e.g. to the EU's Common Agricultural Policy and Trade policies) in relation to these overarching objectives. One proposal requires EU Member States to adopt "healthy food environment plans"—including action on social policies, education, urban planning, zoning and licensing, marketing, public procurement, and beyond—as a condition for continuing to receive agricultural subsidies. Widespread calls for an integrated food policy have been taken up by the European Commission in the "Farm to Fork Strategy" launched in May 2020, which places renewed emphasis on health and nutrition and recognizes the need to look beyond consumer responsibility and build healthy "food environments" (European Commission, 2020).

Note

1 While impacting health through environmental contamination, EDCs are also linked to health impacts as occupational hazards (Impact Channel 1) and contaminated foods (Impact Channel 3).

References

An, R., Shen, J., Bullard, T., Han, Y., & Qiu, D. (2019). A scoping review on economic globalization in relation to the obesity epidemic. *Obesity Reviews*, 21(3).

Anderson, M., & Athreya, B. (2015). Improving the well-being of food system workers. In Global Alliance for the Future of Food (Ed.), *Advancing Health and Well-Being in Food System: Strategic Opportunities for Funders* (pp. 108–127).Butland, B., Jebb, S., Kopelman, P., McPherson, K., Thomas, S., Mardell, J., & Parry, V. (2007). *Tackling obesities: future choice —project report.* Government Office for Science of the United Kingdom, London. Available at: www.safefood.eu/SafeFood/media/SafeFoodLibrary/ Documents/Professional/All-island%20Obesity%20Action%20Forum/foresight-report-full_1.pdf.

Callejon, R.M., Rodriguez-Naranjo, M.I., Ubeda, C., Hornedo-Ortega, R., Garcia-Parrilla, M.C., & Troncoso, A.M. (2015). Reported foodborne outbreaks due to fresh produce in the United States and European Union: trends and causes. *Foodborne Pathogens and Disease.* 12, 32–38. doi:10.1089/fpd.2014.1821.

Carstens, C.K., Salazar, J.K., & Darkoh, C. (2019). Multistate outbreaks of foodborne illness in the United States associated with fresh produce from 2010 to 2017. *Frontiers in Microbiology*, 10, 2667. doi:10.3389/fmicb.2019.02667.

Cavalli, L., Jeebhay, M.F., Marques, F., Mitchell, R., Neis, B., Ngajilo, D., & Watterson, A. (2019). Scoping global aquaculture occupational safety and health. *Journal of Agromedicine: Special Issue from the Fifth International Fishing Industry Safety and Health Conference (IFISH 5)*, 24(4), 391–404. doi:10.1080/1059924X.2019.1655203.

Chan, S.P. (2020) Global economy will suffer for years to come, says OECD. , March 23, BBC News. Available at: www.bbc.com/news/business-52000219.

Doyle, M.P., Erickson, M.C., Alali, W., Cannon, J., Deng, X., Ortega, Y., Smith, M.A., & Zhao, T. (2015). The food industry's current and future role in preventing microbial foodborne illness within the United States. *Clinical Infectious Diseases: An Official Publication of the Infectious Diseases Society of America*, 61(2), 252–259. doi:10.1093/cid/civ253.

European Commission. (2020). A Farm to Fork Strategy for a fair, healthy and environmentally-friendly food system. Brussels, Communication from the Commission to the European Parliament, the Counil, the European Economic and Social Committee and the Committee of the Regions. Available at: https://eur-lex.europa.eu/legal-content/EN/TXT/?uri=CELEX:52020DC0381.

European Food Safety Authority. (2020). National summary reports on pesticide residue analysis performed in 2018. Available at: https://doi.org/10.2903/sp.efsa.2020. EN-1814.

Food and Agriculture Organization of the United Nations. (1996). Rome Declaration on World Food Security and World Food Summit Plan of Action. Available at: www.fao.org/3/w3613e/w3613e00.htm.

Food and Agriculture Organization of the United Nations. (2017). Healthy people depend on healthy food systems: Methodology for valuing the Agriculture and the wider food system Related Costs of Health (MARCH).

Food and Agriculture Organization of the United Nations, International Fund for Agricultural Development, UNICEF, World Food Programme, & World Health Organization. (2019). The State of Food Security and Nutrition in the World 2019. Safeguarding against economic slowdowns and downturns. Rome, FAO. Available at: www.fao.org/3/ca5162en/ca5162en.pdf.

Global Panel on Agriculture and Food Systems for Nutrition. (2016). Food systems and diets: Facing the challenges of the 21st century. Global Panel on Agriculture and Food Systems for Nutrition, London. Available at: http://glopan.org/sites/default/ files/ForesightReport.pdf.

Gore, A.C., Chappell, V.A., Fenton, S.E., Flaws, J.A., Nadal, A., Prins, G.S., Toppari, J., & Zoeller, R.T. (2015). Executive summary to EDC-2: the endocrine society's second scientific statement on endocrine-disrupting chemicals. *Endocrine Reviews*, 36(6), 593–602. doi:10.1210/er.2015-1093.

Gould, L.H., Rosenblum, I., Nicholas, D., Phan, Q., & Jones, T.F. (2015). Contributing factors in restaurant-associated foodborne disease outbreaks, FoodNet Sites, 2006 and 2007. *Journal of Food Protection*, 76(11), 1824–1828. doi:10.4315/0362-028X.JFP-13-037.

Grundy, S.M. (2016). Metabolic syndrome update. *Trends in Cardiovascular Medicine*, 26(4), 364–373. doi:10.1016/j.tcm.2015.10.004.

Gu, B., Sutton, M.A., Chang, S.X., Ge, Y., & Chang, J. (2014). Agricultural ammonia emissions contribute to China's urban air pollution. *Frontiers in Ecology and the Environment*, 12(5), 265–266. doi:10.1890/14.WB.007.

International Diabetes Federation. (2019). *IDF Diabetes Atlas* (9th ed.). Available at: www.diabetesatlas.org.

IPES-Food. (2016). From uniformity to diversity: a paradigm shift from industrial agriculture to diversified agroecological systems. International Panel of Experts on Sustainable Food Systems. Available at: www.ipes-food.org/_img/upload/files/UniformityToDiversity_FULL.pdf.

IPES-Food. (2017). Unravelling the food-health nexus: addressing practices, political economy, and power relations to build healthier food systems. The Global Alliance for the Future of Food and IPES-Food. Available at: https://futureoffood.org/wp-content/uploads/2017/10/FoodHealthNexus_Full-Report_FINAL.pdf.

Kaveeshwar, S. & Cornwall, J. (2014). The current state of diabetes mellitus in India. *Australasian Medical Journal*, 7(1), 45–48. doi:10.4066/AMJ.2014.1979.

Kessler, D. (2020). *Fast Carb, Slow Carbs: The Simple Truth about Food, Weight and Disease*. New York, NY: HarperCollins.

Khetan, S. (2014). *Endocrine Disruptors in the Environment* (1st ed.). Hoboken, NJ: Wiley.

Kirkpatrick, S.I., McIntyre, L., & Potestio, M.L. (2010). Child hunger and long-term adverse consequences for health. *Archives of Pediatrics & Adolescent Medicine*, 164(8), 754–762. doi:10.1001/archpediatrics.2010.117.

Kopittke, P.M., Menzies, N.W., Wang, P., McKenna, B.A., & Lombi, E. (2019). Soil and the intensification of agriculture for global food security. *Environment International*, 132, 105078. doi:10.1016/j.envint.2019.105078.

Larsen, M.H., Dalmasso, M., Ingmer, H., Langsrud, S., Malakauskas, M., Mader, A., Moretro, T., Mozina, S., Rychli, K., Wagner, M., Wallace, R.J., Zentek, J., & Jordan, K. (2014). Persistence of foodborne pathogens and their control in primary and secondary food production chains. *Food Control*, 44, 92–109. doi:10.1016/j.foodcont.2014.03.039.

Lelieveld, J., Evans, J.S., Fnais, M., Giannadaki, D., & Pozzer, A. (2015). The contribution of outdoor air pollution sources to premature mortality on a global scale. *Nature*, 525(7569), 367–371. doi:10.1038/nature15371.

Lu, Y., Song, S., Wang, R., Liu, Z., Meng, J., Sweetman, A.J., Jenkins, A., Ferrier, R.C., Li, H., Luo, W., & Wang, T. (2015). Impacts of soil and water pollution on food safety and health risks in China. *Environment International*, 77, 5–15. doi:10.1016/j.envint.2014.12.010.

Lund, B.M. (2015). Microbiological food safety for vulnerable people. *International Journal of Environmental Research and Public Health*, 12(8), 10117–10132. doi:10.3390/ijerph120810117.

Lunner Kolstrup, C., Kallioniemi, M., Lundqvist, P., Kymäläinen, H., Stallones, L., & Brumby, S. (2013). International perspectives on psychosocial working conditions, mental health, and stress of dairy farm operators. *Journal of Agromedicine: A Global Perspective on Modern Dairy: Occupational Health and Safety Challenges and Opportunities,* 18(3), 244–255. doi:10.1080/1059924X.2013.796903.

Manitz, J., Kneib, T., Schlather, M., Helbing, D., & Brockmann, D. (2014). Origin detection during food-borne disease outbreaks - A case study of the 2011 EHEC/HUS outbreak in Germany. *PLoS Currents,* 6. doi:10.1371/currents.outbreaks. f3fdeb08c5b9de7c09ed9cbcef5f01f2.

McEntire, J. (2013). Foodborne disease: the global movement of food and people. *Infectious disease clinics of North America,* 27(3), 687–693. doi:10.1016/j.idc.2013.05.007.

McIntyre, L., Williams, J.V.A., Lavorato, D.H., & Patten, S. (2013). Depression and suicide ideation in late adolescence and early adulthood are an outcome of child hunger. *Journal of Affective Disorder.* 150(1), 123–129. doi:10.1016/j.jad.2012.11.029.

Monteiro, C.A., Moubarac, J.C., Cannon, G., Ng, S.W., & Popkin, B. (2013). Ultra-processed products are becoming dominant in the global food system. *Obesity Reviews,* 14(S2), 21–28. doi:10.1111/obr.12107.

Moyce, S.C., & Schenker, M. (2018). Migrant workers and their occupational health and safety. *Annual Review of Public Health,* 39(1), 351–365. doi:10.1146/annurev-publhealth-040617-013714.

Neff, R. (Ed.). (2014). *Introduction to the US Food System: Public Health, Environment, and Equity.* Hoboken, NJ: Wiley.

Newell, D.G., Koopmans, M., Verhoef, L., Duizer, E., Aidara-Kane, A., Sprong, H., Opsteegh, M., Langelaar, M., Threfall, J., Scheutz, F., van der Giessen, J., & Kruse, H., (2010). Food-borne diseases—the challenges of 20 years ago still persist while new ones continue to emerge. *International Journal of Food Microbiology,* 139(S1), S3–S15. doi:10.1016/j.ijfoodmicro.2010.01.021.

Painter, J.A., Hoekstra, R.M., Ayers, T., Tauxe, R.V., Braden, C.R., Angulo, F.J., & Griffin, P.M. (2013). Attribution of foodborne illnesses, hospitalizations, and deaths to food commodities by using outbreak data, United States, 1998–2008. *Emerging infectious diseases,* 19(3), 407–415. https://doi.org/10.3201/eid1903.111866.

Paulot, F., & Jacob, D.J. (2014). Hidden cost of U.S. agricultural exports: Particulate matter from ammonia emissions. *Environmental Science & Technology,* 48(2), 903–908. doi:10.1021/es4034793.

Rauber, F., da Costa Louzada, M.L., Steele, E.M., Millett, C., Monteiro, C.A., & Levy, R. B. (2018). Ultra-processed food consumption and chronic non-communicable diseases-related dietary nutrient profile in the UK (2008–2014). *Nutrients,* 10(5), 587. doi:10.3390/nu10050587.

Reffstrup, T.K., Larsen, J.C., & Meyer, O., (2010). Risk assessment of mixtures of pesticides. Current approaches and future strategies. *Regulatory Toxicology and Pharmacology,* 56(2), 174–192.

Robertson, L.J., van der Giessen, J.W.B., Batz, M.B., Kojima, M., & Cahill, S. (2012). Have foodborne parasites finally become a global concern? *Trends in Parasitology,* 29(3), 101–103. doi:10.1016/j.pt.2012.12.00.

Schaible, U.E., & Kaufmann, S.H.E. (2007). Malnutrition and infection: Complex mechanisms and global impacts. *PLoS Medicine,* 4(5), e115. doi:10.1371/journal. pmed.0040115.

Silva, V., Mol, H.G., Zomer, P., Tienstra, M., Ritsema, C.J. and Geissen, V., 2019. Pesticide residues in European agricultural soils–A hidden reality unfolded. *Science of the Total Environment*, 653, 1532–1545. doi:10.1016/j.scitotenv.2018.10.441.

Srour, B., Fezeu, L.K., Kesse-Guyot, E., Allès, B., Méjean, C., Andrianasolo, R.M., Touvier, M. (2019). Ultra-processed food intake and risk of cardiovascular disease: Prospective cohort study (NutriNet-santé). *BMJ (Clinical Research Ed.)*, 365(365), l1451. doi:10.1136/bmj.l1451.

Strawn, L.K., Fortes, E.D., Bihn, E.A., Nightingale, K.K., Gröhn, Y.T., Worobo, R.W., … Bergholz, P.W. (2013). Landscape and meteorological factors affecting prevalence of three food-borne pathogens in fruit and vegetable farms. *Applied and Environmental Microbiology*, 79(2), 588–600. doi:10.1128/AEM.02491-12.

Swinburn, B.A. et al. (2019) the global syndemic of obesity, undernutrition and climate change: The *Lancet* Commission report. *The Lancet*, 393(10173):791–846.

UNICEF, World Health Organization, & World Bank Group. (2018). *Levels and trends in child malnutrition: Key findings of the 2018 edition*. Available at: https://data.unicef. org/wp-content/uploads/2018/05/JME-2018-brochure-.pdf.

U.S. Department of Labor, Bureau of Labor Statistics. (2019). News Release: National Census of Fatal Occupational Injuries in 2018. December 17, 2019.

Vozoris, N.T. & Tarasuk, V.S. (2003). Household food insufficiency is associated with poorer health. *The Journal of Nutrition*, 133(1), 120–126. doi:10.1093/jn/133.1.120.

Wang, Y.C., McPherson, K., Marsh, T., Gortmaker, S.L., & Brown, M. (2011). Obesity 2: Health and economic burden of the projected obesity trends in the USA and the UK. *The Lancet*, 378(9793), 815.

Wielogórska, E., Elliott, C.T., Danaher, M., & Connolly, L. (2015). Endocrine disruptor activity of multiple environmental food chain contaminants. *Toxicology in Vitro*, 29(1), 211–220. doi:10.1016/j.tiv.2014.10.014.

Willett, W.et al. (2019) Food in the Anthropocene: the EAT-Lancet Commission on healthy diets from sustainable food systems. *The Lancet*, 393(10170): 447–492.

World Food Programme. (2020). COVID-19 will double number of people facing food crises unless swift action is taken. Available at: www.wfp.org/news/covid-19-will-double-number-people-facing-food-crises-unless-swift-action-taken.

World Health Organization. (2015). WHO estimates of the global burden of foodborne diseases. Available at: www.who.int/foodsafety/publications/foodborne_disease/fergreport/en.

World Health Organization. (2017). Climate change and human health: biodiversity. Available at: www.who.int/globalchange/ecosystems/biodiversity/en.

World Health Organization. (2018). Non-communicable diseases: key facts. Available at: www.who.int/news-room/fact-sheets/detail/noncommunicable-diseases.

World Health Organization & United Nations Environment Programme. (2013). State of the science of endocrine disrupting chemicals—2012: An assessment of the state of the science of endocrine disruptors prepared by a group of experts for the United Nations Environment Programme (UNEP) and WHO. Available at: www.who.int/ceh/publications/endocrine/en.

Yeni, F., Yavaş, S., Alpas, H., & Soyer, Y. (2016). Most common foodborne pathogens and mycotoxins on fresh produce: A review of recent outbreaks. *Critical Reviews in Food Science and Nutrition*, 56(9), 1532–1544. doi:10.1080/10408398.2013.777021.

Section 3

From the Field

How can True Cost Accounting (TCA) be relevant to the millions of farmers and food producers worldwide? The sheer diversity of farming and food systems globally make this a central challenge and opportunity for TCA as it becomes an established approach, practice, and tool. A common thread through the contributions in this section is the importance of building resilience across farming landscapes to address climate change, biodiversity loss, food insecurity and malnutrition, and sustainable livelihoods. In the multifunctional landscapes described by the authors of these chapters, farmers and food providers are at the heart of this resilience, providing essential services beyond healthy food that result in myriad ecological, social, and community benefits.

The authors provide concrete strategies to address the negative impacts of agriculture by supporting or incentivizing positive change from the ground up. In Chapter 6 Patrick Holden and Adele Jones criticize existing farm assessment, certification, and labelling schemes for failing to provide an accurate understanding of impacts that can be compared across different production practices. In the current system, farmers are penalized for pursuing sustainability certification through higher relative costs and complex assessment and monitoring. Holden and Jones are working to harmonize sustainability assessment tools and certification schemes in the UK and beyond and provide a frank and fresh reflection on this process, its challenges and opportunities. They propose a nested approach where on-farm sustainability assessments join up with high level food system frameworks, including the United Nations Sustainable Development Goals, the Natural Capital Protocol, and the TEEBAgriFood Evaluation Framework.

In Chapter 7, Gábor Figeczky and co-authors use the example of the organic sector as an illustration of how governments can support and incentivize a variety of public goods stemming from agriculture. Instead of subsidizing practices that degrade the environment, a number of regulatory tools—policy, taxation, payments, and programs—could be enlisted to reduce nitrogen runoff and CO_2 emissions, build soil fertility, and enhance biodiversity. They provide multiple examples from around the world where governments at different jurisdictional levels have taken action to support organic practices: for example, South Korea, India, Costa Rica, Switzerland, the USA, the European Union,

Italy, Norway, Denmark, and Mexico have all advanced policies of some kind to limit the negative effects of agriculture and incentivize positive practices. But the authors point to significant structural barriers and vested interests across the system that prevent systemic change.

In Chapter 8 (Transforming the Maize Treadmill), four examples of maize systems—from the USA, Mexico, Malawi, and Zambia—are provided. This chapter describes the multiple agroecological and sociocultural impacts of maize systems in very different contexts, upholding a strong case for transforming these systems. Several key messages emerge from this chapter and the case studies: how infrastructure, programs, and policies entrench certain kinds of production (like corn production in Minnesota), despite evidence of negative impacts; how historic power dynamics shape maize systems, as in the case of Malawi; the central role of smallholder farmers who provide essential evolutionary services related to agricultural biodiversity and ecosystem services, in the case of Mexico; and how changing practices really matter and can result in positive impacts, both economic and environmental, in the case of Zambia.

Finally, Chapter 9, which focuses on soil health in California, sets out an ambitious challenge to refocus on soil health as a central indicator of food system health. Joanna Ory and Alistair Iles describe how changes need to cascade across the entire supply chain to support the necessary shifts on farm. The systemic approach these authors describe is a perfect example of how TCA can be deployed to understand the relationship between supply chain practices, actors, and policies. Seeing farmers as part of a complex system can help us to collectively understand what motivates and incentivizes behaviour change, and the case of almond production in the Central Valley of California is illustrative of how and why farmers make decisions and how these decisions are connected (or not) to broader positive outcomes.

The authors in this section offer insights and opportunities for revaluing the essential services of farmers and food producers through TCA and engaging them as central stakeholders in the food system. Supporting farm-level change requires a nuanced understanding of the economic, social, cultural, and historical context of farmers and food producers.

6 Harmonizing the Measurement of On-Farm Impacts

Patrick Holden and Adele Jones

If the Food System is an Organism, its Farms are the Cells

Imagine the world's food systems as a giant organism, with its component parts, including the farmers, growers, abattoirs, packhouses, processors, distributors, and retailers all contributing to its overall health. In that immensely complex yet interconnected system, the farms are the cells. Building on this metaphor—if the patient Earth is sick and the farms represent the cells fueling the planet's lungs and digestive system—it will clearly be impossible to restore the food system to full vitality without first ensuring that the cells (the farms) are healthy.

Without question, enabling such a transition to farm cellular health is of critical importance. As the leading environmentalist, Dr. Vandana Shiva, observed in a talk she gave in London in 2019, as the vast majority of planet Earth is now covered with farms, the only way to avert irreversible climate change and a likely associated population collapse is to enable a global transition to more healthy, sustainable, and resilient farming methods. Indeed, many countries are including food and land use in their nationally determined contributions (NDCs) as they strive to achieve their long-term climate goals under the United Nations Framework Convention on Climate Change. However these NDCs often reflect purely technical fixes, neglecting the need for holistic and systemic approaches (Leippert et al., 2020) reflecting challenges in mobilizing the agriculture sector and ensuring that farms actually deliver on the changes required.

Thankfully, there is now a growing consensus that in relation to the range of increasingly existential threats to human civilization (including climate change, biodiversity loss, depletion of natural capital, pollution, and growing food insecurity), farming has not only been one of the most significant contributors to the damage, but also potentially holds the key to reversing it.

Barriers to Change

This leads to a key question: if such a consensus now exists, why is the transition not already underway? There are a number of reasons, the primary one being economics. At present, owing to the failure to place a value on the impacts associated with food and farming systems, those that degrade natural, social, and human capital are more profitable than their sustainable equivalents. As a result, the vast majority of the farming community are currently contributing to the problem rather than representing part of the solution.

Another barrier is the lack of a harmonized framework for measuring and valuing on-farm sustainability. The current plethora of overlapping sustainability assessments and certification schemes is time-consuming, costly, and bureaucratic for farmers. It is also frustrating for government agencies, nongovernmental organizations (NGO) and food companies, as well as confusing for consumers, who have no unified means of linking their purchasing power to support sustainable and healthy food production.

As a direct consequence of both of these problems, farmers and food companies have become locked into a cycle of dependence on commodity products, the production of which is damaging on multiple levels. At the same time, consumers, unaware that for every £1 they spend on food they are paying an additional £1 in 'hidden' costs (including through their taxes to clean up polluted waterways for example, or environmental degradation at the cost of generations to come), are forced to continue buying the apparently 'cheap food' which results from the most intensive and harmful production systems (Fitzpatrick and Young, 2017).

Given the severity of the situation we now find ourselves in with climate change and diet-related ill health to name but two issues, it begs the question, how can these key barriers to change be overcome? We believe the answer lies in combination of three key things:

- The emergence of an internationally harmonized framework for measuring and valuing on-farm sustainability.
- The application of True Cost Accounting, including through the introduction of the polluter pays principle and redirection of subsidies to support farming systems which deliver "public good" outcomes.
- An honest, transparent and thriving market for sustainably produced food.

Taken together, these mechanisms could enable the global adoption of farming practices that mitigate against climate change, reduce, and eventually eliminate the pollution of air and water, rebuild soil health, and reinstate biodiversity.

Based on our experience and the work of others, we will begin by addressing the first of the aforementioned barriers and thus a potential solution—the need for a common approach for measuring and valuing on-farm sustainability.

An International Framework for Measuring and Valuing On-Farm Sustainability

In order for the application of True Cost Accounting (TCA) at farm level to be successful, we must first agree on a common language, framework and metrics for measuring the impact of different farming practices. Then, and only then, can we begin to apply the discipline of TCA to monetize these impacts and thus correct the economic distortions described above.

Although this sequence of events might seem obvious now, it has taken us some time to realize the significance of getting this right. As this book has detailed, there are now a number of high-level frameworks for measuring and valuing food system-related natural and social capital impacts, most notably the TEEBAgriFood Evaluation Framework (The Economics of Ecosystem and Biodiversity, 2018). However, as yet, there is little information about how these can be utilized practically at the cellular farm level.

The Sustainable Food Trust (SFT) has had a long history of involvement in the development of TCA (see Box 6.1), but as we became increasingly aware of the need to work through this process, we resolved to focus our attention on catalyzing the emergence of an internationally harmonized framework for measuring on-farm sustainability.

At present, there is a diverse range of overlapping assessment tools and certification/labelling schemes for monitoring and communicating how well a farm is performing. This makes it impossible for consumers, farmers, food businesses, and policymakers to gain an accurate understanding of the comparative impact of products resulting from different methods of production, as well as having a polarizing effect on farming communities due to the "you're either certified or you're not" approach (see reference to organic standards in Box 6.1).

To address this problem and undertake this work, we began by establishing a British farmers and land managers working group to review the current diversity of different frameworks and tools. The composition of the group included a wide range of farming scales, practices, and enterprise types. Unsurprisingly, this included organic farming members, big and small, as well as conventional producers and large estates that had been at the forefront of producing high yields using technically efficient and intensive methods.

Our motives for undertaking this project were simple: as farmers, we were all subjected to multiple annual audits including organic inspections, government assessments, and other certification schemes to enable us to gain public funding and market access. However, none of these assessments ever gave us any indication about whether our farming enterprises were more or less sustainable than the year before! Realizing this, we decided to take things into our own hands and worked together to pull the best and most common elements of all the existing sustainability assessment tools and certification schemes, to eventually arrive at a draft framework of categories, indicators, and metrics that we believed had the potential to be used on any farm, whatever scale and intensity, anywhere in the world (of course while recognizing that there will need to be an element of "bespokeness" for different geographies, climates, and cultures).

In 2017 the group commissioned a thorough gap analysis of the most widely used sustainability assessment tools (Smith et al., 2017) and identified areas of overlap. This exercise further reinforced the case for harmonization, as it revealed to us that there was more than a 60% overlap of data requirements between the different assessment tools and certification schemes. Based on a strengthened resolve and with a sense of naive enthusiasm, we began to realize our audacious aim of developing an internationally harmonized language for on-farm sustainability assessment akin to the language that already exists for financial accounting protocols.

The framework will aid farmer understanding and provide a common language for farm-level sustainability. It should take an inclusive approach, allowing all farmers to make incremental steps towards becoming more resilient, sustainable, regenerative, and ultimately a climate change solution.

The framework could also be used by governments to design future public support schemes, by food companies to aid supply chain transparency, by the finance community as a basis for sustainable investment, and by consumers to better understand the relative sustainability of food products that they purchase. In this way, we can reward those producers who are delivering genuine benefit to the environment and public health and subsequently shift the balance of financial advantage towards more sustainable production on a global scale. The Sustainable Food Trust has been consulting with farmers for over four years to help to design the first iteration of a harmonized framework of assessment. The categories of assessment are:

- Soil
- Water
- Air and climate
- Biodiversity
- Energy and resource use
- Nutrient management
- Plant and crop health
- Animal husbandry
- Social capital
- Human capital
- Productivity

Progress to Date

As we have progressed with this work, the project has increasingly attracted external attention from a wide range of organizations who share an interest in harmonization. These include academic and research institutes, NGOs, certification organizations, food companies and retailers, financial bodies, as well as most recently, policymakers, and government agencies.

At the time of writing this chapter, we are in the midst of conducting a British. Government-funded trial on 30 English farms ranging from arable to mixed cropping and livestock, beef and sheep, pigs, poultry, horticulture, and upland hill farming. The Welsh Government has now also announced that it is its intention to require an annual sustainability audit based on the harmonized framework as a prerequisite for Welsh farmers being able to enter in the new

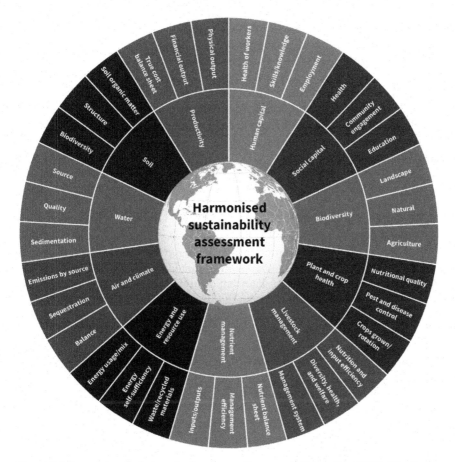

Figure 6.1 Internationally harmonized assessment framework (September 2020). The framework includes 11 categories of assessment with three key indicators under each.

post Brexit farm support scheme. We have also had expressions of interest from several impact investment funds and certification organizations, both in the UK and internationally, who are keen to trial our framework as a means of linking up with others and improving how they measure the change with their farming clients on the ground.

A barrier to harmonized metrics is that many organizations, understandably, work to promote their own, often proprietary frameworks or assessment tools. It is important to state here that at the Sustainable Food Trust, we hold no interest in "owning" the harmonized framework or running a new sustainability assessment scheme. Our ambition is simply to bring all the key players to the table to be able to move towards consensus and then work with organizations actively assessing farms to "superimpose" the framework onto what they are already doing. This way, there can be a great diversity of activities and tools being used for multiple purposes, but the presence of a harmonized framework means that data being collected and communicated is consistent and comparable.

Going forward, the most pressing imperative will be to ensure that a harmonized framework for measuring on-farm sustainability couples naturally with the existing high level food system frameworks, including the United Nations' Sustainable Development Goals, Food and Agriculture Organization's (FAO) Tool for Agroecology Performance Evaluation Tool, the Natural Capital Protocol, and the TEEBAgriFood Evaluation Framework, to mention a few. Once such linkages are place, we can then work to identify, categorize, quantify, and eventually monetize the range of externalities, both positive and negative, on all farms across the world in a way that is meaningful to those with their hands in the soil.

Creating an Enabling Policy Environment for Farmers

The second barrier that we will address is the lack of an enabling policy environment for sustainable food producers. As the harmonization of on-farm assessment process evolves, the focus of attention should then move to governments and government agencies whose interventions will be essential if we are to shift the balance of financial advantage away from the present situation where unsustainable farming pays better than its sustainable counterparts, to a system which rewards the delivery of so called "public goods."

It is a truism that unless we do this, the number of individuals who will be able to practice sustainable agriculture will remain confined to the passionate, the idealistic, the entrepreneurial, and their committed consumer (often wealthy) counterpart, who occupy niche markets and/or are prepared to pay more for more sustainable and healthy food.

"Carrots and Sticks"

Building on a baseline of common data, it will then be possible to develop policy instruments, such as those that fall under the "polluter pays principle," which ensure that in the future, farming practices which deliver environmental and social "goods" are rewarded in proportion to the benefits delivered (carrots), while those that are causing damage become financially accountable for these impacts (sticks).

An example of such a "stick" under the polluter pays headline would be a tax on nitrogen fertilizer. There is a paradox associated with the use of nitrogen: it has become a seemingly essential component of modern agriculture, brought enormous increases in productivity, and allowed the global population to grow food rapidly. But these benefits come with huge negative impacts including the steady degradation of the natural environment and the soil upon which future food productivity depends. The European Nitrogen Assessment—a major exercise involving over 200 scientists across the European Union—established that the negative impact of the use of nitrogen fertilizer could be valued between €35 and €230 billion in 2011 (Sutton and Van Grinsven, 2015). A nitrogen tax would therefore constitute a bold initiative to challenge entrenched behavior by making users financial accountable for the negative impact it causes, driving down its use in favor of practices that deliver public benefit.

One such benefit, or "carrot," would be the maintenance or building of soil carbon. If farmers were to be paid to build organic matter (a proxy measure for soil carbon) as a result of their management practices, it could potentially transform the economics for sustainable farming and bring multiple environmental benefits (KeySoil, 2015). Soil organic carbon makes up approximately 50% of soil organic matter. Degraded soils retain less moisture and are therefore highly vulnerable to droughts. But for every 1% increase in organic matter in soils, the first foot of soil is able to hold an additional 16,500 gallons of water per acre (40,000 gallons per hectare) (Gould, 2015). Of course to achieve this, we need to reach agreement on which practices have the potential to build soil carbon as well as a common metric and the emergence of a reliable and cost effective means of measuring soil carbon outcomes on farm. If this approach were applied globally, it is estimated that soils (already recognized as the second largest carbon sink after the oceans), would re-sequester significant quantities of the CO_2 that have been emitted through the use of fossil fuels over the last 50+ years.

Until now, governments throughout the world have been reluctant to introduce a polluter pays principle for farming, owing to its possible impact on food prices. This is of course a huge issue that must be addressed, yet the failure to do so amounts to environmental vandalism, pollution, and irresponsible asset stripping of finite natural capital. It is our job now to provide them with the evidence and toolkits to be able to make this an obvious choice for future policy development.

A Thriving Market for Sustainable Food

The third barrier to change—and potentially the most powerful lever in driving food system transformation—is the market. Even with corrective policy mechanisms in place, in order for sustainable agriculture to become the default model of choice for farmers all over the world, there must be a thriving and transparent market for the output from such systems, which recognizes and rewards the positive benefits being delivered.

It is imperative that the retail sector now makes itself fully aware of the true cost of the food that it sells and must be proactive in demanding food and farming policies that ensure that its supply-chain partners are producing food that is genuinely sustainable from the perspective of the environment, the farmers, and the rural community.

To take the example of food, the most obvious output from farming, the Sustainable Food Trust believes that we need a new innovative, harmonized, and transparent framework for labelling food products, which draws on the results from an annual sustainability assessment (based on the harmonized framework) and empowers consumers by providing accurate information about the degree of sustainability and provenance of the product. This could take the form of a traffic light system or a sliding scale, but importantly the language, categories, and metrics used must be consistent, comparable, and easily understood.

Of course there are now also new and potentially exciting markets for farmers to start tapping into, including those surrounding the delivery of natural capital or carbon offsets. This, together with the recognition that intensive agriculture is a large greenhouse gas emitter, has given rise to the development of a large number of on-farm carbon calculators. Again, unless there is an internationally harmonized framework for measuring farm-level impact under which these schemes and calculators can all fall, there is a huge risk of repeating the "siloized" mistakes made by the certification organizations all those years ago.

Conclusion

To return to where we started, if we are to restore the global food system "organism" to full health, we need to work from the top down (with governments and the market), from the bottom up (by empowering the millions of farmers who now straddle the majority of the planet's land surface), and in the middle (by empowering citizens to vote for the farming systems that they would like to support through their purchasing powers). It is true that today the patient is sick, but by working within each of these levels, the application of TCA can be used to re-establish farming systems that operate within planetary boundaries and produce food in harmony with nature across the globe.

Box 6.1 A Brief History of True Cost Accounting and Sustainability Assessment From the Farm Up

1970s: The British Soil Association developed organic standards for the production and marketing of sustainably produced food. The objective behind the scheme, which was eventually adopted all over the world, was to enable farmers who were using sustainable production methods to remain financially viable by attracting a price premium for their products in the market place; this was necessary because of the absence of the polluter pays principle which resulted in "conventional" farming being more profitable than its organic equivalent.

Viewed in market terms, the organic project has been successful; the global retail market for organic food now stands at approximately €100 billion (Ouest France, 2020), and the EU is currently in the process of setting a target of 25% of total EU farmland reaching organic production standards by 2030 through its Farm to Fork strategy (European Commission, 2020). Despite these successes, it has now become clear that a partial transition to sustainable farming will not be sufficient to avert runaway climate change and biodiversity loss, with catastrophic impacts on civilization as a whole. In other words, the organic project was treating the symptoms not the cause of the problem—namely the absence of True Cost Accounting.

1980s/90s: Another early pioneer of the economic discipline of True Cost Accounting (TCA) was Professor Jules Pretty, who published a report in 1991 (Conway and Pretty, 1991) for the first time identifying the extent to which the failure to account for pollution and other negative impacts of intensive farming in monetary terms was disadvantaging farming systems which avoided these costs and delivered true social and environmental benefits.

2010–2016: These factors contributed significantly to the decision to launch the Sustainable Food Trust (SFT), which works internationally to accelerate the transition towards more sustainable food and farming systems. As it became increasingly clear that the absence of True Cost Accounting in food and farming was perpetuating a system that was ultimately destructive of planetary and public health, the work of the SFT became increasingly focused on TCA. The SFT organized a number of events, including: 2011, when the Future of Food conference at Georgetown University, Washington, DC, (at which the Prince of Wales gave a memorable speech in which he mentioned perverse subsidies and the absence of the application of the polluter pays principle); 2012, when the Prince of Wales hosted a meeting at his home, Highgrove, which led to the formation of the Global Alliance for the Future of Food at which True Cost Accounting was voted the most important area for their future work; 2013, when conferences were held in Louisville, Kentucky and London, both on the theme of TCA; and 2016, when a conference was held in San Francisco on The True Cost of American Food.

2014: Pavan Sukhdev decided to focus more of his efforts on food and farming. Hitherto, his work had mainly concentrated on the establishment of the Natural Capital Coalition and the foundation of The Economics of Ecosystems and Biodiversity (TEEB). Subsequently this new interest in food and farming coalesced around the TEEBAgriFood project.

2017: The SFT produced a report, *The Hidden Cost of UK Food*, whose author, Richard Young, analyzed the known scientific data on food system externalities, highlighting the reality that for every £1 that is spent in the United Kingdom on food there is another hidden £1, split 50/50 between negative environmental impacts and damage to public health. In summary, the conclusion of the *Hidden Cost* report was that the current food pricing system is dishonest and misleading to the general public.

Since 2012: The Global Alliance for the Future of Food has emerged as the most significant advocate of TCA. Members of the Global Alliance partly funded the work of the TEEBAgriFood research program and more recently seed-funded the True Cost Accounting community of practice—a coalition of the key individuals and organizations with an interest in TCA. It has played a key part in strengthening the discipline of TCA and helping those involved to reach out to the wider global community to incorporate TCA principles into future food systems accounting.

Box 6.2 What Does a Sustainable Farming Operation Look Like?

The truth is, there is no single "correct" answer to this question, but there are some fundamental principles to which we should all be moving, notably producing as much high-quality, nutritious food as possible, while minimizing the use of non-renewable external inputs, including mineral fertilizers and fossil fuel energy, thereby reducing emissions and environmental pollution, while at least maintaining and preferably building soil fertility and other forms of natural capital. Farming systems should be self-sufficient and as resilient against the external "shocks" as possible, in terms of nutrients, seeds, animal feed, and bedding. They should be agriculturally diverse, including the crop varieties and livestock breeds, and as well as preserving "natural" diversity, in terms of wild plants, insects, birds, and animals. Finally, farming systems must provide a reasonable economic return as well as a high quality of life for the farmer and those who surround the farm.

Box 6.3 Box Acknowledgments

In describing the progress that we have made with this initiative, it would be inappropriate not to mention the very important work that has been undertaken by other institutions that have been active in accelerating the collective international understanding of True Cost Accounting at the farm level. Specifically, Nadia El-Hage Scialabba at the FAO spent a number of years developing Sustainability Assessment of Food and Agriculture, which was arguably the first and leading initiative in the field of sustainability metrics harmonization and has informed much of our thinking. The International Federation of Organic Agricultural Movements, whose work over a quarter of a century in bringing together organic standards has in many ways contributed towards the emergence of a global awareness of the necessity for measuring food systems sustainability.

A number of individuals have also made important contributions; notably the Prince of Wales raised the issue of the absence of the polluter pays principle as early as the 1980s, and Christy Brown who coined the term "True Cost Accounting" in 2013. The Global Alliance for the Future of Food also deserves tremendous affirmation, as it has prioritized investing in the discipline and the science of TCA right from the outset.

References

Conway, G.R. & Pretty, J.N. (1991). *Unwelcome Harvest: Agriculture and Pollution.* Washington, DC: Earthscan.

European Commission. (2020). *From Farm to Fork. Our food, our health, our planet our future.*

Fitzpatrick, I. & Young, R. (2017). The Hidden Cost of UK Food. Sustainable Food Trust. Available at: http://sustainablefoodtrust.org/wp-content/uploads/2013/04/HCOF-Report-online-version-1.pdf.

Gould, M.C. (2015). Compost increases the water holding capacity of droughty soils. Michigan State University Extension.

KeySoil. (2015). Profiting from Soil Organic Matter. GY Associates and Rothamsted Research.

Leippert, F., Darmaun, M., Bernoux, M., & Mpheshea, M. (2020). The potential of agroecology to build climate-resilient livelihoods and food systems. Food and Agriculture Organization and Biovision.

Ouest France. (2020). Dans la baie de Saint-Brieuc, le blé noir conquiert les terres.

Smith, L., Mullender, S., & Padel, S. (2017). Sustainability Assessment: The Case for Convergence. The Organic Research Centre.

The Economics of Ecosystems and Biodiversity. (2018). TEEB for Agriculture & Food: Scientific and Economic Foundations. Geneva: United Nations Environment.

7 Incentives to Change

The Experience of the Organic Sector

Gábor Figeczky, Louise Luttikholt, Frank Eyhorn, Adrian Müller, Christian Schader and Federica Varini

The Need for Change

If we as society want to see real change in the world and move towards true sustainability, the real cost of our current food system needs to be made evident—and eventually reflected in the price—so that all participants can take full responsibility for their actions. We need to change how we treat the environment and use natural resources, as well as how we interact with each other when it comes to food and agriculture. Organic agriculture, which has been practised for several decades now, can teach many lessons about how to steer such a transformative process effectively.

Organic Agriculture: Farming with Lower External Costs

Organic agriculture aims at sustaining the health of soils, ecosystems, and people. It relies on ecological processes, biodiversity, and cycles adapted to local conditions, rather than the use of inputs with adverse effects. Organic agriculture combines tradition, innovation, and science to benefit the shared environment and promote fair relationships and good quality of life for all involved. The principles of health, ecology, fairness, and care are the roots from which organic agriculture grows and develops. They express the contribution that organic agriculture can make to the world and a vision to improve all agriculture in a global context. In practice, this translates into reliance on natural ecosystems by mimicking nature and enhancing soil quality, greater biodiversity, and radically reduced impacts from synthetic inputs, such as chemical pesticides, fertilizers, and antibiotics, and the aspiration to improve livelihoods through more fairness and resilience. Organic agriculture has a long history of success and a growing market share now worth $120 billion globally for certified organic alone. In addition, millions of small-holder farmers, particularly in the Global South, also practice these methods, still adhering to the traditional practices of their ancestors. Recognizing the wide range of systems based on the principles of organic agriculture, the organic movements have agreed to embrace organic agriculture in its full diversity (IFOAM—Organics International, 2017).

Building on this, organic agriculture allows for farming with lower external costs and the ability to deliver more positive externalities (van der Werf *et al.*, 2020; Seufert and Ramankutty, 2017) than the prevailing agricultural system.

The potentially lower yield performance of organic versus conventional systems is usually put at the center of the discussion when comparing sustainability per unit of product (Meemken and Qaim, 2018, Seufert, 2018). Current estimates of the yield gap are about 20%, but there are indications that this can be considerably reduced if due focus is laid on diversity in crops and crop rotations, etc. (Ponisio *et al.*, 2015). When comparing the environmental performance of conventional and organic systems on the whole and in relation to the natural environments where they are located, it is clear that high productivity in intensive agriculture is to the detriment of the health of ecosystems and our planet, as it contributes to countless externalities, the costs of which are paid by society. A truly sustainable agriculture and food system is not performing maximally on one indicator (e.g. yields) and badly on all others (e.g. pesticide contamination, nutrient surplus), but is set up to perform relatively well on all sustainability indicators, thus also optimally managing potential trade-offs (e.g. between high yields and transgressing local ecosystem carrying capacities). Thus, the debate on comparing agricultural production systems needs to go beyond mere yield or greenhouse gas footprint comparisons and has to build on encompassing systems comparisons based on total impacts (as shown, e.g., in Müller *et al.*, 2017; van der Werf *et al.*, 2020). TCA can provide a helpful conceptual framework to support such approaches in practice.

True Cost Accounting, an Approach to Transform Food Systems

Modern economies are designed in a way that consumers actually pay, in the end, three or even four times the cost of seemingly cheap food. They pay a lower retail price for the product, and then without realizing it consumers also finance, through tax payments, the mitigation of the negative impacts of production (e.g. to clean up drinking water). Some actors in mainstream food systems are allowed to pollute without having to pay for it and become disproportionately wealthy and powerful. They become politically influential and put pressure on governments to use money from taxpayers to further subsidize unsustainable agriculture through government schemes promoting the purchase of synthetic fertilizers or intensive animal husbandry. In the end, citizens bear the health costs of noncommunicable diseases caused by unhealthy food or environmental pollution. TCA can help governments and consumers understand and identify these "hidden costs," making them visible and quantifiable while at the same time revealing the societal benefits of more sustainable systems such as organic farming (IFOAM—Organics International, 2019).

TCA provides strong arguments for policy reforms that incentivize beneficial practices and systems and disincentivizes harmful ones (Eyhorn *et al.*, 2019) (Figure 7.1).

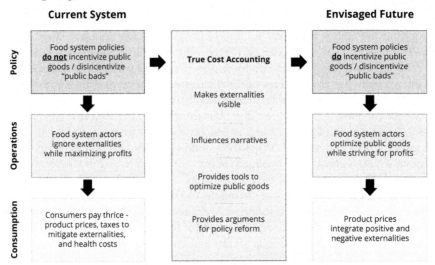

Figure 7.1 The theory of change of TCA.

Some people consider it unethical to value aspects of our world in monetary terms, for example, the beauty of a landscape, the survival of an endangered species or, ultimately, human life. In the organic sector we recognize that the currency of nature is life itself and that many phenomena cannot be expressed through money alone.

Money, however, is universally understood as a way to assign value in societies across the world and as a means to inform decisions on the use of scarce resources. Therefore, notwithstanding its limitations in accurately reflecting societal costs and benefits, TCA is an important tool for change and offers a way to correct some core failures in our current model, where profits are privatized while the rest of the costs are borne by society. TCA makes these external costs, public good provision, and ecosystem services visible to the decision-makers. If used as a basis for regulations or policies, it thus forces operators to account for the externalities that their operations might cause, as they directly influence their private profits.

What to Measure, How to Measure: The Challenges Behind Accounting for Externalities in the Food System

Environmental as well as human costs and benefits add up as products move through the value chain, to the end consumer. For instance, the overall impact of processed food depends on the environmental impacts of the production of its raw ingredients, the transformation processes that these ingredients undergo, as well as the distribution of the final product, including packaging, transportation, storage, etc. and, if relevant, also the impacts of its remains that might become waste.

The true—or full—cost of any given product, activity, enterprise, sector, production model, or system will inevitably be only an approximation, or an incomplete snapshot, limited by a given set of boundaries over a given period of time. The calculations will always show some uncertainty, and results might vary but, importantly, they consistently help to identify whether key practices and their impacts move in a positive or a negative direction; and they can facilitate identification of particular trade-offs and synergies between different aspects along the value chain and between different sustainability indicators.

An extensive amount of scientific literature has explored possible models for evaluating the costs linked to pollution, the environmental impact of certain value chains, as well as for calculating the economic value of the environmental services provided by agriculture. Life Cycle Assessment (LCA) studies focus on the environmental impact of single products measured by impacts per unit of product, disregarding total impacts of a whole production system. Owing to this narrow view, products from intensive production systems tend to perform better in such comparisons. Looking at these systems with a broader perspective, however, accounting for their total impacts in relation to relevant local and regional ecosystem boundaries or also planetary boundaries, we find that their environmental sustainability is lower (Geiger *et al.*, 2010; Gibbs *et al.*, 2009; Meehan *et al.*, 2011). Furthermore, agricultural production is multifunctional, and all outputs besides the core product, such as ecosystem services provision, need to be accounted for when doing an encompassing assessment. LCAs that compare whole farming systems thus need to look at all the outputs produced by agriculture and allocate the environmental impacts to these (Schader *et. al.*, 2012). Thus, previous efforts to account for externalities, such as LCA, have a number of shortcomings that TCA should seek to address.

None of this is simple. In order to reverse the unsustainable course of agriculture, we need to start with the key leverage points, such as nitrogen inputs and pesticide use, and continue to add criteria and refine the methodology continuously over time. A focus on specific practices can serve as a powerful initiator of change that will compel other changes in a similar direction, eventually leading to sustainable agriculture and food systems.

Incentives to Change: How Policies Apply TCA to Transform Food Systems

Accounting for True Costs Through Policies: Creating a Level Playing Field for All Food Production Systems

An examination of how TCA has been implemented in lawmaking, in order to create a more conducive policy environment for organic agriculture, makes it possible to distinguish two main types of measures adopted: incentives (subsidies on agri-environmental measures) and disincentives (taxes). These instruments, given primarily as market support, are justified politically through consistent evidence showing the multiple societal benefits delivered by organic farming in contrast to the detrimental environmental impacts of industrial agriculture. However, governmental support for the organic sector remains too often a marginal intervention for a niche market, while unsustainable practices, such as the use of chemical inputs in the name of achieving food security, or intensive animal husbandry, remain heavily subsidized. The overall agricultural policy package is therefore incoherent. In the next section we provide some concrete examples of policy measures from different contexts.

Incentives Through Agri-Environmental Measures

Income support for organic producers is generally included under policy interventions dedicated to support agri-environmental measures. As previously argued, the political reasoning for subsidizing organic agriculture resides in the multiple socioeconomic and environmental benefits that this type of production delivers. These positive externalities are partly remunerated by the market itself in the form of a premium price, as well as through specific subsidy programs. These subsidies are usually given to organic farmers or farmers in conversion to organic in the form of a fixed amount per hectare or as a reward for voluntary agri-environmental measures implemented. Such forms of support are meant to foster a wider adoption of organic agriculture among farmers. Subsidies given during the conversion period are often higher than those received for maintaining a business under organic production. This is due to the fact that during the conversion period, the farmer bears the additional costs of organic production without yet receiving the benefit of the premium prices for their products.

In Europe and the USA, but also in certain Latin American and Asian countries, the organic sector has benefited in the last two decades from different forms of subsidy redistribution.

In Asia, **South Korea** is one of the first states to have implemented direct payments to organic producers to support their income. Starting in 1999, in agreement with the Environment-Friendly Agriculture Promotion Act, farmers certified as performing environmentally friendly agriculture are rewarded with direct payments which differ according to the certification category (organic production gets higher payments compared to no-pesticides or low pesticides), type of crop, and the area cultivated. The total budget earmarked to this type of payments was approximately €188.5 million between 1999 and 2012, €23 million in 2013, €19 million in 2014, and €23 million in 2015 (Choi, 2015).

Since 2015 **India** has had a program which includes direct payments dedicated to small-scale producers performing organic farming. The Paramparagat Krishi Vikas Yojana (PKVY) program, running under the National Mission on Sustainable Agriculture, supports small producers that organize themselves in clusters under organic production. The subsidy covers a variety of costs, such as input purchase, harvesting, transportation of costs, marketing and also a direct payment based on area. The total amount allocated for the scheme in the period 2015–18 was €118 million, and the government estimates that under this scheme 237,820 hectares of land were converted into organic farming land, and 394,550 farmers benefited from the support. Although the Indian government provides support to organic farming through different programs, the reality is that the majority of government support is allocated to conventionally grown cash crops under monocropping conditions. This is especially true if one looks at the support for agricultural inputs: in 2017–18 the Indian government allocated approximately €9 billion to provide subsidies for synthetic fertilizers. In comparison, the two dedicated programs for organic farming—PKVY and MOVCDENR[1]—were supported over the period 2015–18 with approximately €168 million. In 2018 India had almost 2 million hectares under organic production, which is equal to about 1% organic share of its agricultural land (Willer *et al.*, 2020).

In Central America, **Costa Rica** represents an example of a country that has also implemented area payments for conversion. Since 2007 this measure supports small and medium-sized organic farmers allocating the subsidy according to the type of crop over a period of three years. Interestingly, such a measure is funded by a taxation scheme imposed on fuel. In 2018, for instance, the government disbursed a budget of €424,000 for the implementation of this measure.

In **Switzerland**, direct payments are a central element of national agricultural policy, and farmers are eligible to apply for direct payment only if they comply with certain environmental requirements. These cross-compliance criteria include a set aside area of at least 7% of the utilized agricultural area, reduction of soil erosion and nutrient run-off, and basic requirements for good agricultural practices. Until 2014 a distinction was made between general direct payments and environmentally friendly payments. However, as the Swiss agricultural policy known as AP14–17 has now been extended until 2021, this distinction is not in place anymore, and seven types of contributions have been introduced: maintenance of the cultural landscape, ensuring food supply (which include support for farming systems in unfavorable areas), biodiversity, landscape quality, production systems (which includes support to organic agriculture), and efficient use of resources and transition contributions (to ensure a socially sustainable transition from the old to the new policy). Farmers are eligible to receive these types of payments only if they comply with some stringent environmental performance requirements (PER) that cover domains such as animal welfare, biodiversity, nitrogen surpluses, soil management, and phytosanitary products. As a result of this reform, the funds are used more specifically in favor of production systems delivering services of public interest which are not automatically remunerated by the market (i.e. improvement of animal welfare and promotion of biodiversity). In 2018 organic farming reached 156,098 hectares (15% of the total agricultural land), an increase of more than 10,000 hectares (+7%) compared with the previous year. In total, more than €51 million was paid out in 2018 for the promotion of organic farming—that is, €4.6

million more than in 2017, which equals approximately 2% of the total expenditure on direct payments in the country. However, payments are cumulative and organic farms are eligible for further Ecological Direct Payments (Swiss Federal Office for Agriculture, 2004).

In the **USA**, there are a number of programs at the United States Department of Agriculture that support the organic sector. For example, the voluntary Environmental Quality Incentives Program (EQIP), introduced in 1985, provides technical and financial support to farmers engaging in conservation practices. Organic farmers have a dedicated sub-program within EQIP aiming at accompanying farmers in the conversion phase. This successful program ran with a budget of $1 billion reaching more than 6,800 farms in 2016. Additional programs have since been designed and implemented, some on a state rather than federal basis (see Chapter 9, "Fostering Healthy Soils in California: Farmer Motivations and Barriers"). Other programs are not as well funded, unfortunately, such as the Organic Transitions Program (which provides support for producers wishing to transition to organic farming, funded recently at only $4 million) and the Organic Agriculture and Research Extension Initiative (funded recently at only $20 million).

Almost all member states of the **European Union** (EU) offer a dedicated measure for area conversion and/or maintenance payments for organic producers included among other agri-environmental measures under Pillar 2 of the Common Agricultural Policy (CAP). The CAP, which represents almost 40% percent of the total EU budget, is indeed the most powerful policy instrument shaping EU food systems since 1962. The CAP cycle 2014–2020 made organic farming more prominent, corroborating its value as production practice able to deliver on sustainability targets in the agri-food sector under its two main components of the policy, Pillar 1 and 2. Approximately half of the budget is earmarked to Pillar 1, which provides support to income for farmers in the form of direct payments. Since 2015 Member States must use at least 30% of their national direct-payment allocations to fund the greening components introduced to provide additional support to offset the cost of delivering environmental public goods not remunerated by the market; organic farms automatically receive such additional support without having to fulfill any additional requirements.

Public support dedicated through the CAP in recent decades has contributed to the development of the organic sector at a communitarian level, with the share of organic farming area in total utilized agricultural area (UAA) in 2018 reaching almost 8% (Willer *et al.*, 2020). None the less, despite being the most significant type of financial support to organic agriculture in Europe, the CAP 2014–2020 earmarked to the organic sector a mere 1.5% (€6.3 billion) of current EU agricultural spending (Stolze *et al.*, 2016). In a dedicated report, the European Court of Auditors (2017) highlighted how the current CAP still presents important shortcomings when addressing environmental degradation and climate change challenges, as well as economic sustainability issues related to the agri-food sector. The main shortcoming is that so far most of those payments are provided as income support for fulfilling the policy's minimum conditions and are based on the amount of land that a farmer owns without being conditionally linked to result-orientated and quantifiable targets. In the attempt to overcome and improve the sustainability performance of the policy, the European Commission is working on a new CAP proposal for 2021–2027, with a view to implementing it starting from 2022. The most relevant innovation is introducing more margin for local adaptation at Member State level while at the same time making it compulsory for each country to adopt the Green Architecture, which introduces the possibility for producers to voluntarily apply for environmental measures, included not only for the Rural Development (Pillar 2) but also for direct payments (Pillar 1).

Besides these instruments based on compensating farmers for the income forgone owing to the implementation of organic practices, several other policy measures have been implemented worldwide to sustain the development of the organic sector. This is done by fostering production, consumption, or by creating an enabling environment for the development of the sector. In 2017 IFOAM—Organics International published extensive research which looked at public support provided to organic agriculture in over 80 countries, spanning from support to inputs development to free organic certification or promotion of public food procurement prioritizing organic produce (IFOAM—Organics International, 2017).

Moving Towards Environmental Payments Tied to Result-Orientated Objectives

Although the basic policy justification behind these subsidies is to reward environmental and societal benefits, the calculation methods used by governments to determine the level of payments for various types of production or environmental measures adopted have so far not been based on defining a value for those positive externalities, but rather on compensating the additional costs and income foregone from farming organically or adopting agri-environmental measures.

For this reason, such policy incentives still remain a somewhat imperfect tool to apply TCA into policymaking, as payments remain mainly constrained by the "extra cost or loss of income" as dictated by the World Trade Organization (WTO)[2], which was introduced to respect the principle that agri-environmental payments should not create trade distortions (Burton and Schwarz, 2013). This requirement has been, for instance, strictly observed by the European Commission and its Member States when designing its agri-environmental scheme, preventing the adoption of innovative environmental mechanisms better equipped to link environmental performance to the level of payment.

Allocating payments according to the degree of achievement against predetermined sustainability targets requires complex models and analytical frameworks, which have not yet been fully integrated into lawmaking. So far, only the Tuscany Region in its Rural Development Programme for the period of 2014–20 includes a unique attempt to bring in a holistic evaluation when comparing organic farms with conventional systems (see Box 7.1).

The ultimate goal should be to encourage farmers to adopt a selected variety of agri-environmental measures tailored to the pedo-climatic area where they are operating. Thus, policymakers could focus their interventions and support on local needs like decreasing soil erosion or nitrogen leaching.

Box 7.1 Defining Area Payment for Organic Farming in the Tuscany Region, Italy

EC Regulation 2078/92 institutionalized the promotion of organic farming in the European Union (EU) as a result of the explicit beneficial impact at an environmental and landscape level. The current calculation of the payments level for any agro-environmental scheme in the EU, however, is defined by the EAFRD[3] regulation (1305/2013) (European Union, 2013) as a compensation for all or part of the additional costs and income foregone resulting from the commitments made by the beneficiary. Likewise, the payments related to Measure 11, which grants support for maintenance or conversion to organic farming as part of the second pillar of the current Common Agricultural Policy, also follow this rule. The additional costs and forgone income are defined by Member States based on parameters such as differences in yield, production costs, prices, and transaction costs. Therefore, this system, despite being motivated by the acknowledgment that "market failure" should be compensated, does not recognize or address the cost or benefits related to the environmental or social dimension linked to the performance of organic farming systems.

In the model used by the Tuscany Region to define the level of payment under Measure 11, the foregone income and the value of the ecosystem services provided are calculated using a set of modelling frameworks developed in collaboration with the University of Florence (Pacini *et al.*, 2015). This model includes a great variety of ecological (i.e., biodiversity, nitrogen leaching, soil erosion, pesticides risks, etc.) and financial indicators (i.e., gross margin, expenses, depreciation, etc.) that allows comparison of the economic and environmental performance of organic and conventional systems. The model adopted by the Tuscany Region is limited to area payments only for arable crops such as cereals, leguminous, and oleaginous crops and mixed farms with livestock and arable crops.

Even though this model applies to a limited type of farms, the Tuscany Region's Rural Development Programme calculates the direct payment for Measure 11 as an estimation of the ecosystem services delivered by organic farms in a specific pedo-climatic area, through the economic and environmental evaluation of a complex model.

As a result, the framework defines a level of payment that presents an increased efficacy compared to the classic calculation based only on the cost-benefit balance.

The problem of improved efficiency and effectiveness of agri-environment payments is indeed a complex one, as finding methods to increase effectiveness and efficiency in real life conditions is challenging. So far, we are lacking models and scientific methodology able to validate on a farm and field scale the benefits while simultaneously including farmer perspective and behavior, production, and pedo-climatic conditions and a range of environmental potential impacts such as biodiversity decrease, pesticide risk, nitrogen pollution, soil erosion, and conservation of ecological infrastructures. The model in the Tuscany Rural Development Plan is meant to address precisely this gap at least for a limited typology of farms.

However, according to Pacini *et al.* (2015):

...further optimization of the cost-effectiveness of intended policy measures could result from an identification of the efficiency of resource use/production of ecosystems services of a given farm type as compared to the efficiency of an ideal farm.

If this would be achieved, the payments dedicated to organic farming could be fine-tuned to their efficiency performance, ensuring a continuous improvement of farming practices and stronger environmental achievements.

Disincentives Through Taxation

Rewarding positive externalities while minimizing the impact of the negative ones should always be the dual objective of consistent and conducive policy-making supporting an agroecological transition. This should be translated into a coherent policy approach that provides support for producers implementing best practices and also increases legal requirements (i.e., in the area of ecosystem conservation or environmental protection) and levies environmental taxes to disincentivize unsustainable practices.

In the agricultural sector, there are some examples of how economic instruments try to place the economic consequence of pollution and health harm on those responsible for it, thus internalizing the negative externalities. Environmental taxes, or taxes and levies on pollution, are one of these examples and they can be implemented to, for example, discourage nitrate run-off and leaching from farming activities or to reduce greenhouse gas emissions. Taxes and levies can be also applied on inputs such as chemical fertilizers and pesticides.

Several studies (Böcker and Finger, 2016; IFOAM EU, 2018; Slunge and Alpizar, 2019) highlight the effectiveness of taxes on pesticides, if coupled with an additional set of supporting measures that can be included in National Action Plans for an agroecological transition. These should be implemented as highly differentiated tax schemes, which are calculated based on the damage caused by a certain input on the environment. These taxes can be even more effective and create more incentives if they remove taxation or offer lower tax rates for those products that are less hazardous (i.e., plant protection products accepted in the production standards for organic agriculture, as is done in France and Italy with a lower value-added tax [VAT]). The target of these taxes can be two-fold: on the one hand, taxes on chemical pesticides might aim at producing revenue, while on the other hand, they can target a decrease in adoption of harmful substances.

In Europe several countries, including Norway, France, and Denmark, have been adopting taxation schemes on pesticides with different degrees of success.

In 1988 **Norway** introduced its first tax on pesticides. The country has now a differentiated tax, based on seven categories with a focus on the risks for human health and the environment. All pesticides for professional use are tested according to several criteria and then categorized in a low, medium, or high risk. Products allowed in organic farming are exempted from the tax. An impact assessment of the scheme revealed a decrease in the number of detected residues in the water.

Denmark introduced its first tax on pesticides in 1982, which back then was only for impacting consumption from households. After several reforms that extended the taxation schemes to agricultural products as well, the current taxation program, enforced since 2013, is considered to have played a major role in reaching the governmental goal of reducing the use of pesticides by 40% in the period 2013–15 (See Box 7.2 for the case study).

In **India,** from 2005 onwards, the Sikkim state government gradually phased out subsidies on chemical fertilizers and pesticides. This made inputs costlier in order to embrace other strategies to ensure sustainable soils and plant protection. This measure was coupled with a gradual closing-down of the selling points of synthetic inputs and a levy restricting their import. As a final step, the Sikkim Agricultural, Horticultural Inputs and Livestock Feed Regulation Act, 2014[4] ratified the ban on the import of synthetic agricultural inputs and selling any of such substances in the state. These measures were combined with programs to support farmers in producing their own plant protection products and fertilizers on the farm.

In 2014 the government of **Mexico**, in an attempt to address the high amount of pollutants and contamination derived from inputs used in agriculture, imposed an excise tax on pesticides according to the level of their toxicity. The government collected more than €83 million through this scheme between 2012 and 2017. There are plans to earmark these funds to increase efforts of alternative pesticide control and promote more agroecological practices.

Box 7.2 Pesticides Taxation in Denmark

Pedersen and Nielsen (2017) undertook a comprehensive assessment of the performance of pesticides policies in Denmark in the period 1986–2015. According to their analysis, Denmark, which has been implementing different pesticide taxation schemes since 1982, can be regarded as the European pioneer in terms of policy intervention tackling the reduction of pesticides use. It is also important to highlight that these policy interventions were promoted and enforced by a large spectrum of political parties owing to the strong Danish tradition of ensuring protection to their precious groundwater reservoirs, which are used as a source of untreated drinking water and therefore need to be protected from agricultural runoff (Pedersen and Nielsen, 2017).

The previous tax scheme, which enforced an ad valorem tax on retail prices of pesticides between 1996 and 2012, proved to be inefficient to reach the targeted reduction of pesticide use to 50%, as it did not have a built-in incentive for farmers to choose the least hazardous pesticides. In 2013 a new scheme reformed the taxation policy on pesticides introducing a better proxy to evaluate the effect of the scheme (Pedersen and Nielsen, 2017). This new indicator, the Pesticide Load (PL), gives information about three dimensions of the use of a certain pesticide, comprising sub-indicators for the risk on human health as a result of exposure of the operator, the ecotoxicology on non-target organisms, and the persistence and accumulation in the environment. In this new scheme, the tax on pesticides became a differentiated pesticides tax, as the tax fee is calculated for every single pesticide through an individual assessment that takes into account the different aspects mentioned above. According to this tax design, higher taxation rates apply to more harmful pesticides, and in general the tax rates were more than doubled compared with the previous scheme, making Denmark one of the countries with the highest taxation rate for pesticides worldwide. The objective of the new taxation scheme was to reduce the PL by 40% by the end of 2015 compared with that estimated in 2011 (Ministry of Environment and Food, 2017; Pedersen et al., 2020).

The tax revenues are mainly redistributed to the agricultural sector, reimbursing farmers through a reduction of taxes on land. This constitutes an incentive for reducing the use of pesticides, as producers would then have the benefit of low pesticide tax and lower land tax (Pedersen and Nielsen, 2017). In 2017 the tax revenue amounted to approximately €71 million (Pedersen et al., 2020).

A recent assessment done by the Danish Environmental Protection Agency shows that this new taxation scheme achieved indeed a substantial reduction in PL from before the tax introduction until the season 2016/17 within the range of 12–27% (depending on baseline year) according to farmers' registered pesticide use (Pedersen et al., 2020), going up to 44% if we considered sales figures for the period 2011–17 (Pedersen et al., 2020). The difference between use and sales figures are due mainly to hoarding behavior that happened before the introduction of the new tax and which might still cause a reduction in sales today (Pedersen et al., 2020). Despite the substantial reduction, the objective of reaching 40–50% load decrease, as was estimated in *ex ante* analyses prior to the tax introduction, has not been achieved yet (Pedersen et al., 2020).

The demand of pesticides among European farmers has proved to be rather inelastic (Pedersen *et al.*, 2020). Over the years the different policies on pesticides have struggled to reach their objectives as a consequence of incorrect *ex ante* policy analysis. The expected result of the policy did not take into consideration that some farmers might respond to such taxes in a way that economically does not make sense. These farmers, instead of focusing on the production costs, aim primarily at increasing yields over all other dimensions and are less concerned about pesticide prices.

In synergies with its taxation schemes on pesticides, Denmark has been implementing since 1986 different National Pesticides Action plans aiming at reducing the quantity of approved pesticides used in the country, as requested also by the Directive 2009/128/EC, article 4 (European Commission, 2016). Beside the taxation scheme, these plans have introduced over the years different sets of measures, such as support to research, stricter approval procedure for pesticides, mandatory spraying certificates for professional users of pesticides, creation of buffer zones around watercourses and lakes, an increase in the share of organic farmed land, and advice to farmers on reduction of pesticide use (Pedersen and Nielsen, 2017).

Less effective and more difficult to implement is indirect taxation on specific food products, for instance by differentiating VAT on products issued by organic farming and those issued by industrial agriculture enterprises. Applying different VAT rates that favor organic products over non-organic ones is likely to be perceived as against the principle of neutrality, which ensures that similar products should be taxed in the same way to ensure fair competition. In reality, even though certification provides a tool to distinguish between organic and non-organic products, it is still arguable that the properties of the final product (i.e., organoleptic ones) are actually dissimilar.

It is, however, important to mention that, as the economic impact of environmental taxes, such as those on chemical pesticides owing to the low-price elasticity of pesticides, are likely to be paid for by end-consumers, with higher impacts on the lower-income consumers, governments should couple this type of intervention with re-distribution of tax revenue or reduced taxes on, for example, labor.

The Rugged Road Ahead

TCA can help to achieve a better understanding of the sustainability benefits and threats coming from different agricultural actions, practices, and systems and contribute to better decisions. In order to truly benefit from this potential, a number of changes are needed.

We suggest first to focus on approaches that are robust—that is, their beneficial consequences do not depend much on the exact cost estimates used, address the most pressing problems, and do not result in trade-offs and inconsistencies. TCA can be used, for example, to establish a tax on key pollutants from agriculture, namely nitrogen, pesticides, and greenhouse gases, as well as to support practices that promote soil fertility and biodiversity.

Nitrogen is problematic when applied in levels beyond the amount that can be continuously recycled in a circular system. Thus, taxes on nitrogen sources external to some regional ecosystem boundaries are a measure that could be established, complemented with support payments for practices that build on reduced nitrogen inputs, such as grass-based feed harvested largely on-farm, etc.

Second, taxes on carbon dioxide (CO_2) emissions from fossil fuels could be established. If a nitrogen tax is implemented, it is less necessary to establish a general carbon tax in agriculture that also works on nitrous oxide (N_2O) and methane (CH_4). N_2O is covered by the nitrogen tax, as less nitrogen fertilizer input in farming results in lower nitrous oxide emissions from soils. Taxing CH_4 would put ruminants at a disadvantage, although they play a central role in sustainable food systems by converting feed from non-arable grassland areas into food, which is not possible otherwise. To work on CH_4 emissions, support for CH_4-mitigating measures in herd management (e.g., increasing the number of lactations) and manure application as well as housing could be provided. For pesticides, the tax level could be linked to some gross estimate of their damage potential. Some impacts, however, could be deemed unacceptable, so that bans need to be established and enforced.

Regarding soil fertility and biodiversity, support payments could work best. For soil fertility, it can be linked to soil-organic carbon, without applying carbon credits. For biodiversity, a number of practices and systemic changes could be supported, with some conditions on regional and networked landscapes, as successful biodiversity support often depends on connected corridors of a certain size. TCA can give a basic argument for the usefulness of these policies and payments and a first gross estimate of required tax and support levels.

In general, governments should support only agriculture and food systems that deliver on the United Nations' Sustainable Development Goals, counterbalancing the powerful vested interests that global and national agribusiness corporations and commodity groups represent.

Notes

1 Mission Organic Value Chain Development for North East Region (MOVCDNER) is a value chain-based organic farming scheme that began in 2015 and was implemented in the northeastern states of India under the Ministry of Development of the North Eastern region. The mission aims to support the creation of producers' organizations, on-farm and off-farm organic inputs production, certification, post-harvest matters, processing, and marketing.
2 Annex 2, Section 12 of the Agreement on Agriculture. www.wto.org/english/res_e/ publications_e/ai17_e/agriculture_ann2_jur.pdf.

3 European Agricultural Fund for Rural Development.
4 Available at: www.lawsofindia.org/pdf/sikkim/2014/2014Sikkim10.pdf

References

Böcker, T. and Finger, R. (2016). European pesticide tax schemes in comparison: An analysis of experiences and developments. *Sustainability (Switzerland)*, 8(4), 1–22. doi:10.3390/su8040378.

Burton, R.J.F. and Schwarz, G. (2013). Result-oriented agri-environmental schemes in Europe and their potential for promoting behavioural change. *Land Use Policy*, 30(1), 628–641. doi:10.1016/j.landusepol.2012.05.002.

Choi, S. (2015). *Agriculture in Korea*. Naju: Korea Rural Economic Institute.

European Court of Auditors. (2017). Greening: a more complex income support scheme, not yet environmentally effective. EU Court Audit. 287, 1977–2017.

European Union. (2013). REGULATION (EU) No 1306/2013 OF THE EUROPEAN PARLIAMENT AND OF THE COUNCIL of 17 December 2013 on the financing, management and monitoring of the common agricultural policy. *Official Journal of the European Union*, 1305, 549–607.

Eyhorn, F., Müller, A., Reganold, J.P., Frison, E., Herren, H.R., Luttikholt, L., Mueller, A., Sanders, J., Scialabba, N.E., Seufert, V., & Smith, P. (2019). Sustainability in global agriculture driven by organic farming. *Nature Sustainability*, 2, 253–255. doi:10.1038/s41893-019-0266-6.

Geiger, F., Bengtsson, J., Berendse, F., Weisser, W.W., Emmerson, M., Morales, M.B., Ceryngier, P., Liira, J., Tscharntke, T., Winqvist, C., Eggers, S., Bommarco, R., Pärt, T., Bretagnolle, V., Plantegenest, M., Clement, L.W., Dennis, C., Palmer, C., Oñate, J. J., … Inchausti, P. (2010). Persistent negative effects of pesticides on bio-diversity and biological control potential on European farmland. *Basic and Applied Ecology*, 11(2), 97–105. doi:10.1016/j.baae.2009.12.001.

Gibbs, K.E., MacKey, R.L., & Currie, D.J. (2009). Human land use, agriculture, pesti-cides and losses of imperiled species. *Diversity and Distributions*, 15(2), 242–253. doi:10.1111/j.1472-4642.2008.00543.x.

IFOAM EU. (2018). Taxation as a tool towards true cost accounting. Available at: www.organicseurope.bio/content/uploads/2020/06/ifoameu_final_study_on_taxation_as_a_tool_towards_true_cost_accounting.pdf?dd.

IFOAM—Organics International. (2017). *The Full Diversity of Organic Agriculture: What We Call Organic*. Available at: www.ifoam.bio/sites/default/files/2020-03/position_full_diversity_of_oa.pdf.

IFOAM—Organics International. (2019). *Full Cost Accounting to Transform Agriculture and Food Systems*. Available at: www.organicseurope.bio/content/uploads/2020/06/Full-cost-accounting.pdf.

Meehan, T.D., Werling, B.P., Landis, D.A., & Gratton, C. (2011). Agricultural land-scape simplification and insecticide use in the Midwestern United States. *Proceedings of the National Academy of Sciences of the United States of America*, 108(28), 11500–11505. doi:10.1073/pnas.1100751108.

Meemken, E.M., & Qaim, M. (2018). Organic Agriculture, Food Security, and the Environment. *Annual Review of Resource Economics*, 10(1), 39–63. doi:10.1146/annurev-resource-100517-023252.

Ministry of Environment and Food. (2017). *Danish National Actionplan on Pesticides 2017–2021*.

Pacini, G.C., Merante, P., Lazzerini, G., & Van Passel, S. (2015). Increasing the cost-effectiveness of EU agri-environment policy measures through evaluation of farm and field-level environmental and economic performance. *Agricultural Systems*, 136, 70–78. doi:10.1016/j.agsy.2015.02.004.

Pedersen, A.B. & Nielsen, H.Ø. (2017). Effectiveness of Pesticide Policies. *Environmental Pest Management*, December, 297–324. doi:10.1002/9781119255574.ch13.

Pedersen, A.B., Nielsen, H.Ø., & Daugbjerg, C. (2020). Environmental policy mixes and target group heterogeneity: analysing Danish farmers' responses to the pesticide taxes. *Journal of Environmental Policy and Planning*, 22(5), 608–619. doi:10.1080/1523908X.2020.1806047.

Ponisio, L.C., M'Gonigle, L.K., Mace, K.C., Palomino, J., de Valpine, P., & Kremen, C. (2015). Diversification practices reduce organic to conventional yield gap. *Proceedings of the Royal Society, B*, 282. 20141396. doi:10.1098/rspb.2014.1396.

Schader, C., Stolze, M., & Gattinger, A. (2012). Environmental performance of organic farming. In: Boye, J. I., Arcand, Y. (Eds.), Green Technologies in Food Production and Processing. *Food Engineering Series*, 183–210. https://doi.org/10.1007/978-1-4614-1587-9.

Seufert, V. (2018). Comparing yields: Organic versus conventional agriculture. In P. Ferranti, E.M. Berry, & J.R. Anderson (Eds.). *Encyclopedia of Food Security and Sustainability: Volume 3: Sustainable Food Systems and Agriculture, 3*, 196–208. doi:10.1016/B978-0-08-100596-5.22027-1.

Seufert, V. & Ramankutty, N. (2017). Many shades of gray—the context-dependent performance of organic agriculture. *Science Advances*, 3(3). https://doi.org/10.1126/sciadv.1602638.

Slunge, D. & Alpizar, F. (2019). Market-based instruments for managing hazardous chemicals: A review of the literature and future research agenda. *Sustainability (Switzerland)*, 11(16). doi:10.3390/su11164344.

Stolze, M., Sanders, J., Kasperczyk, N., Madsen, G., & Meredith, S. (2016). *CAP 2014–2020: Organic farming and the prospects for stimulating public goods.* Brussels: IFOAM EU.

Swiss Federal Office for Agriculture. (2004). *Swiss Agricultural Policy.* Federal Department of Economic Affairs.

van der Werf, H.M.G., Knudsen, M.T. & Cederberg, C. (2020). Towards better representation of organic agriculture in life cycle assessment. *Nature Sustainability.* doi:10.1038/s41893-020-0489-6.

Willer H., Schlatter B., Trávníček, J., Kemper, L., & Lernoud, J. (2020). *The World of Organic Agriculture - Statistics and Emerging Trends 2020.* Research Institute of Organic Agriculture (FiBL), Frick, and IFOAM – Organics International. doi:10.4324/9781849775991.

8 Transforming the Maize Treadmill

Understanding Social, Economic, and Ecological Impacts

Francisca Acevedo Gasman, Lauren E. Baker, Mauricio R. Bellon, Caroline Burgeff, Alicia Mastretta-Yanes, Rainer Nerger, Harpinder Sandhu, Esmeralda G. Urquiza-Haas, Stephanie White and Gyde Wollesen

Introduction

Maize was domesticated in Mexico around 9,000 years ago. Some 5,000 years later, maize spread from its center of origin in Mesoamerica to the rest of the American continent and subsequently to the rest of the world (Vigouroux *et al.*, 2011). In the early 1930s modern varieties began to be developed. These replaced almost all of the locally adapted traditional varieties, or landraces, of maize in the USA in little more than three decades (Duvick, 2001). In Europe, modern varieties were introduced at the end of World War II, completely modifying the traditional agricultural systems typified by high biodiversity and low external inputs (Duvick, 2001, 2005). This agricultural transformation, or "green revolution," has had profound consequences for Earth's biodiversity and ecosystem services (Bommarco *et al.*, 2013), agricultural systems (Garbach *et al.*, 2014; Power, 2010; Zhang *et al.*, 2007) and diets (Popkin, 1993; Baker, 2013).

Maize has been at the heart of socio-economic-ecological food systems transformations. Easily adapting to diverse climatic and agricultural conditions, as well as agro-industrial processes, maize is a central ingredient in high-fructose corn syrup (driving a transition to calorie-dense, nutrient-poor diets), animal feed (driving the meat industrial complex), biofuel, and plastics, among others. Monoculture corn production is the most widely planted crop in the world, driving land use change with significant climate impacts. At the same time, diverse smallholder maize systems continue to thrive in Mexico and beyond, demonstrating the resilience of these agricultural practices and their associated cultural and culinary traditions (Baker, 2013).

How do we understand, evaluate and account for the dynamics across diverse maize systems and their impacts over time? Recognizing the multiple costs and benefits provided by these systems is a first step in assessing their value and promoting strategies to mitigate negative externalities and strengthen positive benefits. This chapter highlights four TCA applications undertaken in very different contexts in Mexico, the USA, Malawi, and Zambia.[1] These studies help us to understand the challenges and potential of using TCA to address food system externalities, both positive and negative. The approaches and methodologies vary significantly, but all shed light on the utility and limitations of TCA. Most importantly, they help us better to understand maize systems, their contributions, their histories, their impacts, and the ways that they are embedded in broader ecological, political, and economic systems. Key insights from these kinds of assessments can inform the policies, programs, and practices that shape agricultural systems worldwide. The authors of this chapter provide case studies summarizing their research results and then offer some high-level reflections about what was learned across the different cases. The first case study—an assessment of maize by colleagues at the National Commission for the Knowledge and Use of Biodiversity in Mexico—reveals the importance of smallholder agricultural systems for agricultural biodiversity. Through the cultivation of native varieties, traditional smallholders in Mexico and other parts of the world provide important evolutionary services to maize production worldwide. These smallholder systems promote genetic diversity by cultivating maize under a wide range of environmental, socio-cultural and economic conditions and by selecting certain traits to adapt maize to these different conditions or to satisfy cultural preferences. This study provides insights into how these evolutionary services can be further enhanced by supporting traditional agricultural practices and maintaining heterogeneous landscapes linked to cultural and ecological resources.

No country has benefited more from the industrialization of maize production than the USA, where corn is a crop of significant economic importance, adding $48.5 billion annually to the US economy. To understand the US maize system, Harpinder Sandhu and team visited Minnesota where genetically modified (GM) and conventionally grown (hybrid) corn dominate the landscape, and organic corn is grown on only a fraction of the land. The second case study presented in this chapter compares the "true cost" of conventional GM corn and organic corn production in Minnesota, revealing how agricultural policy and programs "lock in" certain production practices, despite the clear benefits of organic corn systems.

The third case study is based on maize in Malawi. The notion that maize is central to food security in Malawi is a widely held view. To date, however, "maize-led development" has produced disappointing outcomes. Stephanie White provides an overview of the historical, political, and environmental context of the maize agri-food system in Malawi and uses the TEEBAgriFood framework to analyze maize in relation to three distinct parameters: input stocks and flows, fertility stocks and flows, and maize stocks and flows. This case study reveals the costs of remaining beholden to a maize-centric agri-food system, as well as the factors that keep this system in place, despite calls for agricultural diversification.

Finally, we draw on a study of three maize cropping systems in Zambia. These three systems have different characteristics—monocropping, smallholder mixed, and larger-scale rotation. Each of these systems is carefully documented and analyzed to understand the implications of specific practices and provide guidance to farmers, businesses, and policymakers. This analysis helps us to understand the experience of farmers and how their farming systems can be improved to enhance soil health, livelihoods, and mitigate environmental impacts.

To conclude, we offer six reflections and challenges that emerge from a collective analysis of these four very different TCA studies. The case studies strive to describe the broader political, ecological, social, and cultural systems in which they are embedded, illustrating the challenge of systemic framing. Because the studies are so different, there is a challenge related to finding a common vocabulary for TCA. The studies balance an imperative to address the historical forces that shape food systems while focusing on direct and indirect costs. All the studies, in one way or another, illustrate the potential benefits provided by agrobiodiversity managed by smallholders, and how these benefits are narrowed in monocropped maize systems. The authors of this chapter noted the challenge of making meaningful comparisons across systems and studies, owing to the different approaches and methodologies used. All studies highlight the challenge of engaging decision-makers—farmers, business leaders, and policymakers—to reflect on research findings and adapt policies and practices accordingly. Finally, the authors note the importance of providing appropriate guidance for TCA studies to ensure their relevance and applicability to decision-making.

Case Study 1: Maize in Mexico: Ecosystem Services Provided by Agrobiodiversity and Traditional Small Holders

Francisca Acevedo Gasman, Mauricio R. Bellon, Caroline Burgeff, Alicia Mastretta-Yanes, and Esmeralda G. Urquiza-Haas.[2]

Traditional smallholder agriculture has generally been considered a low-productivity system. Only recently has the role of traditional farmers as key contributors to local and regional food security been estimated and documented (Ricciardi *et al.*, 2019; Bellon *et al.*, 2018; Graeub *et al.*, 2016). A greater proportion of food and nutrients that feed people living in the most populated regions of the world is produced by small and medium-sized farms (Pengue and Gemmill-Herren, 2018). Small farms represent 93.3% of all farms in the Asia-Pacific region, 89.6% of those in Africa, 80.1% in Latin America and the Caribbean, 76.8% in North America, and 88.5% in Europe (Grain, 2014). Despite them representing 92.3% of all farms worldwide, they hold only 24.7% of agricultural land (Grain, 2014). The importance of smallholder agriculture, however, reaches far across time and space and, as discussed here using the case of Mexico, might also provide multiple benefits at the genetic, species, and landscape levels. The recognition of the multiple benefits provided by agrobiodiversity managed by smallholders at these three levels is the first step in assessing the value of these systems and promoting strategies to scale up some of these principles among more intensive agricultural units.

The Genetic Level

Today, most maize grown worldwide is cultivated in intensive systems, destined for livestock, sweetener, and oil industries, as well as for the production of ethanol and other non-edible products (see review in Comisión Nacional para el Conocimiento y Uso de la Biodiversidad, 2017). A great proportion of this production comes from modern varieties (2017), whose highly uniform genetic background makes them vulnerable to pests and pathogens causing severe crop losses (National Research Council, 1974, 1975). For instance, in the 1970s, the Southern Leaf Corn Blight (*Bipolaris maydis;* also known as *Helminthosporium maydis*) resulted in the loss of 15% of the maize crop in USA and generated economic losses of about $1 billion dollars (Bruns, 2017). The fungus was able to spread rapidly and infect large areas because most hybrid maize at that time was uniform for a condition called cytoplasmic male sterility and happened to be susceptible to this fungus. Between 2012 and 2015 losses of maize yield owing to diverse root rots, seedling blights, and plant-parasitic nematodes in the USA and Ontario, Canada, were calculated at over $27 billion (Mueller *et al.*, 2016). Genetic resources from landraces and wild relatives contributed to the recovery of pre-existing production levels in cases like these (Redden *et al.*, 2015; Maxted *et al.*, 1997).Currently, most of the genetic diversity of maize is harbored in Mesoamerica (Bedoya et al., 2018), the center of origin and domestication of this crop (Pickersgill, 2007; Matsuoka *et al.*, 2002; Vavilov, 1926). Centers of domestication and diversity of crops represent key regions for the food security of present and future human societies (Brush, 1995) and are generally characterized by a long history of use of these species and a high varietal diversity that is fostered by traditional smallholders and co-occurrence with their ancestors and wild relatives (Vavilov, 1926, 1951; Vavilov *et al.*, 1992).

Traditional smallholders around the globe continue to grow maize, from sea level to 3,800 meters, including arid regions and regions with more than 11,000 mm of rainfall per year, and from 42° latitude South to 50° latitude North (Timothy *et al.*, 1988), using seed from the previous cycle, subjecting maize populations to a continuous evolution under domestication. The genetic base that confers maize its ability to prosper in such diverse environmental conditions is the result of its cultivation during millennia under this wide range of environmental conditions (Ruiz Corral et al., 2008).

Considering the area cultivated with maize by traditional farmers in Mexico, it is estimated that about 1.38 trillion genetically different individual plants are subjected to both natural and artificial selection, under heterogeneous cultural, biological, and environmental conditions. From this, a breeding population of about 5.24 billion maize mother plants transmit their variation to the next generation each cycle, which contributes to both preserving rare alleles and to generate new potentially adaptive variations (Bellon *et al.*, 2018). This "evolutionary service" constitutes a public good offered by traditional farmers in Mexico (Bellon *et al.*, 2018) and other regions of the world (Bedoya *et al.*, 2017; Bracco *et al.*, 2009; Carvalho *et al.*, 2004; Hartings *et al.*, 2008; Kumar *et al.*, 2015; Qi-Lun *et al.*, 2008) and to society in general. In Mexico, the genetic diversity of cultivated maize is also constantly broadening through the influx of genes from wild relatives like teosinte, which is tolerated or actively promoted by traditional farmers (Rojas-Barrera *et al.*, 2019; Hufford *et al.*, 2013; Matsuoka *et al.*, 2002; Wilkes, 1977).

Traditional farmers cultivating landraces in Mexico and other regions of the world do not only maintain their genetic diversity (i.e. evolutionary services), but also promote the identification and fixation of adaptive traits through the selection of the desired phenotypes. Local landraces are preferred by traditional smallholders in areas where they are traditionally grown, which tend to be areas with poorer soils or suboptimal climate conditions (Comisión Nacional para el Conocimiento y Uso de la Biodiversidad, 2017). Some maize landraces thrive in a wide range of environments, like the *Tuxpeño* which is to be found in 19 of the 28 climatic types present in Mexico. Others like the *Blando de Sonora* and *Azul* landraces were only found in three climatic types (Comisión Nacional para el Conocimiento y Uso de la Biodiversidad, 2011). In line with this, landraces have shown specific adaptations to nitrogen poor soils (e.g. *Rojo, Piedra Blanca, Llano*: Van Deynze et al., 2018), resistance to drought (e.g. *Nan-Tel, Conejo, Ratón, Cónico Norteño* (population Zac-58), *Chalqueño* (population MICH-21: Avendaño et al., 2005; Muñoz, 2003; Wellhausen et al., 1951) and to different pest and pathogens (e.g. *Zapalote chico, Tuxpeño, Olotón*: Bellon and Risopolous, 2001; Muñoz, 2003; Ramírez et al., 2005; Widstrom *et al.*, 2003). Besides their adaptability (i.e. resistance to biotic and abiotic stresses), landraces are also appreciated for their precocity and their food/forage quality (Comisión Nacional para el Conocimiento y Uso de la Biodiversidad, 2011; González-Amaro, 2016; Muñoz, 2003; Perales, 1996; Rodríguez *et al.*, 2007; Sociedad Mexicana de Fitogenética 2007, 2009). Hence, maize landrace adaptations provide or complement ecosystem services of pest and disease control, weather and water regulation, soil fertility, and cultural services derived from their use.

The Species Level

In Mesoamerica, maize is cultivated along with beans (*Phaseolus vulgaris*) and squash (*Cucurbita pepo, C. ficifolia, C. maxima, C. mixta and C. moschata*), also known as the "three sisters," and a varying number of domesticated and semi-domesticated species in a system called *milpa* (Gómez-Pompa, 1987; Hernández-Xolocotzi *et al.*, 1995; Terán and Rasmussen, 2009). These species not only balance the nutritional quality/quantity of the milpa-derived diet, but also represent species that provide ecosystem services for the agricultural system itself (Amador and Gliessman, 1990; Table 1).

The review by Hajjar *et al.* (2008) indicated that ecosystem services (e.g. pollination, CO_2 sequestration, pest and disease control, improved soil fertility, and reduced soil erosion) provided by crop genetic diversity, are mediated by an increase in the number of functional traits and complex interactions occurring in diverse agroecosystems. In the case of the Mesoamerican *milpa,* the benefits provided by intercropping of the 'three sisters' are attributed to their distinct root architectures which allow them to use potentially complementary water and nutrient acquisition strategies (Postma and Lynch, 2012). The practice of intercropping maize with legumes (and often squash crops) has become a global heritage, implemented by traditional farmers in Mexico, but also by traditional smallholders as well as large-scale agricultural systems around the world (see review in Comisión Nacional para el Conocimiento y Uso de la Biodiversidad, 2017).

Nutrient cycling services, which are facilitated by the intercropping of maize and beans, are enhanced owing to the fixation of aerial nitrogen by *Rhizobium* bacteria present in the roots of beans, increasing the availability of nitrogen for maize (Van Berkum *et al.*, 1996), and of phosphorus for both maize and bean (Latati *et al.*, 2013). Legumes are also used for weed (Caamal-Maldonado *et al.*, 2001) and pest (Altieri *et al.*, 1977) control in traditional agroecosystems. Squash has been found to suppress the growth of weeds and to avoid the loss of soil moisture (Chacon and Gliessman, 1982; Ghanbari *et al.*, 2010). Weeds, some of which are used and tolerated in the agricultural fields (Vibrans, 2016), may serve to reduce the presence of crop pests through different mechanisms: as alternative food sources for herbivorous arthropods or beneficial insects (predations, parasitoids, pollinators), and as a source of diseases for pests (see review in Capinera, 2005; Norris and Kogan, 2005; Wisler and Norris, 2005).

The Landscape Level

The size (e.g. small) and aims (e.g. semi-subsistence) of traditional agricultural systems tend to promote more diverse landscapes through the maintenance of surrounding native vegetation, the presence of home gardens, and the use of multiple species as a life strategy (Bellon *et al.*, 2020; Neulinger *et al.*, 2013; Palacios *et al.*, 2013). The heterogeneous landscapes that these rural strategies promote, not only favor the provision of ecosystem services at the local scale such as pollination, regulation of pests and diseases, and water provision, but could potentially do so at a larger scale (Landis, 2017) (Table 8.1).

The exploratory maize study sought to identify and value both the dependencies and impacts of two contrasting maize systems, from and on, various ecosystem services in Mexico, Ecuador, and the USA (Comisión Nacional para el Conocimiento y Uso de la Biodiversidad, 2017). Here, we decided to highlight the contribution of traditional smallholder farming systems to agro-biodiversity at the genetic, species, and landscape level. Understanding and recognizing these multiple benefits is the first step in assessing their value and

Table 8.1 Ecosystem services provided by intercropping of associated crops and semi-domesticated species in maize-centered agroecosystems and landscapes

Crop/Landscape element	Mechanism	Ecosystem service	Source
Bean	Fix atmospheric nitrogen through its association with bacteria from the *Rhizobium* genus	Nutrient cycling	Caamal-Maldonado et al., 2001; Bilalis et al., 2010; Van Berkum et al., 1996
	Reduces weed density	Provision	
	Increases maize yield		Ebel et al., 2017; Tsubo et al., 2003
Squash	Prevents soil erosion, and increases carbon capture and storage	Nutrient cycling	Caldeira et al., 2004
	Inhibits the growth of weeds through photosynthetically active radiation interception	Provision	Fujiyoshi et al., 2007; Ghanbari et al., 2010
Beans and Squash	Increase of expected yield of maize, bean and squash in monoculture	Provision	Ebel et al., 2017; Postma and Lynch, 2012; Zhang et al., 2014
Weeds	Interfere with the ability of insects to locate crop plants, provide alternative food sources for pests, harbor insect diseases, and attract beneficial insects	Pest control	Altieri et al., 1977; Penagos et al., 2003
Intercropping	Decrease in water loss and increase in water use efficiency due to early high leaf area index	Water provision	Ogindo and Walker, 2005
Landscape complexity, diversified agricultural systems and polycultures	Associated with a lower density of pests, increase of natural enemies, higher mortality of herbivore insects and less damage to crops	Pest and disease control	Kremen and Miles, 2012; Letourneau et al. 2011; Tonhasca and Byrne, 1994
Maintenance of natural habitats around agricultural fields	Provide food and refuge for natural enemies of agricultural pests	Pest and disease control	Landis et al., 2000; Power, 2010; Tillman et al., 2012; Tscharntke et al. 2005;
Use of native vegetation as hedgerows and maintenance of natural habitats	Provide food, refuge and reproduction sites for pollinator species	Pollination	Garbach et al. 2014; Power, 2010; review in Potts et al., 2010 and Garibaldi et al., 2011
Maintenance of riparian vegetation	Reduces water runoff and soil erosion Removes nutrients and sediments before they leach into superficial water bodies	Regulation	Swinton et al., 2007

can support the development of strategies to enhance biodiversity and ecosystem services not only in smallholder systems, but also in more intensive agricultural systems.

Case Study 2: Corn in Minnesota: Comparing Conventional GM and Organic Systems

Harpinder Sandhu[3]

In Minnesota, corn adds $4.5 billion annually to the local economy. Genetically modified (GM) corn dominates the landscape, with organic corn grown in a fraction of the area. GM corn is grown primarily for ethanol production and a by-product from the ethanol processing process is used for animal feed. It is typically grown in rotation with soybean as a monoculture. Organic corn is primarily used directly as animal feed and is grown in mixed farming systems. To provide some context related to the scale of GM and organic production, in the USA, approximately 88% of the corn grown is GM, whereas certified organic corn represents only 0.02%.

In order to examine corn-based farming systems in the US context, two contrasting management systems—GM corn and organic corn—were selected for a study in Minnesota. The TEEBAgriFood food systems evaluation framework (described in Chapter 4) was applied to these two corn production systems in Minnesota to reveal impacts and dependencies on produced, social, human (including health), and natural capital and evaluate the hidden costs and benefits of corn production (Sandhu *et al.*, 2020).

Corn production in the US intersects with a number of critical environmental issues. Corn-based ethanol production has increased the demand for corn and hence increased associated environmental impacts without a clear reduction in the carbon intensity of fuel. Moreover, corn produced for animal feed is much less efficient in terms of producing human food calories per unit area than corn produced for direct human consumption. Large amounts of fertilizers and herbicides are used in GM production systems. This increases the cost of production and reduces net returns. In addition, there is continuous exportation of nitrate, phosphorus, and sediments from farmland to watersheds and ravines in the Mississippi River basin, which leads to the hypoxic zone in the Gulf of Mexico.

A number of policy and program drivers have resulted in the expansion of corn systems in the state of Minnesota and across the USA. These include the Renewable Fuel Standard, which created favorable market prices for corn. However, these farming systems are being modified to include sustainable practices or Best Management Practices, such as inclusion of cover crops, minimum, or strip tillage to minimize soil degradation and prevent loss of nutrients from the system. Some of the practices are promoted by the Environmental Quality Incentive Program and Conservation Stewardship Programs.

The study revealed that corn production is driven by economic factors—produced capital—and is supported by extensive social networks in Minnesota. Regarding human capital, there are significant health costs associated with GM corn production. The health costs are based on the production side of the corn value chain. These do not include capital costs incurred in the public health system, loss of economic productivity, and loss of taxes and Gross Domestic Product. They might, to a certain extent, include the individual medical expenditures associated with the health impacts of living near corn farms (Sandhu *et al.*, 2020). Natural capital—externalities related to climate change, water quality, air quality, and soil quality—is impacted negatively by GM corn production in Minnesota. At a macro level, the study revealed that each bushel of GM corn (at a market price of $3.05 per bushel) generates negative environmental externalities of $0.37 (+$0.02) and $0.88 in health costs (see Table 8.2).

Corn is a dominant crop in Minnesota and is vital for the agricultural economy. About 24,000 corn farmers generate more than $4.5 billion for the state economy. Net returns for farmers are higher in organic corn systems. GM corn yield is higher than organic corn systems, but net returns are lower owing to high variable costs of agrochemicals and lower market price. The total environmental costs (including impacts on climate change, water quality owing to nitrate and phosphorus load, air quality, and soil quality) associated with GM corn production are $71.60 per acre or $557.65 million annually in Minnesota, and this does not include environmental costs associated with the transport, processing, and consumption. Uncertainty and spatial heterogeneity cause this estimate to vary greatly. There are high health costs associated with

Table 8.2 Summary of Health and Environmental Cost $/bushel of Corn Under Two Production Systems

	GM corn	Organic corn
Market price ($/bushel)	3.05	7.46
Environmental costs associated with fertilizer use ($/bushel)	0.37	Not quantified owing to lack of data on organic farms
Environmental costs associated with energy use ($/bushel)	0.02	0.03
Health cost ($/bushel)	0.88	0.00 Although there is some suggested evidence for reduced adverse association of organic corn production with general health, quantifying the health costs requires data on exact location and planted area of organic corn farms. This was not available.

GM corn production. Total annual health costs associated with corn production in Minnesota are \$1.3 billion (\$233 per capita) or \$171 per acre (for 7.6 million acres of harvested corn in Minnesota in 2017). GM and hybrid corn production systems notably use large amounts of ammonium and nitrate fertilizers and herbicides. Fertilizers, herbicides, and dust from corn systems have been associated with different types of cancer (affecting digestive and reproductive organs and blood) and respiratory diseases. Considering that organic corn production refrains from chemical usage, it is assumed that these systems' agri-environment has a neutral impact on health. The study demonstrated that general health of individuals decreases by 0.67% with GM corn production, totaling annual non-financial health costs of corn in Minnesota at some \$1.3 billion.

Outcomes from this study can be used to make decisions about production systems and practices that can improve all four capitals, and policymakers can use this information to incentivize systems and practices that enhance social, environmental, and economic well-being of farmers and society at large. This would require a major shift in US agricultural and energy policies that currently favor GM corn systems.

This multidimensional assessment has helped understanding of key impacts and dependencies and the true costs and benefits of two corn production systems in Minnesota. In order to apply this understanding to effect change there is a need to understand how farmers and policymakers adopt new information and a need to work in consultation with these decision-makers to develop pathways for change. True Cost Accounting (TCA) needs to become a relevant decision-making tool at the farm level and needs to be integrated into policymaking.

Case Study 3: Maize-led Development in Malawi

Stephanie White[4]

Efforts to transform the farm sector in low-income countries (LICs) have persisted in one form or another since the inception of the "Green Revolution" in the 1950s, an approach to farming, food security, and economic progress that rested on the assumption that the basic problem in agri-food systems development was low agricultural productivity. Agriculture, so the argument goes, can serve as an "engine of modernization and growth," provided that yields can be raised, economies of scale achieved, and global markets accessed. At the farm level, the planned transformation of agriculture in LICs was carried out through increased use of modern inputs, which includes hybrid "improved" seeds, chemical pesticides, fertilizers, and insecticides; irrigation and mechanization; land consolidation; and integration of farmers into global markets, which has implications for what crops should be grown. Seventy years after its inception, this basic approach to agricultural development and food security continues to dominate policy interventions and recommendations.

The technologies, policies, and other investments associated with the Green Revolution dramatically boosted aggregate output per person, while income and population growth, policy liberalization, foreign direct investment, and other globalization processes drove exponential growth and consolidation of the retail food sector (Hawkes, 2006, 2018; Reardon *et al.*, 2009, 2018; Thompson and Scoones, 2009). Particularly over the past three decades, financial and trade liberalization encouraged transnational food and beverage corporations to colonize local value chains in many low- and medium-income countries, replacing them with food exchange processes that are spatially long and vertically integrated, that is, supermarkets (Anand et al., 2015; Hawkes, 2006; Reardon *et al.*, 2018; Stuckler *et al.*, 2012). This structural transformation is indicative of economic globalization and therefore reads as economic development and progress towards an "advanced" food system (Reardon *et al.*, 2018). However, the link between agricultural productivity, commercialization, and improvements in well-being are widely assumed, although common development indicators tell a mixed story. They also tell an incomplete story as many outcomes, such as soil degradation, pest build-up, price volatility, or other contextual factors, such as climate change, are left out of analyses and/or treated as separate, unrelated problems.

Transitioning away from agriculture that is preoccupied with yields and governed by the notion of competitive markets, towards one that aims towards sustainable food security requires different frames. Most TEEBAgriFood work has focused on environmental sustainability, but in order for agriculture to be sustainable and *just,* frames must also acknowledge historical injustice and power relationships that are at the foundation of development food policy and practice.

In Malawi, for example, maize is the preferred staple and foundation of the agri-food system. It occupies at least 60% of cultivated land and is farmed by 97% of farming households on very small tracts of land, ranging in size from 0.5 hectares to 1.5 hectares. It makes up 60–70% of total food intake and 48% of protein consumption. Average yields are around 1.2 MT/ha, which is lower than the average for Africa (1.8 MT/ha), also considered far below the average potential. Western economists have referred to maize as "a ray of hope" for Africa's food security crisis, while policymakers and donors assert that raising maize productivity and improving the performance of maize input and output markets is an essential condition for achieving food security. Consequently, huge investments have gone towards raising maize yields. At an international level, public and private agricultural research organizations devote millions of dollars every year to developing improved varieties and cropping techniques.

The association of maize security with food security has its roots in colonialism (Kampanje-Phiri, 2016; McCann, 2001). From about 1912 onwards, the British promoted maize as a foundation for food security and used it as a vehicle to exert control over agricultural production and distribution. Kampanje-Phiri (2016) explains how legislation introduced during the colonial period asserted control over what smallholder farmers grew, rationalized by the British as a necessary measure "because of African improvidence" (Ng'ong'ola, 1986, pp. 244).

Following independence in 1964, Kamuzu Banda, who presided over Malawi from 1964–1994, continued to use maize-based food security as a means of exerting control, but in ways linked more tightly to Malawian culture. Beginning in the 1980s, Banda's government began to promote hybrid maize and fertilizer use among smallholders. The parastatal marketing board Agricultural Development and Marketing Corporation turned its attention to the smallholder sector, distributing subsidized fertilizers, marketing farmers' grain, and transporting grain to food-deficit areas during the hungry season. It was this combination of practices that marked the onset of continuous maize mono-cropping and land tilling. Since 2004, the Malawian government's central policy to bring about maize self-sufficiency is the Farm Input Subsidy Programme (FISP), which seeks to provide around half of farm households with fertilizer and improved seeds at varying subsidized costs. At one point, FISP was celebrated as the Malawi Miracle, but in recent years it is frequently maligned as expensive and inefficient, though many still credit it with creating macro-level food security.

A closer examination guided by the TEEBAgriFood framework reveals that maize-led development has produced disappointing outcomes. Despite notable (reported) increases in average national maize yields, human development indicators have scarcely budged and, in some cases, are deteriorating. High volatility continues to characterize maize markets, diets are poorly diversified, malnutrition among children remains high, and poverty levels have increased in recent years (International Monetary Fund, 2017; Mazunda, 2013; Mockshell and Zeller, 2016; Schiesarie,). In addition, environmental resource stocks such as agrobiodiversity and soil fertility, which are particularly critical to smallholder farmers who are not able to easily access purchased inputs, are deteriorating due to the continuous cropping of hybrid maize (Bezner Kerr and Patel, 2014). Climate change is expected to have widely variable impacts that exacerbate uncertainty and extremes. Changes to rainfall distribution are uncertain, but no models project increased precipitation. In the short term, climate change could benefit maize production, but increased maize production could worsen soil degradation and deforestation.

At the heart of maize-centricity is the persistent narrative that maize security and food security are the same thing. In addition to technical interventions and programmatic investments, there is a basic and fundamental need for a national conversation to challenge that narrative and to engage the population to identify alternatives. In addition to the core "maize security equals food security" narrative, multiple dependencies keep the maize agri-food system locked in. An alternative path forward will include investments and other forms of support to the public extension system and local food exchange practices, institutions, and processes; use of decision-making frames governed less by imperatives to "modernize" and participate in a global food system, and more by socio-ecological well-being and agri-food system resilience in the face of climate change; and immediate transition to regenerative agriculture practices that reduce reliance on imported nutrients. The application of the TEEBAgriFood framework to research protocols could help to develop alternative strategies in collaboration with farmers and other food system actors in Malawi. Among the various geographies and ecologies in Malawi, what alternative food crops could help to transition away from maize? What is their cost of production? How do food security calculations change? What existing food practices could be supported to improve income-earning opportunities to diversify not only the maize production system, but other areas of food exchange, processing, and retailing? What are the hidden costs of other potential staples in relation to the daily realities of average Malawians?

Knowledge creation is a political and ecological process. Any metric system comes with embedded values about what matters, and like any metric system, TEEBAgriFood may be used to further particular interests. Moreover, the TEEBAgriFood evaluation framework is not immune from being used in overly technocentric ways and excluding non-experts from decision-making. To bring about more sustainable and *just* agri-food systems, engagement of marginalized and poor communities should be integral and profound.

Case Study 4: True Cost Assessment of Different Maize Cropping Systems in Zambia

Rainer Nerger and Gyde Wollesen[5]

This assessment was carried out by Hivos Zambia and involved the analysis of three farming systems and ten representative farms.

- Cropping System 1: Small-scale farms producing only maize (monocrop system)
- Cropping System 2: Small-scale farms applying a mixed farm system (small-scale rotation)
- Cropping System 3: Larger-scale farms with a mixed cropping system (large-scale rotation)

The three systems were different in a number of ways. The farms in Cropping System 1 did not have irrigation and were left fallow about half of the year. Comparatively small quantities of fertilizer (synthetic and manure) were applied in these systems, resulting in comparably low yields. Some of the crop residues were left in the field. The farms in Cropping System 2 cultivated maize in rotation with vegetables such as cabbage and onions. Irrigation was used, and slightly higher amounts of synthetic fertilizers were applied, and yield levels were significantly higher compared with Cropping System 1. Some of the farmers in this system applied small amounts of compost and manure, some planted cover crops, and in other cases crop residues were either left on the field or were burnt. In Cropping System 3, the larger-scale farms cultivated maize in rotation with leguminous crops such as field beans or soybeans. Various irrigation systems were used, and comparably high amounts of synthetic fertilizers were applied, resulting in comparably good yields. Some of the assessed farms applied small quantities of compost and manure, but only one farm used cover crops and left crop residues in the field.

To provide a more qualitative description of the farms, two farmers are described below. Mrs Chingambu's farm was part of cropping system one. Over the past 20 years, her best harvest decreased from 2.2 tonnes per hectare to 1.0 tonne per hectare. The seeds that she used were not high quality and not selling at the best possible price. Mrs Chingambu's maize was recently affected by diseases and infections and her plants appeared unhealthy at various stages. Mrs Chingambu has been losing revenue and intends to shift her farming to exclusively livestock. This is a result of the debt that she has accumulated buying fertilizers and chemicals, which she is unable to pay for because of a lack of profit. In contrast is Mr and Mrs Moyo's farm. This farm was part of Cropping System 2 and rotated maize with legume cover crops. Frequent weeding, the addition of animal and green manures, and intercropping with beans resulted in high soil fertility and good yields. Mr and Mrs Moyo planted natural hedges around their maize to discourage pests. They regularly slashed the green manure crop and pruned the legume shrubs, leaving the residues to cover the soil. The family is harvesting about 2.1 tonnes per hectare.

TCA examined the environmental implications and externalities of farming practices across the three cropping systems. Findings illustrate that the real cost of maize production systems is on average 2 to 2.5 times higher than what is actually being paid for in the marketplace. In other words, maize is produced at the expense of future production potential. More sustainable farming practices, as demonstrated by Cropping System 2, can have an almost neutral environmental cost that becomes a net benefit after 5–7 years of using better practices. The major limitation of the TCA calculations in this study is the small number of farms studied.

Table 8.1 shows the environmental costs if maize is more sustainably grown. This includes the following sustainable farming practices: 1) using the crop residues for composting or mulch instead of spreading or burning them; 2) reduce tillage; and 3) systematically use cover crops to loosen the soil, suppress weeds, keep the soil moist and cool, and fix atmospheric nitrogen into a form in the soil that is usable by crops. If these practices are applied, then the environmental costs can be reduced significantly per hectare and year.

Compared with the business-as-usual (baseline) scenario, the significant difference of the sustainable alternative model is that carbon is sequestered, and soil was not eroded but built up. These impacts lead to an increasingly resilient farming system, making better use of nutrients and water, reducing the risk of increasing production costs and crop failures. Ultimately, this moves farmers from a vicious to a virtuous production cycle.

The assessment results showed that the true cost of maize production is 2 to 2.5 times higher than the current market price. However, if sustainable production practices are promoted and incentivized the farmer's risk is reduced and the resilience of the maize-based farming system increases. The currently

Table 8.3 Results of True Cost Accounting of Different Maize Production Systems[6]

	Cropping System 3		Cropping System 1		Cropping System 2		
	Large-scale mixed system		*Small-scale single maize system*		*Small-scale mixed system*		
	Baseline: large-scale maize rotation	*Scenario: large-scale maize rotation*	*Baseline: maize monocrop system*	*Scenario: maize monocrop system*	*Baseline: small-scale maize rotation*	*Scenario 1: small-scale maize rotation*	*Scenario 2: small-scale maize rotation (+7 years)*
n	2	2	3	3	5	5	5
GHG emissions	€-329.99	€-109.90	€-122.12	€-83.88	€-313.80	€-139.26	€-146.48
C–Sequestration	€10.66	€48.83	€13.08	€14.34	€26.69	€207.40	€400.60
Water pollution	€-109.58	€-14.58	€-109.58	€-14.58	€-109.58	€-14.58	€-14.58
Water use	€-264.38	€-237.94			€-264.38	€-237.94	€-237.94
Erosion	€-323.06	€-49.04	€-323.06	€-49.04	€-323.06	€-49.04	€-49.04
Soil build-up	€1.91	€87.43	€2.34	€2.57	€4.78	€37.14	€71.74
Biodiversity	€-12.89		€-12.89		€-12.89		
Total/ha	€-1,027.33	€-275.20	€-552.23	€-130.59	€-992.24	€-196.27	€24.31
Total/kg	€-0.40	€-0.14	€-0.71	€-0.24	€-0.17	€-0.03	€0.03
Add. scenario cost/ ha: cover crops & compost		€100.00		€100.00		€100.00	€100.00
Current cost of production/ha	€670.00	€670.00	€480.00	€480.00	€575.00	€575.00	€575.00
True cost of production/ha	€1,697.33	€1,045.20	€1,032.23	€710.59	€1,567.24	€871.27	€650.69
% true/current cost/ ha	253%	156%	215%	148%	273%	152%	113%
Real cost: How many times higher?	2.5	1.6	2.2	1.5	2.7	1.5	1.1

hidden costs will be lower. In turn, the volatility of food prices could be reduced. Otherwise, the steeply rising cost of living could cause increased poverty and with it social unrest.

To achieve resilience, Zambian maize farmers should be incentivized by policymakers to adopt a balanced crop rotation and to reduce fallow fields through the cultivation of cover crops. These practices are unevenly used by farmers and should be implemented systematically. Cover crops fix nitrogen, keep the soil moist and cool, suppress weeds, build-up root biomass, and increase soil organic matter and microbial life. If appropriate deep rooting cover crop varieties are selected, tillage could be further reduced as the cover crop roots will loosen the soil. Composting crop residues transforms the decomposing plant material into humus, which provides growing plants with the required nutrients. If these practices are promoted and incentivized through the right enabling policy environment and via agricultural extension services, Zambian maize producers will be able to disrupt the vicious cycle of more and more fertilizer requirements per year at increasing fertilizer prices.

Over the past 50 years Zambia's agricultural system has turned into a maize-based monoculture system which is not resilient to climate change. Maize productivity and suitability in Zambia is estimated to decrease between now and 2050 (Ramirez-Cabral et al., 2017; International Center for Tropical Agriculture and World Bank, 2017). Agriculture is the main driver of deforestation (International Center for Tropical Agriculture and World Bank, 2017), and these pressures will increase deforestation and thus foster climate change. This study shows that a resilient maize production is possible using the described farming practices, minimizing environmental externalities and improving the livelihoods of Zambian smallholder farmers. The results of this study are being used to support farmers to adopt new practices and to advocate for policy change.

Cross Cutting Reflections and Challenges

The four case studies above applied TCA to better understand the dynamics of the systems they were assessing, to measure and value dependencies and impacts, and to inform decision-making. In the case of Comisión Nacional para el Conocimiento y Uso de la Biodiversidad (CONABIO), the assessment led to further work supporting smallholder farmers and the local value networks that enhance and strengthen agricultural biodiversity. In the case of corn in Minnesota and maize in Malawi, the study results were shared with local farmers and/or stakeholders through discussions about using TCA as a tool for more holistic agricultural and food systems assessments. In the case of Zambia, the study was used to advocate for policy changes and provide a rationale for sustainable production practices. In all cases, these "proof of concept" TCA applications revealed important lessons for strengthening and mainstreaming TCA in food systems. Collective analysis of these TCA studies has surfaced the following reflections and challenges.

1) The Challenges of Systemic Framing and a Common TCA Language

While TCA and the TEEBAgriFood Evaluation Framework allows for scaling (up and down), it is challenging to determine the Framework's boundaries across scales. The exercise of providing a description of a system—a key element of the process—is often confused as the application itself. Several questions emerge: How can a whole system be captured? The Framework provides a checklist for assessing the system, but how detailed does this need to be? What is the best way to describe and represent nested systems? A common approach to describing food systems would support greater cross-disciplinary understanding of each application from the outset, as well as providing suitable and stable entry points for systems analysis and discussion, whatever the focus of the specific application. Integrated systemic thinking is the underlying driver for the Framework yet poses significant conceptual and practical challenges. The examples from Mexico, Minnesota, Malawi, and Zambia illustrate these challenges well, as they were only able to capture some systems dynamics due to the scope of work.

The very different approaches taken by the study teams reveals the challenge of finding a common TCA vocabulary. All studies outline the impacts, costs, and benefits of different systems, but because of the different scales examined, and the different methodologies used, these studies lack coherence as a set. This kind of cross application coherence might not be necessary, but a common vocabulary or TCA language would enhance the legitimacy and utility of the approach.

2) Balancing the Imperative to Address History and Power that Shape Food Systems While Focusing on Direct and Indirect Costs

The studies cited in this chapter recognize the importance of context and history and how food systems are shaped by these dynamics. For example, the maize in Malawi study describes contemporary maize production as an outcome of colonial policies and programs. The Zambia assessment focused only on production practices, but acknowledges the broader system dynamics that influence these practices. Indeed, it is difficult to approach any moment in time without describing the broader context and history. In thinking about capitals and flows, and trying to quantify these, entrenchment and the dependencies that keep the system in place loom large. Coursing through these systems are power relations that also need to be acknowledged. Where do TCA assessments allow for this analysis?

3) The Multiple Benefits Provided by Agrobiodiversity Managed by Smallholders

CONABIO's study illustrates the important and often invisible contribution smallholder farmers make to agricultural biodiversity. If these contributions were more widely recognized and upheld, farmers could be better recognized as stewards of biocultural landscapes and, in particular, the dynamic agricultural biodiversity that the global food system depends upon. The concept of evolutionary services amplified by CONABIO could be translated into other contexts to support smallholders to stay on the land, to thrive in rural areas, and to contribute important ecological and cultural services essential to global biodiversity and climate priorities. CONABIO's findings seem particularly relevant to the Malawi and Zambia research, and this kind of cross study analysis and reflection can bring different perspectives together, creating new opportunities to promote food systems transformation.

4) The Challenge of Making Meaningful Comparisons Across Systems

The corn in Minnesota and Zambia studies reflect the difficulty of comparing two different systems at vastly different scales, with different data available. How do we make meaningful comparisons? How do we build alignment across the issues of each study or across a series of studies? Study leads and reviewers noted both a lack of data and too much data. Where do we find the right data? What original research needs to be undertaken to support the studies? How do we distil data into key metrics? How do we determine what metrics are meaningful to stakeholders? What is the difference between true cost and true value? These are socially situated questions, and clearly depend on the goals of the comparison or assessment. We use the language of accounting but often the results do not reflect costing and accounting and the answers sought are far broader.

5) The Challenge of Engaging Farmers and Decision-Makers

The intention of these studies is to influence both policy and practice. Results need to be presented in a way that is useful both for policymakers and practitioners, especially farmers. Stakeholder engagement is a central part of TCA's theory of change, but this has not been reflected in all of the early "proof of concept" studies. What does long-term stakeholder engagement around TCA for food systems look like? In the Malawi, Zambia, and Minnesota studies there was a modest attempt to engage local stakeholders and decision-makers. As White (2019) concludes "Knowledge creation is a political and ecological process…To bring about more sustainable and *just* agri-food systems, engagement of marginalized and poor communities should be integral and profound."

6) The Importance of Providing Appropriate Guidance for TCA Studies

What guidance resources can be developed to support study leads? These guidance resources need to reflect the multidimensional aspects of the applications—from study definition and boundaries, to stakeholder engagement, to methodological approaches. How can the community of actors interested in TCA and TEEBAgriFood work together to develop and refine these guidance documents and suite of tools?

As a central actor in food and agriculture systems, maize/corn is worthy of deeper study and consideration. Indeed, the societal, economic, and ecological transformations of food systems will depend on the deep transformation of key commodity value chains. These studies illustrate the value of understanding food system impacts holistically and point to the importance of political economy—or political *ecology*—to understand key systems dynamics and linkages. TCA is both an approach and a tool that supports more holistic understanding of food and agriculture systems and should be mobilized to inform decision-making from the farm level onward.

Notes

1 The case studies are drawn from the following publications:

    1) Comisión Nacional para el Conocimiento y Uso de la Biodiversidad. (2017). *Ecosystems and agro-biodiversity across small and large-scale maize production systems.* TEEB Agriculture & Food, UNEP, Geneva.

    2) Sandhu, H., Scialabba, N.E., Warner, C. *et al.* (2020). Evaluating the holistic costs and benefits of corn production systems in Minnesota, *US Scientific Reports, 10(3922).*

    3) White, S. (2019). *A TEEBAgriFood Analysis of the Malawi Maize Agri-food System.* Global Alliance for the Future of Food.

    4) Bandel, T. & Nerger, R. (2018). *The True Cost of Maize Production in Zambia's Central Province.* Hivos, Soil and More, 2nd ed.

2 Comisión Nacional para el Conocimiento y Uso de la Biodiversidad.
3 University of South Australia.
4 Michigan Department of Health and Human Services, formerly at MSU's Global Center for Food System Innovation.
5 Soil & More Impacts GmbH.
6 This assessment is based on the guidelines of the Natural Capital Protocol (Natural Capital Coalition, 2016) using the monetization factors suggested by the Food and Agricultural Organization of the United Nations (2014).

References

Amador, M.F. & Gliessman, S.R. (1990). An ecological approach to reducing external inputs through the use of intercropping. *Agroecology*, 146–159. New York, NY: Springer.

Altieri, M.A., Van Schoonhoven, A., & Doll, J. (1977). The ecological role of weeds in insect pest management systems: a review illustrated by bean (*Phaseolus vulgaris*) cropping systems. *Pans*, 23(2), 195–205.

Anand, S.S.*et al.* (2015). Food consumption and its impact on cardiovascular disease: importance of solutions focused on the globalized food system: a report from the workshop convened by the World Heart Federation. *Journal of the American College of Cardiology*, 66(14), 1590–1614.

Baker, L. (2013). *Corn Meets Maize: Food Movements and Markets in Mexico.* Washington, DC: Rowman & Littlefield.

Bandel, T. & Nerger, R. (2018). *The True Cost of Maize Production in Zambia's Central Province.* Hivos, Soil and More, iied.

Bedoya, C.A., Dreisigacker, S., Hearne, S., Franco, J., Mir, C., Prasanna, B.M., Taba, S., Charcosset, A., & Warburton, M.L. (2017). Genetic diversity and population structure of native maize populations in Latin America and the Caribbean. *PLoS ONE*, 12(4), e0173488.

Bellon, M. R. & Risopoulos, J. (2001). Small-scale farmers expand the benefits of improved maize germplasm: A case study from Chiapas, Mexico. *World Development*, 29(5), 799–811.

Bellon, M. R., Mastretta-Yanes, A., Ponce-Mendoza, A., Ortiz-Santamaría, D., Oliveros-Galindo, O., Perales, H., Acevedo, & F., Sarukhán, J. (2018). Evolutionary and food supply implications of ongoing maize domestication by Mexican campesinos. *Proceedings of the Royal Society B: Biological Sciences*, 285(1885), 20181049.

Bellon, M. R., Kotu, B. H., Azzarri, C., & Caracciolo, F. (2020). To diversify or not to diversify, that is the question. Pursuing agricultural development for smallholder farmers in marginal areas of Ghana. *World Development*, 125, 104682.

Bezner Kerr, R. & Patel, R. (2014). Food security in Malawi: disputed diagnoses, different prescriptions. *Food Security and Development* (pp. 221–245). Abingdon: Routledge.

Bilalis, D., Papastylianou, P., Konstantas, A., Patsiali, S., Karkanis, A., & Efthimiadou, A. (2010). Weed-suppressive effects of maize–legume intercropping in organic farming. *International Journal of Pest Management*, 56(2), 173–181.

Bommarco, R., Kleijn, D., & Potts, S.G. (2013). Ecological intensification: harnessing ecosystem services for food security. *Trends in Ecology and Evolution*, 28(4): 230–238.

Bracco, M., Lia, V.V., Gottlieb, A.M., Hernández, J.C., & Poggio, L. (2009). Genetic diversity in maize landraces from indigenous settlements of Northeastern Argentina. *Genetica*, 135(1), 39–49.

Bruns, H.A. (2017). Southern corn leaf blight: a story worth retelling. *Agronomy Journal*, 109(4), 1218–1224.

Brush, S.B. (1995). In situ conservation of landraces in centers of crop diversity. *Crop Science*, 35(2): 346–354.

Caamal-Maldonado, J.A., Jiménez-Osornio, J.J., Torres-Barragán, A., & Anaya, A.L. (2001). The use of allelopathic legume cover and mulch species for weed control in cropping systems. *Agronomy Journal*, 93, 27–36.

Capinera, J. L. (2005). Relationships between insect pests and weeds: an evolutionary perspective. *Weed Science*, 53(6), 892–901.

Carvalho, V.P., Ruas, C.F., Ferreira, J.M., Moreira, R.M., & Ruas, P.M. (2004). Genetic diversity among maize (*Zea mays L.*) landraces assessed by RAPD markers. *Genetics and Molecular Biology*, 27(2), 228–236.

Chacon, J.C. & Gliessman, S.R. (1982). Use of the "non-weed" concept in traditional tropical agroecosystems of south-eastern Mexico. *Agro-Ecosystems*, 8(1), 1–11.

Comisión Nacional para el Conocimiento y Uso de la Biodiversidad. (2011). Proyecto Global "Recopilación, generación, actualización y análisis de información acerca de la diversidad genética de maíces y sus parientes silvestres en México". Mexico City.

Comisión Nacional para el Conocimiento y Uso de la Biodiversidad. 2017. Ecosystems and agro-biodiversity across small and large-scale maize production systems, feeder study to the "TEEB for Agriculture and Food". Available at: www.teebweb.org/agrifoodarchive/maize.

Duvick, D.N. (2001). Biotechnology in the 1930s: the development of hybrid maize. *Nature Reviews Genetics*, 2(1), 69–74.

Duvick, D.N. (2005). The contribution of breeding to yield advances in maize (Zea mays L.). *Advances in Agronomy*, 86, 83–145.

Ebel, R., Cárdenas, J.G.P., Miranda, F.S., & González, J.C. (2017). Organic milpa: yields of maize, beans, and squash in mono-and polycropping systems. *Terra Latinoamericana*, 35(2), 149–160.

Food and Agriculture Organization of the United Nations. (2014). *Food Wastage Footprint: Full-Cost Accounting.*

Fujiyoshi, P.T., Gliessman, S.R., & Langenheim, J.H. (2007). Factors in the suppression of weeds by squash interplanted in corn. *Weed Biology and Management*, 7(2), 105–114.

Garbach, K., Milder, J.C., Montenegro, M., Karp, D.S., DeClerck, F.A.J. (2014). Biodiversity and ecosystem services in agroecosystems. *Encyclopaedia of Agriculture and Food Systems*, 2, 21–40.

Garibaldi, L.A., Steffan-Dewenter, I., Kremen, C., Morales, J.M., Bommarco, R., Cunningham, S.A., Carvalheiro, L.G., Chacoff, N.P., Dudenhöffer, J.H., Greenleaf, S.S., Holzschuh, A., Isaacs, R., Krewenka, K., Mandelik, Y., Mayfield, M.M., Morandin, L.A., Potts, S.G,…. Klein, A.M. (2011). Stability of pollination services decreases with isolation from natural areas despite honey bee visits. *Ecology Letters*, 14(10), 1062–1072.

Ghanbari, A., Dahmardeh, M., Siahsar, B.A., & Ramroudi, M. (2010). Effect of maize (*Zea mays L.*)-cowpea (*Vigna unguiculata L.*) intercropping on light distribution, soil temperature and soil moisture in arid environment. Journal of Food, *Agriculture and Environment*, 8(1), 102–108.

Gómez-Pompa, A. (1987). On maya silviculture. *Mexican Studies/Estudios Mexicanos*, 3(1), 1–17.

González-Amaro, R.M. (2016). *Usos locales y preferencias de consumo como factores de la diversidad del maíz nativo de Oaxaca.* Available at: www.conabio.gob.mx/institucion/proyectos/resultados/LE011_Anexo_11_Tesis.pdf.

Graeub, B.E., Chappell, M.J., Wittman, H., Ledermann, S., Kerr, R.B., & Gemmill-Herren, B. (2016). The state of family farms in the world. *World development*, 87, 1–15.

Grain. (2014). *Hungry for land: small farmers feed the world with less than a quarter of all farmland.* Retrieved October 2020. Available at: www.grain.org/article/entries/4929-hungry-for-land-small-farmers-feed-the-world-with-less-than-a-quarter-of-all-farmland.

Hajjar, R., Jarvis, D.I., & Gemmill-Herren, B. (2008). The utility of crop genetic diversity in maintaining ecosystem services. *Agriculture, Ecosystems & Environment*, 123(4), 261–270.

Hartings, H., Berardo, N., Mazzinelli, G.F., Valoti, P., Verderio, A., & Motto, M. (2008). Assessment of genetic diversity and relationships among maize (*Zea mays L.*) Italian landraces by morphological traits and AFLP profiling. *Theoretical and Applied Genetics*, 117(6), 831.

Hawkes, C. (2006). Uneven dietary development: linking the policies and processes of globalization with the nutrition transition, obesity and diet-related chronic diseases. *Globalization and Health*, 2(1), 4.

Hawkes, C. (2018). Globalization and the Nutrition Transition: A Case Study (10-1). *Case Studies in Food Policy for Developing Countries: Policies for Health, Nutrition, Food Consumption, and Poverty*, 3, 113.

Hernández-Xolocotzi, E., Bello Baltazar, E, & Levy Tacher, S. (Eds.) (1995) *La Milpa en Yucatán, un sistema de producción agrícola tradicional.* Tomos I y 2. Colegio de Postgraduados, México.

Hufford, M.B., Lubinksy, P., Pyhäjärvi, T., Devengenzo, M.T., Ellstrand, N.C., & Ross-Ibarra, J. (2013). The genomic signature of crop-wild introgression in maize. *PLoS Genetics* 9(5).

International Monetary Fund. (2017). Malawi, Economic Development Document. IMF Country Report. No. 17/184. Washington, DC.

International Center for Tropical Agriculture and World Bank. (2017). Climate-Smart Agriculture in Zambia. CSA Country Profiles for Africa Series.

Kampanje-Phiri, J.J. (2016). *The Ways of Maize: Food, Poverty, Policy, and the Politics of Meaning in Malawi.* LAP Lambert Academic Publishing.

Kremen, C. & Miles, A. (2012). Ecosystem services in biologically diversified versus conventional farming systems: benefits, externalities, and trade-offs. *Ecology and Society,* 17(4).

Kumar, A., Kumari, J., Rana, J.C., Chaudhary, D.P., Kumar, R., Singh, H., & Dutta, M. (2015). Diversity among maize landraces in North West Himalayan region of India assessed by agro-morphological and quality traits. *Indian Journal of Genetics and Plant Breeding,* 75(2), 188–195.

Landis, D.A. (2017). Designing agricultural landscapes for biodiversity-based ecosystem services. *Basic and Applied Ecology,* 18, 1–12.

Landis, D.A., Wratten, S.D., Gurr, & G.M. (2000). Habitat management to conserve natural enemies of arthropod pests in agriculture. *Annual Review of Entomology,* 45(1), 175–201.

Latati, M., Pansu, M., Drevon, J.J., & Ounane, S.M. (2013). Advantage of intercropping maize (*Zea mays L.*) and common bean (*Phaseolus vulgaris L.*) on yield and nitrogen uptake in Northeast Algeria. *International Journal of Research in Applied Sciences,* 1, 1–7.

Letourneau, D.K., Armbrecht, I., Rivera, B.S., Lerma, J.M., Carmona, E.J., Daza, M. C., Escobar, S., Galindo, V., Gutiérrez, C., Duque-López, S., López-Mejía, J., Acosta-Rangel, A.M., Herrera-Rangel, J., Rivera, L., Saavedra, C.A., Torres, A.M., & Reyes-Trujillo, A. (2011). Does plant diversity benefit agroecosystems? A synthetic review. *Ecological Applications,* 21(1), 9–21.

Matsuoka, Y., Vigouroux, Y., Goodman, M.M., Sanchez, J., Buckler, E., & Doebley, J. (2002). A single domestication for maize shown by multilocus microsatellite genotyping. *Proceedings of the National Academy of Sciences,* 99(9) 6080–6084.

Maxted, N., Ford-Lloyd, B.V., & Hawkes, J.G. (1997). *Plant Genetic Conservation. The In Situ Approach.* Dordrecht: Springer Netherlands.

Mazunda, J. (2013). Budget allocation, maize yield performance, and food security outcomes under Malawi's farm input subsidy programme (No. 17), International Food Policy Research Institute.

McCann, J. (2001). Maize and grace: history, corn, and Africa's new landscapes, 1500–1999, *Comparative Studies in Society and History,* 43(2), 246–272.

Mueller, D.S., Wise, K.A., Sisson, A.J., Allen, T.W., Bergstrom, G.C., Bosley, D.B., Bradley, C.A., Borders, K.D., Byamukama, E., Chilvers, M.I., Collins, A., Faske, T.R., Friskop, A.I., Heiniger, R.W., Hollier, C.A., Hooker, C.A., Isakeit, T., Zackson-Ziems, T. A. ... & Collins, A. (2016). Corn yield loss estimates due to diseases in the United States and Ontario, Canada from 2012 to 2015. *Plant Health Progress,* 17(3), 211–222.

Muñoz, O.A. (2003). *Centli-maíz.* Colegio de Postgraduados. Montecillo, Texcoco.

National Research Council. (1974). Genetic Vulnerability of Major Crops. Committee on Genetic Vulnerability of Major Crops Agricultural Board. National Academy of Sciences.

National Research Council. (1975). *Annual report, Fiscal years 1973 and 1974. 94th Congress, 1st Session. Senate Document No. 94–41*. National Academy of Sciences, National Academy of Engineering Institute of Medicine, National Research Council.

Natural Capital Coalition. (2016). *Natural capital protocol.* Available at: http://natural capitalcoalition.org/protocol/:ICAEW.

Neulinger, K., Vogl, C.R., & Alayón-Gamboa, J.A. (2013). Plant species and their uses in homegardens of migrant Maya and Mestizo smallholder farmers in Calakmul, Campeche, Mexico. *Journal of Ethnobiology*, 33(1), 105–124.

Ng'ong'ola, C. (1986). Malawi's agricultural economy and the evolution of legislation on the production and marketing of peasant economic crops. *Journal of Southern African Studies*, 12(2), 240–262.

Norris, R.F. & Kogan, M. (2005). Ecology of interactions between weeds and arthropods, *Annual Review of Entomology*, 50, 479–503.

Ogindo, H.O. & Walker, S. (2005). Comparison of measured changes in seasonal soil water content by rainfed maize-bean intercrop and component cropping systems in a semi-arid region of southern Africa. *Physics and Chemistry of the Earth, Parts A/B/C*, 30(11–16),799–808.

Palacios, M.R., Huber-Sannwald, E., Barrios, L.G., de Paz, F.P., Hernández, J.C., Mendoza, M.D.G.G. (2013). Landscape diversity in a rural territory: Emerging land use mosaics coupled to livelihood diversification. *Land Use Policy*, 30(1), 814–824.

Penagos, D.I., Magallanes, R., Valle, J., Cisneros, J., Martinez, A.M., Goulson, D., Chapman, J.W., Caballero, P., Cave, R.D., & Williams, T. (2003). Effect of weeds on insect pests of maize and their natural enemies in southern Mexico. *International Journal of Pest Management*, 49(2), 155–161.

Pengue, M. & Gemmill-Herren, B. (2018). Eco-agrifood systems: Today's realities and tomorrow's challenges. In: *TEEB for Agriculture and Food: Scientific and Economic Foundations*. UN Environment.

Perales, H. (1996). *Conservation and evolution of maize in Amecameca and Cuautla valleys of México*. University of California.

Pickersgill, B. (2007). Domestication of plants in the Americas: Insights from Mendelian and molecular genetics. *Annals of Botany*, 100, 925–940.

Popkin, B. (1993). Nutritional Patterns and Transitions. *Population and Development Review*, 19(1), 138–157.

Postma, J.A. & Lynch, J. P. (2012). Complementarity in root architecture for nutrient uptake in ancient maize/bean and maize/bean/squash polycultures. *Annals of Botany*, 110(2), 521–534.

Potts, S.G., Biesmeijer, J.C., Kremen, C., Neumann, P., Schweiger, O., & Kunin, W.E. (2010). Global pollinator declines: trends, impacts and drivers. *Trends in Ecology & Evolution*, 25(6), 345–353.

Power, A.G. (2010). Ecosystem services and agriculture: tradeoffs and synergies. *Philosophical Transactions of the Royal Society B: Biological Sciences*, 365(1554), 2959–2971.

Qi-Lun, Y., Ping, F., Ke-Cheng, K., & Guang-Tang, P. (2008). Genetic diversity based on SSR markers in maize (*Zea mays L.*) landraces from Wuling mountain region in China. *Journal of Genetics*, 87(3), 287–291.

Ramirez-Cabral, N.Y., Kumar, L., & Shabani, F. (2017). Global alterations in areas of suitability for maize production from climate change and using a mechanistic species distribution model (CLIMEX). *Scientific Reports*, 7(1), 1–13.

Reardon, T., Barrett, C. B., Berdegué, J. A., & Swinnen, J.F. (2009). Agrifood industry transformation and small farmers in developing countries. *World Development*, 37(11), 1717–1727.

Reardon, T., Echeverria, R., Berdegué, J., Minten, B., Liverpool-Tasie, S., Tschirley, D., & Zilberman, D. (2018). Rapid transformation of food systems in developing regions: Highlighting the role of agricultural research & innovations. *Agricultural Systems*.

Redden, R., Yadav, S.S., Maxted, N., Ehsan, D.M., Guarino, L., & Smith, P. (2015). *Crop Wild Relatives and Climate Change*. Hoboken, NJ: Wiley Blackwell.

Ricciardi, V., Ramankutty N., Mehrabi, Z., Jarvis, L., Chookolingo, B. (2019). How much of the world's food do smallholders produce? *Global Food Security*, 17, 64–72.

Rodríguez, L., Veles, J., Gómez, R., Figueroa, J.D., & Gaytán, M. (2007). Physico-chemical and thermal properties of maize varieties and their relation to the dry and wet milling performance. *Cereal Food World*, 52(4), A62.

Rojas-Barrera, I.C., Wegier, A., González, J.D.J.S., Owens, G.L., Rieseberg, L.H., & Piñero, D. (2019). Contemporary evolution of maize landraces and their wild relatives influenced by gene flow with modern maize varieties. *Proceedings of the National Academy of Sciences*, 116(42), 21302–21311.

Ruiz Corral, J.A., Durán Puga, N., Sanchez Gonzalez, J.D.J., Ron Parra, J., González Eguiarte, D.R., Holland, J.B., & Medina García, G. (2008). Climatic adaptation and ecological descriptors of 42 Mexican maize races. *Crop Science*, 48(4), 1502–1512.

Sandhu, H., Scialabba, N.E., Warner, C, Behzadnejad, F., Koehane, K., Houston, R., & Fujiwara, D. (2020). Evaluating the holistic costs and benefits of corn production systems in Minnesota, *US Scientific Reports*, 10, 3922.

Schiesari, C., Mockshell, J., & Zeller, M. (2016). Farm input subsidy program in Malawi: the rationale behind the policy. MPRA Paper No. 81409.

Sociedad Mexicana de Fitogenética. (2007). *Primera reunión de mejoradores de variedades criollas de maíz en México. Memoria* (pp. 237). Guadalajara, Jalisco.

Sociedad Mexicana de Fitogenética. (2009). *Tercera reunión nacional para el mejoramiento, conservación y uso de maíces criollos. Memoria de resúmenes* (pp. 101). Celaya, Guanajuato.

Stuckler, D., McKee, M., Ebrahim, S., & Basu, S. (2012). Manufacturing epidemics: the role of global producers in increased consumption of unhealthy commodities including processed foods, alcohol, and tobacco. *PLoS Medicine*, 9(6),e1001235.

Swinton, S.M., Lupi, F., Robertson, G.P., & Hamilton, S.K. (2007). Ecosystem services and agriculture: cultivating agricultural ecosystems for diverse benefits. *Ecological Economics*, 64(2), 245–252.

Terán, S. & Rasmussen, C. (2009). *La milpa de los mayas*. (2nd ed). Universidad Autónoma de México-Universidad de Oriente, Mérida.

Thompson, J., & Scoones, I. (2009). Addressing the dynamics of agri-food systems: an emerging agenda for social science research. *Environmental Science & Policy*, 12(4), 386–397.

Tillman, P. G., Smith, H., and Holland, J. (2012). Cover crops and related methods for enhancing agricultural diversity and conservation biocontrol: Successful case studies. In: Gurr, G.M., Wratten, S.D., Snyder, W.E., & Read, D.M.Y. *(Eds.) Biodiversity and Insect Pests: Key Issues for Sustainable Management* (pp. 309–327). London: John Wiley & Sons.

Timothy, D.H., Harvey, P.H., & Dowswell, C. R. (1988). *Development and spread of improved maize varieties and hybrids in developing countries*. Bureau for Science and Technology, Agency for International Development, Washington, DC.

Tonhasca Jr, A. & Byrne, D.N. (1994). The effects of crop diversification on herbivorous insects: a meta-analysis approach. *Ecological Entomology*, 19(3), 239–244.

Tscharntke, T., Klein, A.M., Kruess, A., Steffan-Dewenter, I., & Thies, C. (2005). Landscape perspectives on agricultural intensification and biodiversity–ecosystem service management. *Ecology Letters*, 8(8), 857–874.

Tsubo, M., Mukhala, E., Ogindo, H.O., & Walker, S. (2003). Productivity of maize-bean intercropping in a semi-arid region of South Africa. *Water SA*, 29(4), 381–388.

Van Berkum, P., Beyene, D., & Eardly, B.D. (1996). *Phylogenetic relationships among Rhizobium species nodulating the common bean (Phaseolus vulgaris L.). International Journal of Systematic and Evolutionary Microbiology*, 46(1), 240–244.

Vavilov, N.I. (1926). *Studies on the Origin of Cultivated Plants, Institute of Applied Botany and Plant Breeding, Leningrad, USSR*. Institute of Applied Botany and Plant Breeding.

Vavilov, N.I. (1951). Phytogeographic basis of plant breeding. The origin, variation immunity and breeding of cultivated plants. *Chronica Botanica*, 13, 1–366.

Vavilov, N.I., Vavilov, M.I., Vavlov, N. Í., & Dorofeev, V.F. (1992). *Origin and Geography of Cultivated Plants*. Cambridge: Cambridge University Press.

Vibrans, H. (2016). Ethnobotany of Mexican weeds. In Lira, R., Casas, A., & Blancas, J. (Eds.). *Ethnobotany of Mexico: Interactions of people and plants in Mesoamerica* (pp. 287–317). New York, NY: Springer.

Vigouroux, Y., Barnaud, A., Scarcelli, N., & Thuillet, A.C. (2011). Biodiversity, evolution and adaptation of cultivated crops. *Comptes Rendus Biologies*, 334(5–6),450–457.

White, S. (2019). *A TEEBAgriFood Analysis of the Malawi Maize Agri-food System*. Global Alliance for the Future of Food.

Wilkes, H.G. (1977). Hybridization of maize and teosinte, in Mexico and Guatemala and the improvement of maize. *Economic Botany*, 254–293.

Wisler, G.C. & Norris, R.F. (2005). Interactions between weeds and cultivated plants as related to management of plant pathogens. *Weed Science*, 53(6), 914–917.

Zhang, W., Ricketts, T.H., Kremen, C., Carney, K., & Swinton, S. (2007). Ecosystem services and dis-services to agriculture. *Ecological Economics*, 64(2), 253–260.

Zhang, C., Postma, J.A., York, L.M., & Lynch, J.P. (2014). Root foraging elicits niche complementarity-dependent yield advantage in the ancient 'three sisters'(maize/bean/squash) polyculture. *Annals of Botany*, 114(8), 1719–1733.

9 Fostering Healthy Soils in California

Farmer Motivations and Barriers

Joanna Ory and Alastair Iles

Introduction

As the farmer walks down the tree row, she brushes past the cover crop that she planted several months ago. The truck that comes to pick up the bee boxes after the almond bloom is not due for two weeks, but the white almond blossoms are already falling to the orchard floor. She planted mustard and clover seeds so that bees would have food even after the almond bloom. Also, the mustard tap root will open the soil and let water soak in. When the cover crop grows taller, and the time nears for almond harvest, the vegetation will get mowed, and the plant material will break down, returning organic matter to the soil. This is a farmer with soil health on the top of her mind. What does it cost her, to follow these regenerative practices? More importantly, what does it cost us, if she does not?

Across the USA, many farmers face declining soil quality. The true costs of soil loss from farms include substantial water degradation and toxic exposure from nutrient and pesticide runoff. Pursuing soil health offers many benefits to farmers, including more fertile soil, increased productivity, higher crop quality, and other environmental and economic gains (Blanco-Canqui et al., 2015). Each season, farmers make decisions about whether to use soil-building practices like cover cropping, rotating crops, or using compost. This chapter looks at how almond farmers in California consider adopting soil health practices in an industry that emphasizes productivity and efficiency. What incentives and costs do farmers take into account? What motivations and barriers influence their ability to act? We discuss how use of True Cost Accounting (TCA) might help to change farmer behavior.

Transitioning to Healthy Soils

Only a minority of US producers have adopted soil health practices (Soil Health Institute, 2019). Less than 5% of intensive vegetable farmers in the Central Coast in California use cover crops (Brennan, 2017). Such techniques are often not adopted by farmers because of market preferences, knowledge gaps, and agronomic, environmental, and policy barriers (Carlisle, 2016).

Carlisle suggests that perceived long-term benefits, farmer knowledge and training, and stable land tenure are major factors that influence farmers to adopt soil health practices.

To make transitions, farmers, food companies, and policymakers need to understand the true environmental, social, and health costs associated with their production. This data alone will not necessarily persuade individual farmers to switch, because farmers might not experience the negative impacts that they cause. For example, farmers might not be aware that the erosion from their farms can cause water quality problems downstream. They also face significant barriers built into the economic and technological structures of their industry.

In this chapter, we demonstrate that many farmers are not implementing healthy soils practices owing to the perceived costs of trying to implement them, alongside an incomplete view of the external impacts caused by unsustainable farming practices. We highlight how farmers often miss out on soil health benefits because they are locked into production schedules, food safety standards, or "efficient" orchard management practices. Some of these barriers could be overcome by changing the dominant industrial supply chain, but this can be difficult to achieve. At present, farmers need to be extremely motivated to pursue improved soil health beyond making the minimal changes that the supply chain permits. Some innovative farmers are using more demanding practices because they are very committed to sustainability and soil quality. They are also sometimes isolated in their farming community, not knowing other farmers who are experimenting with soil health practices and lacking research in organic systems.

Case Study of Almonds in California

More than 80% of the world's almonds are grown in California. Almonds are the third most valuable agricultural commodity in California, amounting to $5.5 billion in 2018 (California Department of Food and Agriculture, 2020a). Of this production, 67% is exported to other countries (Almond Board, 2019). Almonds are a permanent tree crop that usually spans 20 to 30 years of production before an orchard is removed. Since the 1950s, orchards have become larger, significantly mechanized, and less diverse in their varieties. Orchards are primarily a no-till system, which limits the ability of farmers to incorporate compost or other materials into the soil, whereas most annual vegetable cropping involves tilling or working with the soil every season.

A TCA of almond production includes examination of the impacts related to soil health challenges and mismanagement. Many farmers in the Central Valley of California have experienced drought and water scarcity, which have led to salt accumulation in the soil. Farmers have largely transitioned to high-efficiency drip irrigation, which allows for water use efficiency but limits the ability to grow cover crops, use compost, and nurture microbial life. Dust pollution from clearing orchard floors and during the harvest results in poorer air quality and public health in communities close to the orchards. Loss of habitat has undermined bee health and biodiversity. Farmers are evaluating the best options for disposing of old trees once the trees have reached the end of their productive life. Traditionally, farmers have burned trees on site or at cogeneration plants, but some are beginning to grind them up and incorporate the trees back into the soil—a process called orchard recycling.

For our case study, we interviewed 17 almond orchardists in the Central Valley, as well as industry and agricultural extension personnel. The growers we interviewed represent a spectrum of agricultural production types: from very large (thousands of acres) to small (less than ten acres), organic and regenerative to conventional farms, and market channels ranging from wholesale to direct-to-consumer.

Putting Soil Health into Practice

To build soil health, almond farmers might use techniques such as cover cropping, letting native vegetation grow between tree rows, applying compost, reducing equipment passes in the field to reduce compaction and dust, recycling old orchards, and integrating animals into the orchard. All of the growers we interviewed said that maintaining or improving soil health was important to them. Yet orchardists differed greatly in their adoption of soil-building practices, with some not taking any steps and others experimenting with many practices. Both motivations and barriers influenced the degree of adoption, and orchardists who were more prone to the adoption of soil health practices displayed more consciousness of the true costs of not using sustainable soil health practices.

The key motivations for implementing soil health practices that farmers cited can be grouped into three main categories: (1) environmental improvement; (2) yield and profitability; and (3) soil quality (Bergtold *et al.*, 2017; Reimer *et al.*, 2012). Regarding motivations for environmental improvement, one orchardist described their own true cost accounting related to healthy soil management, "You're looking at a whole systems approach. It's not just healthy soil, but healthy soil equals healthy plants, healthy animals, healthy humans, and a healthy environment for our water and our air." Farmers who care about environmental improvement often invoke ecosystem benefits and a keen awareness of the true costs of poor management on dust control and air quality, bee health, or carbon sequestration.

Those orchardists who emphasize yield and profitability refer to nutrient availability, disease prevention, and tree health as key outcomes of improved soil health. One interviewee noted: "Where compost is incorporated you can visibly see differences in the trees... They just look more vibrant. Greener. I mean the soil's healthier and so the trees are just happier. If you can be on a program doing it year after year keeping that soil balanced healthy—the trees will respond to it." Such orchardists link healthy trees with higher levels of production. This production-orientated set of motivations was more common among growers that used fewer soil health practices or used practices to remedy problem areas of the orchard but not throughout their land.

Cover Crops in Particular

In orchards with no cover crop, the inter-row spaces are mowed and sprayed with herbicide to control weeds. The soil is bare, with little plant matter to enrich the soil or provide habitat. Compare this desolate orchard with one planted with clovers, mustard, and grass as a cover crop mix. There are flowers and lush ground cover. Bees, worms, and microbial life have a habitat. Water can percolate into the ground instead of pooling on the surface.

Cover cropping in California typically occurs at a very low rate compared with other US regions, largely because of the intensive commercial production system that we examine below. Planting a cover crop in an almond orchard involves planting seeds (often combinations of legumes, brassicas, and grass seeds) in the spaces between the tree rows and either relying on rain or sprinkler irrigation to water them. If the almonds are harvested off of the ground (which is the standard practice), the cover crop is typically grown until the spring, and then it is mowed and usually sprayed with herbicide to ensure a clean orchard floor for harvest.

True Cost Accounting of Cover Crops

True Cost Accounting (TCA) reveals many costs to the environment from not using cover crops, including ground compaction, water runoff and pollution off the farm, and reduced habitat (See Table 9.1 for a more in depth view of TCA for cover cropping). According to scientific research, cover crops offer many environmental and ecosystem benefits, including improved soil structure that inhibits erosion, more organic matter that encourages soil microbes (which in

Table 9.1 True Cost Accounting For Cover Crops in Almonds

Type of cost	Specific costs (-) and benefits (+) related to cover cropping in almonds
Environmental	Bee health (+), erosion control (+), water infiltration (+), increased soil microbiology (+), increased ecosystem biodiversity (+), water quality improvement (+), carbon storage (+), reduced green house gas emissions (+), herbicide use for termination (-)
Social	Dust reduction (+), landscape beautification (+), identification with environmental ethic (+), worker well-being (+ or - depending on use of herbicide)
Economic	Implementation of practice costs (time, labor, equipment, fuel, seeds) (-), weed reduction (+), cost savings from diminished synthetic fertilizer use (+), impact on yield and production cost (+ or -), long-term benefit of practice for soil health (+), future impact costs in terms of soil health degradation of not using the practice (-), possible fee from huller for debris in final raw product (-), changing irrigation practices to accommodate practice (-)

turn boost tree root performance), improved nutrient cycling, reduced fertilizer use, and readier absorption of water into soil in an often-dry environment (e.g., Basche *et al.*, 2014; Baas *et al.*, 2015; Delate *et al.*, 2015; Hu *et al.*, 2015; Daryanto *et al.*, 2018). In terms of farm performance, there can be substantially less runoff from fertilizers, fewer nitrogen emissions (which is important in the Central Valley owing to high prevailing air pollution there), and higher soil carbon sequestration. These multiple benefits are difficult to quantify in monetary terms.

Most orchardists interviewed had favorable views on cover cropping. They tended to focus on improved soil quality and orchard management. Often, orchardists would note multiple goals, especially to increase water absorption, improve bee health, increase microbial diversity, and provide dust control. Their goals also varied depending on the lifecycle phase of an orchard, whether it was too young to produce nuts, or it was mature, or it was close to reaching "old age."

Table 9.1 demonstrates the types of cost (and benefit) data that are necessary to develop the full picture for TCA for cover crops in almonds. The cost of not using practices should include the benefits the farmers—and the region—are missing out on in terms of natural, social, and human capital. Having specific evaluations for the different costs and benefits would be a useful decision-making tool for farmers and extension staff who provide recommendations to farmers.

Motivations for Adoption

True costs of not using cover crops, like ground compaction and water drainage problems, were prominent in motivating farmers to utilize cover cropping. Off-farm costs, like reduced carbon storage, were generally further removed from the motivations of most of the farmers we interviewed.

Many farmers said that they use cover crops to target areas of their orchards that have problematic drainage or erosion. One orchardist said, "We wouldn't necessarily think to put a cover crop on everything. We want cover crop on the areas where there are infiltration problems. You would put a cover crop with mustard seed in the ground to reduce compaction and standing water." Another orchardist explained, "For fields where we have some water penetration issues, we will plant [a cover crop] to try to help it, and we think it helps. It is an added cost but we feel like there's some benefits for problematic fields." A third farmer said, "I might plant a cover crop on one area where I have a nightmare of an erosion issue in part of the field." Cover cropping as a "fix" for orchard problems suggests that farmers are more motivated to invest time and money when they can see clear benefits from increased water savings and improved yields.

Organic farmers use cover crops to supply nutrients to trees. One organic orchardist mentioned, "Nitrogen is the deficit around here, so I use a mix that has root nodules that do the nitrogen fixation. I use some clover, vetch, beans and peas." Other farmers emphasized the importance of a cover crop for increasing soil microbe activity. For example, one grower stated: "Soil diversity more than anything [is the reason to cover crop]. To help the microbes with a place to survive longer." Not all farmers use cover crops—if they do at all—in their mature orchards. Several growers primarily planted cover crops in their young, non-nut producing orchards to enhance microbe communities, but once the orchard matures, they do not plant cover crops again.

One particularly interesting goal is offering bee forage during times when the almond trees are not in bloom. While bee hives are imported into the Central Valley on a vast scale, bees are locally active year-round, including in winter. Imported bees also need help to survive. One orchardist said, "Let's give the bees something not only to feed off when they arrive here because they get here before almonds start blooming. They need something to feed on after bloom season, the two weeks the beekeepers are trying to get everything out." The co-benefits of a cover crop for soil health *and* bee health underscore the multiple ecosystem services that some farming practices can provide.

Many orchardists who use cover cropping find that it helps with dust control. Because orchard floors tend to be kept bare to allow mechanized harvest, and exposed soils dry out in the Central Valley climate, dust during the harvest season is a major public health concern throughout the Central Valley. One orchardist remarked that the dust issue is something that needs to be seriously addressed by the almond industry. "The amount of dust we produce is almost embarrassing at times. We need to get better at it." Another farmer mentioned that "when it does not rain, the upper one inch and a half becomes dust right away because it is burned by the sun. Cover-cropping is great for dust control and for the workers because otherwise they can't see with the dust." The Almond Board (2018)—a business group that promotes this industry—has a goal of reducing harvest dust by 50% by 2025. The Almond Board outlines a suite of steps that growers can take to reduce harvest dust, yet cover cropping is not currently included.

Most of the farmers linked better soil health back to healthier trees, increased yields, and greater farm revenues. Orchardists who greatly prioritize production are more willing to abandon practices if these do not clearly generate income. One orchardist commented, "Ultimately it has to come back to yield. If we don't improve our yield [we won't continue a practice]." This production oriented set of motivations was more common among growers that used fewer soil health practices or used practices to remedy problem areas of the orchard but not throughout their land.

Barriers to Using Cover Crops

Many farmers we interviewed did not practice cover cropping themselves or did so only on a very limited part of their operation. We found that major barriers to orchardists using cover crops include (1) concerns about water use;

(2) incompatibility of cover crops with production schedules and equipment; (3) costs of implementing cover crops; (4) food safety requirements; (5) lack of market support; and (6) an incomplete view of the true costs associated with not cover cropping. We focus on a few examples here. These barriers must be factored in when considering how TCA might help farmers to transition to soil health.

Planting cover crops is not cost-free: orchardists must pay for seed, the equipment, labor, and fuel to plant the seed, the cost of water to irrigate the cover crop (if it is not rainfed), and the time and fuel (and possibly herbicide) needed to terminate the cover crop before harvest. Farmers must learn which mixes of cover crops work best for their land and how to grow cover crops (which could be different for an orchardist who is primarily used to tree care).

Farmers often said that they did not want to plant cover crops owing to lack of rain/irrigation water to grow the cover crop. Many parts of the Central Valley experience low rainfall, and farmers depend on California's vast water infrastructure or their own groundwater to irrigate crops. The issue of water unavailability is evident for farmers who use only drip irrigation, and do not have the option of using sprinkler irrigation to water the cover crop. The drip irrigation situation means that growers must rely on rain to sustain a cover crop, which is difficult during years with little precipitation. One orchardist said that he would not plant a bag of cover crop seed for the 2019 season because he did not want to waste the seed, owing to drought.

Overcoming this barrier, which is both technological (the type of irrigation system) and environmental (availability of rain) does not have an easy solution. For example, one farmer said: "One of the things we would love to do here and it just has not worked, we spend money and it has not worked, is grow a cover crop. The rains have not come at the times when we needed them. It just hasn't worked out. It's cost us money for really no gain." However, some farmers have been experimenting with different seed mixes that might not require as much water.

The industrial production system and supply chain is another barrier. Many orchards rely on contracts with buyers, where farmers are paid to produce high yields from the orchards without much concern for production methods or sustainability impacts. One family farmer commented: "The corporate farms are perfect and clean and use lots of chemicals. The people who take care of farms, the people that they're growing for, they do whatever they have to do as contract work for that investment group. They want to keep their jobs, and their job is to grow the nuts and they will do that with a lot of inputs." Conventional orchardists tend to stick to production schedules and to worry about cover crops interfering with the almond harvest. For example, a cover crop must be terminated and cleared in time to allow harvest machines to be used. Cover crops that are mowed can leave behind debris that pose food safety problems if nuts are gathered from the ground. For many orchardists, it is much easier and cheaper to use pesticides rather than mow.

Finally, few market incentives exist for orchardists to use cover crops in the industrial supply chain. Almonds are generally pooled together at the processing plant where they are shelled and hulled. Many buyers have little interest in—and input into—the production practices used at the farm level, and little contact with actual producers. Practices that protect soil health are often not valued or even discussed in the supply chain. One grower involved in the wholesale market explained: "My almonds end up in the Almond Complex. I don't think the buyers are asking so much [about sustainability], like are these organic or regenerative? I don't think they're selling to people who are asking for that. This is not the industry that is asking for specific types of almonds."

TCA could be used as a tool within the supply chain to evaluate the sustainability of the almond industry. Creating closer links between consumers, buyers, and producers would be a step in strengthening transparency about sustainability issues within the supply chain. The Almond Board currently runs a sustainability program for farmers to perform self-assessments. Including a TCA tool as part of this sustainability program—and increasing buyer awareness and demand for participation in the program—could help the supply chain to focus on the costs and benefits both on and off the farm for different practices.

Sustainability initiatives that take into account the true costs of agricultural production are becoming more common within food and farming industries. While initiatives like bee- or bird-friendly production are growing in popularity, soil health is not usually a major aspect of sustainability programs. For example, almond milk producers work with farms on different sustainability measures, particularly water efficiency and bee health, although soil health is not part of their initiatives. However, some food companies are starting to show interest in regenerative agriculture and working with farmers to supply regenerative almonds. One interviewee stated, "We work with a (specific food company, name withheld) because they want to find somebody that they can source almonds from that's raising almonds regeneratively. That's very much a concern for them. It's not for everybody, though." Increasing awareness of the externalities related to soil management among food buyers could put more pressure on growers to adopt practices that support soil health.

Valuing Soil Health Through Innovative Policy

Several US Department of Agriculture and California government agri-conservation programs have contributed to the growing uptake of health practices through incentives and technical advice. These programs include the Environmental Quality Incentive Program and the Conservation Innovation Program at the US Department of Agriculture's Natural Resources Conservation Service. These programs are under-resourced and routinely over-subscribed. Many farmers in California might not gain access to the resources. It is also unclear whether and how the programs influence farmers to diversify and to successfully protect soils. In response, in 2017 California created the Healthy Soils Program, administered by the California Department of Food and Agriculture (CDFA), which seeks to incentivize farmers to pursue healthy soils.

This program offers farmers up to three years of funding to introduce one or more soil health practices on land where the practices have previously not been used (California Department of Food and Agriculture, 2020b). A key factor in

CDFA's determination of whether a practice is eligible is whether it stores carbon. To date, hundreds of farmers, including a few almond orchardists, have received grants. For the 2020 grant round, the CDFA awarded over $22 million in grants. The individual grants are up to $100,000 per farm. Interviewees saw financial incentives as a way to minimize the risks of using practices on large acreage. A measure of program success will be whether farmers continue to use the practices after their grants end. Will they become more willing to invest in soil health?

While incentive programs like the HSP could expand adoption of soil health practices, the inclusion of TCA tools for farmers would strengthen these programs. The HSP would improve if it not only considered greenhouse gas emission reduction benefits when valuing different agricultural practices, but also included protecting water quality, reducing soil erosion, and maintaining healthy agro-ecosystems.

Conclusion

Increasing soil health practices calls for many changes to the current agricultural system. Enabling farmers to see and act on the long-term benefits of soil health for their land is a necessary step. TCA offers a valuable tool for farmers and others to understand the ways in which adopting or not adopting certain practices affects their profitability, environment, and communities. To further strengthen this insight, TCA should include the missed benefits from being unable to carry out practices due to economic, technological, and production barriers. To overcome current policy and market barriers, policymakers, buyers, and consumers also need to value soil health practices much more.

Funding Acknowledgment

Thank you to the Berkeley Food Institute, USDA AFRI Grant #2019-67019-29537, NSF Award # 1824871, and Patagonia, Inc. for supporting this research.

References

Almond Board of California. (2018). Almond Orchard 2025: Reducing harvest dust. Available at: www.almonds.com/sites/default/files/content/attachments/Almond%20Orchard%202025%20Reducing%20Dust%20at%20Harvest.pdf.

Almond Board of California. (2019). Almond Almanac 2019. Available at: www.almonds.com/sites/default/files/2020-04/Almanac_2019_Web.pdf.

Baas, D., Robertson, G., Miller, S., & Millar, N. (2015). *Effects of cover crops on nitrous oxide emissions, nitrogen availability, and carbon accumulation in organic versus conventionally managed systems*. Final report for ORG project 2011–04952. CRIS Abstracts.

Basche, A., Miguez, F., Kaspar, T., & Castellano, M. (2014). Do cover crops increase or decrease nitrous oxide emissions? A meta-analysis. *Journal of Soil and Water Conservation*, 69(6), 471–482. https://doi.org/10.2489/jswc.69.6.471.

Bergtold, J.S., Ramsey, S., Maddy, L., and Williams, J.R. (2017). A review of economic considerations for cover crops as a conservation practice. *Renewable Agriculture and Food Systems*. 34, 1–15. https://doi.org/10.1017/S1742170517000278.

Blanco-Canqui, H., Shaver, T.M., Lindquist, J.L., Shapiro, C.A., Elmore, R.W., Francis, C.A., Hergert, G.W. (2015). Cover crops and ecosystem services: insights from studies in temperate soils. *Agronomy Journal*, 107(6), 2449–2474. https://doi.org/10.2134/agronj15.0086.

Brennan, E. (2017). Can We Grow Organic or Conventional Vegetables Sustainably Without Cover Crops? *HortTechnology, 27(2)*, 151–161. https://doi.org/10.21273/HORTTECH03358-16.

California Department of Food and Agriculture. (2020a). California Agricultural Production Statistics. Available at: www.cdfa.ca.gov/statistics.

California Department of Food and Agriculture. (2020). Healthy Soils Program. Available at: www.cdfa.ca.gov/oefi/healthysoils.

Carlisle, L. (2016). Factors Influencing Farmer Adoption of Soil Health Practices in the United States: A Narrative Review. *Agroecology and Sustainable Food Systems*, 40(6), 583–613. https://doi.org/10.1080/21683565.2016.1156596.

Daryanto, S., Fu, B., Wang, L. Jacinthe, P., & Zhao, W. (2018). Quantitative synthesis on the ecosystem services of cover crops. *Earth-Science Reviews*, 185, 357–373. doi:10.1016/j.earscirev.2018.06.013.

Delate, K., Cambardella, C., & Chase, C. (2015). Effects of cover crops, soil amendments, and reduced tillage on carbon sequestration and soil health in a long term vegetable system. Final Report for ORG project 2010–03956. CRIS Abstracts.

Hu, S., Shi, W., Meijer, A., & Reddy, G. (2015). Evaluating the potential of winter cover crops for carbon sequestration in degraded soils transitioning to organic production. Project proposal and final report for ORG project 2010–04008. CRIS Abstracts.

Reimer, A, Thompson, A., & Prokopy, L. (2012). The multi-dimensional nature of environmental attitudes among farmers in Indiana: Implications for conservation adoption. *Agriculture and Human Values*, 29(1), 29–40.

Soil Health Institute. (2019). Progress Report: Adoption of Soil Health Systems Based on Data from the 2017 US Census of Agriculture. Available at: https://soilhealthinstitute.org/wp-content/uploads/2019/07/Soil-Health-Census-Report.pdf.

Section 4

For the Public Good

Food is the one necessity for life that is not as comprehensively managed in the public interest as other essentials for life, such as water or energy. For the most part, the food system is fundamentally driven via various levels of commerce. There are certainly international agreements, national policies, and programs that are deployed at the level of local government, all of which influence the agriculture and food economy. But a holistic, agroecologically orientated frame is for the most part, not palpable throughout government. Modern governing doctrines in this area are a palimpsest: amending policies painted over existing policies in so many layers of decades that the structural issues remain the same—particularly those that cause inequities, such as the US health inequities pointed out in Chapter 10 of this section—The Real Cost of Unhealthy Diets—as well as the five channels of health impacts described earlier in Chapter 5.

Current inroads to structural reform call for a strategic, surgically precise reconstruction of systemic policies relating to food, perhaps in acknowledgment of the challenge of changing national policy in national economies that are deeply entrenched in industrialization. The ideas for True Cost Accounting (TCA) for the public good illustrated in this section represent incrementally sound yet transformative ideas. The authors in this section provide information on programs and policy suggestions ranging from local levels of government to the international: Chapter 11 describes the growing participation of a number US school districts in a program with a holistic values rubric which rates and ranks their large-scale food purchases according to TCA-aligned value categories, including local economies, environmental sustainability, a valued workforce, animal welfare, and nutritional health; Chapter 12 sets out a potential course for embedding TCA in US administrative decision-making; while Chapter 13 expands the TCA policy frame to explore the many multilateral venues from which to develop and embed true cost policies in international policy.

The chapters in this section are not the only examples of TCA in policy or practice, to be sure; in earlier chapters in this volume we have learned of TCA applications underway in Egypt, Ecuador, Bhutan, Vanuatu, New Zealand, South Korea, Italy, Norway, Denmark, India, Mexico, Malawi, Zambia, and the UK. What the chapters in this section illuminate are paths forward for public policy and the imminent potential of these opportunities.

10 The Real Cost of Unhealthy Diets

Sarah Reinhardt, Rebecca Boehm and Ricardo Salvador

Introduction

Understanding the health consequences and costs of the foods we eat is a key component of True Cost Accounting (TCA). A growing body of research makes it clear that we pay for the cost of poor diets many times over: in addition to the social and environmental costs of food described elsewhere in this book, consumers are paying once for the foods they buy at grocery stores, restaurants, and other food service establishments, and again for the health and medical care that they might require to treat diet-related diseases. Other costs still are associated with reduced quality of living and lost productivity owing to sickness and early death. These costs are borne not just by individuals, but also by the public sector, which subsidizes health care and nutrition programs for millions of people, and by society as a whole. This chapter focuses on the USA, where few people meet recommendations for a healthy diet, and many live with one or more diet-related diseases. Here we summarize research describing what we spend on food in the USA, how much we spend treating diseases associated with poor diets, and policy solutions aimed at reducing these costs and improving overall health and quality of life.

Describing the US Diet: A Long Way from Health

Since 1980 the federal government has outlined the broad contours of a healthy diet via the *Dietary Guidelines for Americans*—the nation's leading set of science-based nutrition recommendations for disease prevention and health promotion. An important purpose of the guidelines, updated every five years, is to inform federal nutrition programs such as the Supplemental Nutrition Assistance Program (SNAP), the Special Supplemental Nutrition Program for Women, Infants, and Children (WIC), and the National School Lunch Program (NSLP) and School Breakfast Program (SBP) that serve millions of children, parents, seniors, and veterans each year, many of whom are low-income and at nutritional risk.

The recommendations contained in the *Dietary Guidelines* are far from the reality of what most people eat on a daily basis. Although the healthfulness of diets has modestly improved for both adult and youth populations in recent decades, overall diet quality remains low, and disparities persist between segments of the population (Liu *et al.*, 2020a; Wang *et al.*, 2014; Zhang et al., 2018). Compared with diets recommended by the *Dietary Guidelines*, most diets fall short of meeting the daily recommended amounts of fruits, vegetables, and whole grains, while containing excess amounts of refined grains and some meats, as well as added sugar, sodium, and total calories (Centers for Disease Control and Prevention, 2016). The typical US diet also contains many highly processed foods, which can increase the risk of overweight and obesity, metabolic syndrome, hypertension, and other markers for disease (Liu *et al.*, 2020b).

The Multidimensional Costs of US Diets

How Much Do Households Spend on Food?

On average, US households[1] spend $378 per month on food eaten at home and $286 on food eaten in places like cafeterias, restaurants, and other food service establishments, for a total of $664 per month—up to a third of which is spent on food that is wasted (Conrad, 2020). Food purchases account for nearly 13% of monthly spending in the average US household (Bureau of Labor Statistics, 2020). However, the proportion of household income spent on food varies by income level: the poorest one-fifth of households spend 35.1% of their income on food and tend to have lower diet quality, while the richest one-fifth spend only 8.2% of household income on food and tend to have higher diet quality (Economic Research Service, 2020; Hiza *et al.*, 2013). Meanwhile, average daily food costs based on self-reported dietary intake have been estimated at approximately $5.80 per person per day, adjusted to a standard 2,000 calorie diet (Figure 10.1) (Fulgoni and Drewnowski, 2019; Rehm *et al.*, 2015). These costs reflect only the price of food consumed and do not account for food waste.

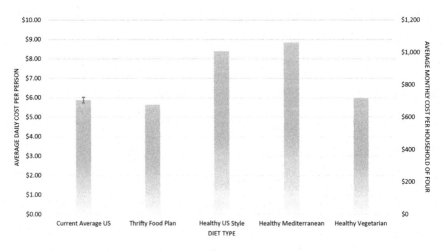

Figure 10.1 Estimated costs of current and model healthy US diets.

What Is the Cost of a Healthy Diet?

Healthier diets are typically viewed as more expensive, particularly when cost is measured per calorie (Beydoun *et al.*, 2015; Rehm *et al.*, 2015). Of course, a healthy diet can be achieved in many ways, and these variations could be accompanied by different price tags. A recent study estimated that the standard healthy "US-Style" diet recommended by the Dietary Guidelines would cost $8.27 per person per day, while a healthy Mediterranean diet would cost $8.73 and a healthy vegetarian diet $5.90, based on a standard 2,000 calorie diet (see Figure 10.1) (Fulgoni and Drewnowski, 2019). In contrast, if measured on a unit of edible weight (e.g. 100 grams) or average portion (e.g. one cup) basis, many healthy foods cost *less* than unhealthy foods. This is explained in part by the fact that many nutritious foods rich in vitamins and minerals, such as fruits and vegetables, are lower in calories per average serving size or equivalent weight (Carlson and Frazao, 2012).

The US Department of Agriculture's Economic Research Service produces estimates of the cost of four model diets that meet nutritional needs: the thrifty food plan, low-cost food plan, moderate-cost food plan, and liberal food plan (Carlson *et al.*, 2007). According to these estimates, the current cost of following the lowest-cost diet (thrifty food plan) for a family of four is between $591 and $679 per month (see Figure 10.1), while the highest-cost diet (liberal food plan) is between $1,153 and $1,350 (US Department of Agriculture, 2020). However, critics of the thrifty food plan point out that it relies on unrealistic assumptions about food availability, affordability, and preparation time and lacks the dietary diversity recommended by the *Dietary Guidelines* (Davis and You, 2010; Food Research and Action Center, 2012; Rose, 2007). As a result, many families that qualify for public food assistance find that their monthly SNAP[2] benefit allotments, which are based on the thrifty food plan, fall short of helping them afford a healthy diet (Mulik and Haynes-Maslow, 2017).

Healthcare and Related Costs of Diet–Related Disease

Poor diets are among the leading causes of disease and death in the USA, accounting for an estimated 18–26% of all deaths nationwide (Afshin *et al.*, 2019; Wang *et al.*, 2019). Low intake of whole grains, nuts and seeds, fruits, vegetables, and omega-3 fats from seafood, together with high intake of sodium, processed meats, and sugar-sweetened beverages, have been identified as leading dietary factors associated with greater risk of death and disease, as measured by disability-adjusted life years (DALYs) (Afshin *et al.*, 2019; Micha *et al.*, 2017). DALYs are often used to represent healthy years of life lost owing to illness affecting quality of life and premature death (World Health Organization, n.d.).

A vast majority of US deaths associated with dietary risk factors (84%) are due to cardiovascular diseases, followed by cancer, diabetes, and other diseases (The US Burden of Disease Collaborators *et al.*, 2018). Meanwhile, the annual costs specific to diet-related cardiovascular diseases and type 2 diabetes, known collectively as cardiometabolic disease, have been estimated at $50.4 billion across the US population, or $301 per person aged 35 and older (Jardim *et al.*, 2019). Research has found that meeting dietary recommendations for fruits and vegetables could save more than $32 billion in medical costs per year owing to reductions in cardiovascular disease, while decreasing sugar-sweetened beverage intake by one serving daily could save $16 billion through reductions in type 2 diabetes (Reinhardt, 2019).

Diet-related diseases also lead to lost productivity, as many working-age adults with chronic illnesses are prevented from working owing to disability and lost years of life. Poor diets account for one in nine DALYs lost, and about half of all DALYs lost due to cardiovascular disease and type 2 diabetes in the USA (Afshin *et al.*, 2019). The annual cost of lost productivity from all cardiovascular disease, not just those cases attributable to dietary factors, is estimated at $153 billion, while the total cost of productivity losses from all type 2 diabetes is $90 billion (Benjamin *et al.*, 2019; Reinhardt, 2019). Much of these productivity losses could be avoided with improvements in diet. It has been estimated that meeting fruit and vegetable recommendations could recoup $20 billion in productivity costs by preventing cases of cardiovascular disease and reducing intake of sugar-sweetened beverages by one serving daily could recoup $6 billion via reductions in type 2 diabetes (Reinhardt, 2019).

People living with diet-related diseases might also be more susceptible to communicable diseases, such as viral infections, and to health risks posed by climate change. For example, those with poor nutritional status and underlying health conditions such as type 2 diabetes may be more susceptible to diseases such as that caused by the 2019 novel coronavirus disease (COVID-19) which has necessitated trillions in federal spending to support national healthcare systems and economic relief (Naja and Hamadeh, 2020; Richardson *et al.*, 2020; Riddle *et al.*, 2020). People with existing medical conditions have also been found to be at greater risk of illness and death when faced with climate change impacts such as extreme heat and weather events, poor air quality, and other environmental stressors (Ebi *et al.*, 2017, pp. 8; Environmental Protection Agency, 2016).

Although it is impossible to put a price on human life, economists have developed methods for estimating how much a life free of certain illnesses or injuries is worth, or the "value of a statistical life." Using this method, researchers found that a 10% reduction in deaths from diet-related cardiovascular diseases, cancer, and diabetes would be worth $14.8 trillion (measured in 2020) (Murphy and Topel, 2005). More recently, a study applying disease-specific outcomes from this model found that reducing cardiovascular risk by 12.8% if the US population met federal fruit and vegetable consumption recommendations would be worth $10 trillion per year (measured in 2019), while reducing type 2 diabetes risk by 7.3% via reductions in sugar-sweetened beverage intake would be worth $470 billion per year (measured in 2019) (Reinhardt, 2019).

US Policy Solutions to Address Poor Diets

Solutions responding to the public health crisis caused by poor diets have emerged at all levels of society: in federal, state, and local policies; at institutions such as colleges, universities, and hospitals; within the private sector; across health care professions; and within communities.

This chapter focuses on public policy, as the government spends more than $1.5 trillion each year on health care and nutrition programs serving millions of families, kids, and seniors (Centers for Medicare and Medicaid Services, 2018; Economic Research Service, 2019a). Effective policymaking should center the needs of those historically marginalized and at highest risk, including low-income populations and many communities of color who have faced a history of exclusion and structural violence, often at the hands of government itself (Elsheikh and Barhoum, 2013; Haynes-Maslow and Stillerman, 2016). Some of the policies described below have drawn criticism for failing to do so, thereby perpetuating systems that have done harm.

Encouraging Healthy Eating through Education and Food Environment Improvements

Federal nutrition programs, including SNAP, WIC, NSLP, and SBP, serve one in four Americans every year (US Department of Agriculture, n.d.-b). Given their broad reach, these programs have been recognized as powerful points of intervention for promoting healthy diets. Yet despite legislation dictating that nutrition programs should align with federal dietary guidance, there are few mechanisms to ensure accountability, meaning that additional policies might be needed to bring programs into compliance with evidence-based nutrition standards (National Nutrition Monitoring and Related Research Act of 1990, 1990). For example, the *Healthy, Hunger-Free Kids Act of 2010* was a landmark policy that improved the healthfulness of school meals, resulting in increased consumption of fruits and vegetables by students and less refined grains, sodium, and empty calories in meals—all without increasing plate waste or decreasing school revenue (Cohen *et al.*, 2014; Fox and Gearan, 2019). Other policy, systems, and environmental changes are increasingly being used within nutrition programs to shape food environments that support healthy food choices—for example, by developing healthy retail programs, community and school gardens, or institutional wellness policies (Honeycutt *et al.*, 2015; Story *et al.*, 2008).

Many nutrition programs also offer direct education. Supplemental Nutrition Assistance Program Education, known as SNAP-Ed, serves nearly 5 million people at 60,000 sites across the country using a diverse array of education programs and initiatives (Gleason *et al.*, 2018; Naja-Riese *et al.*, 2019). Although cost-benefit evaluations of evidence-based nutrition education programs are challenging, owing in part to wide variation in program design and implementation, it has been estimated that every dollar spent on nutrition education could yield between $2.66 to $17.04 in health care cost savings (Joy et al., 2006; Rajgopal et al., 2002).

Increasing Household Purchasing Power and Healthy Food Access

Numerous programs have been introduced to improve the purchasing power and diet quality of low-income SNAP and WIC participants by using financial incentives to encourage purchase of healthy foods. For example, the 2014 Farm Bill first authorized a program now known as the Gus Schumacher Nutrition Incentive Program that subsidizes SNAP participants' purchases of fruits and vegetables at farmers markets and retail stores (Parks *et al.*, 2019). Other forms of subsidies aiming to make the relative price of fruits and vegetables less expensive than other foods have also been shown to increase fruit and vegetable purchases and consumption among low-income consumers (Olsho *et al.*, 2016).

Combinations of restrictions and incentives for food purchases made by SNAP participants have also been proposed as a way to improve diet quality. One study evaluating the potential health and economic impacts of a 30% fruit and vegetable SNAP incentive paired with sugar-sweetened beverage restrictions found that nearly 94,000 cases of cardiovascular disease would be prevented, nearly 46,000 quality-adjusted life years gained, and $4.33 billion saved in healthcare costs over a period of five years. Pairing incentives for fruits, vegetables, nuts, whole grains, fish, and plant-based oils with restrictions on sugar-sweetened beverages, junk foods, and processed meat could achieve even greater gains (Mozaffarian *et al.*, 2018). Although potentially effective, this approach has received criticism for its implications for participant dignity and autonomy (Schwartz, 2017). It should also be noted that these programs impact only those participating in federal nutrition programs, and issues of diet quality persist even in higher income groups in the USA.

Other federal programs developed to improve healthy food access include the Healthy Food Financing Initiative, which helps to fund healthy food retail and food enterprise projects in underserved areas; the Farmers Market and Local Food Promotion Programs, which fund projects such as farmers markets, community supported agriculture programs, and farm-to-institution sales; and the Farm to School program, which helps farmers and food producers to sell directly to schools. Such programs can offer the dual benefit of improving healthy food access while building community wealth and supporting local and regional food economies (US Department of Agriculture, n.d.-a, n.d.-c; US Department of Health and Human Services, 2019).

Taxes and Subsidies: Leveraging Food Prices to Shift Purchases

Price regulation is a public health strategy that has been effective in curbing the purchase of alcohol and cigarettes. Cities nationwide have begun employing this strategy to reduce the purchase of sugar-sweetened beverages and support healthy food purchases (Chaloupka *et al.*, 2019).

Sugar-Sweetened Beverage Taxes

Sugar-sweetened beverages, including soft drinks, fruit drinks, sweetened coffee and tea, and sports and energy drinks, are associated with weight gain and increased risk of type 2 diabetes, cardiovascular disease, and dental caries, among other diseases (Bleich and Vercammen, 2018; Imamura *et al.*, 2015; Malik and Hu, 2019). The World Health Organization (WHO) has recommended the taxation of sugar-sweetened beverages as one tool to address obesity and non-communicable diseases (World Health Organization, 2017). Evidence suggests that such taxes can effectively reduce purchases of sugar-sweetened beverages, although their effects have varied across the USA and might diminish when consumers can avoid the tax by purchasing beverages in adjacent cities or counties (M.M. Lee *et al.*, 2019; Seiler *et al.*, 2018). One study found that adopting a national tax of $0.01 per ounce could generate as much as $23.6 billion in ten-year obesity-related health care savings and generate an additional $12.5 billion in annual revenue; a second study estimated $53.2 billion in cost savings across a lifetime owing to cardiovascular disease and diabetes prevention, and an additional $80.4 billion generated in revenue (Y. Lee *et al.*, 2020; Long *et al.*, 2015). The potential health benefits of such taxes are twofold: populations benefit first through consuming fewer sugar-sweetened beverages, and second through the reinvestment of tax revenue in other public health measures.

Food as Medicine

Although many health professionals offer nutrition education, it can be difficult for patients to follow their recommendations if healthy foods remain unaffordable. Produce prescription programs, which were first authorized as a pilot program in the 2014 Farm Bill, allow health care providers to offer fruit and vegetable vouchers to low-income and nutritionally at-risk patients (Economic Research Service, 2019b). Preliminary research has shown that such programs can increase fruit and vegetable intake among low-income children and adults, and also increase fruit and vegetable variety and reduce body mass index (BMI) and hemoglobin A1C levels among some low-income adult populations (Bryce *et al.*, 2017; Cavanagh *et al.*, 2017; Ridberg *et al.*, 2019).

Programs such as Medicare, which provides health insurance primarily to individuals over 65, and Medicaid, which serves primarily low-income households, could also provide incentives for healthier food as a means of improving health and generating medical cost savings. A recent study found that a 30% subsidy for fruits and vegetables offered to Medicaid and Medicare participants could prevent nearly 2 million cardiovascular events and save $39.7 billion in healthcare, while a 30% subsidy for whole grains, nuts and seeds, seafoods, and plant oils, could prevent nearly 3.3 million cardiovascular events and 120,000 cases of diabetes, saving more than $100 billion in healthcare costs (Y. Lee *et al.*, 2019). Some private health insurers have also begun incentivizing behaviors that improve diet and health among clients—a practice that could eventually find support in federal regulation (John Hancock Insurance, 2020).

Improved Access to Healthcare and Preventive Services

Expanding access to healthcare and health insurance, particularly among low-income households, improves self-assessed health, reduces out of pocket spending on healthcare, and increases use of preventive services, including nutrition counseling and education (Bhattarai *et al.*, 2013; Cunningham, 2008; Hu *et al.*, 2016; Nikpay *et al.*, 2016; Simon *et al.*, 2017; Sommers *et al.*, 2016). Research has now begun to explore if expanding access to free or low-cost health care could also improve health behaviors such as food choices and diet. Early results are promising: following the adoption of the *2010 Patient Protection and Affordable Care Act*, which provided free healthcare to 15 million people by offering states the option to expand Medicaid, carbonated soft drink purchases declined in states that chose to expand Medicaid versus those that did not (Cotti *et al.*, 2019; He *et al.*, 2020).

Regulation of Food Marketing

Extensive marketing of unhealthy foods poses a major challenge to improving diet quality nationwide (Sadeghirad *et al.*, 2016). The food industry spends nearly $11 billion each year on television advertisements alone, and companies heavily target advertisements for unhealthy foods such as fast food, candy, sugary drinks, and snacks to Hispanic and Black consumers, and to children and adolescents in particular (Harris *et al.*, 2019, 2020). The WHO recommends regulating food and beverage marketing, which has been shown to effectively decrease sales of unhealthy foods. However, the USA has yet to adopt such strategies on a broad scale (Kraak *et al.*, 2016). As voluntary and self-regulatory initiatives have proven to be largely ineffective, health experts and leading advocacy groups continue to recommend federal regulation of unhealthy food marketing to children (American Heart Association, 2019; Graff *et al.*, 2012).

Barriers to Policy and Systems Change Supporting Healthier Diets

Barriers to achieving a healthy diet are numerous. For many, challenges include the real or perceived costs of nutritious foods, poor geographic access or inadequate transportation, lack of time or skills needed to prepare healthy foods, and deeply ingrained preferences, habits, and cultural norms (French *et al.*, 2019; Monsivais et al., 2014; Seguin *et al.*, 2014; Ver Ploeg *et al.*, 2017). Meanwhile, the food manufacturing industry extensively markets an abundance of highly processed or "ultra-processed" foods, which contribute to poor diets (Askari *et al.*, 2020; Rico-Campà et al., 2019). In addition to being perceived as more palatable and more affordable, these foods are nearly ubiquitous in the US food landscape and compete with healthier options for a place in our diets.

Importantly, the ability to eat healthfully is inextricably linked to larger social forces, including poverty and racism. Systemic racism has functioned for decades to keep many people of color living in poverty and in neighborhoods inundated with fast food and lacking in quality healthy food, while being disproportionately exposed to junk food marketing (Cooksey-Stowers *et al.*, 2017; Harris *et al.*, 2019; Mitchell and Franco, 2018; Ohri-Vachaspati *et al.*, 2015). The decades of discriminatory policy and practice that shaped these conditions have been described as "food apartheid." (Bower *et al.*, 2014; Brones, 2018). As described in other chapters, many of the people who work to grow, transport, sell, and serve the food that we eat are unable to afford healthy diets themselves and might lack access to health care and paid sick days (Food Chain Workers Alliance and Solidarity Research Cooperative, 2016).

There are also many challenges to achieving systems change through public policy. Chief among them is political influence wielded by the food and beverage industry. In 2019 this industry spent $24.6 million lobbying the United States Congress on food and beverage policy (Center for Responsive Politics, n.d.). As a result, the federal government has long been reluctant to challenge or counter industry interests, often investing in individual behavior change strategies such as nutrition education instead. Although behavior change can be an effective approach, particularly when coupled with other interventions targeting the food environments where people live, work, and play, focusing exclusively on behavior change is unlikely to generate substantial improvements in dietary quality for all populations (US Department of Agriculture and US Department of Health and Human Services, 2015).

Outlook: The Future of Food and Health

Understanding the health consequences and costs of the foods that we eat is key to capturing the true costs of the food system. As a result, the concepts and current research described in this chapter are critical to informing the framework and methodology of TCA. As the body of research on the relationships between diet, health, and health care expands, it must also become more interdisciplinary in nature, drawing on social and environmental sciences to paint a more complex picture of the true costs of the food system.

Ultimately, the utility of TCA is contingent on whether all actors in the food system—particularly those that have effectively externalized health costs associated with poor diets—can be held to account. Currently, much of this burden is borne by the federal government and by the individual. The food industry, meanwhile, has largely evaded responsibility for the public health consequences of its business models. More broadly, the political and economic institutions of the USA have yet to answer for a long history of policies and practices that have resulted in widespread social and economic inequality.

The future of food and health in the USA will be determined at a fork in the road. Down one road is business as usual: the social, economic, and environmental impacts of the food system that burden many are traded for the benefit of the few. Down another, a vastly different picture: policy prioritizes public health and wellness, bringing into balance the true costs of the food system and holding stakeholders accountable for the roles that they play. When faced with these futures, we hope that TCA will be a valuable tool for decision-making that puts healthy eating in reach for all people.

Notes

1 Average number of total people in household (consumer unit) is 2.5.
2 Supplemental Nutrition Assistance Program, administered by the United States Department of Agriculture.

References

Afshin, A.*et al.* (2019). Health effects of dietary risks in 195 countries, 1990–2017: A systematic analysis for the Global Burden of Disease Study 2017. *The Lancet*, 393 (10184), 1958–1972. https://doi.org/10.1016/S0140-6736.

American Heart Association. (2019). Unhealthy and Unregulated: Food Advertising and Marketing to Children. Available at: www.heart.org/-/media/files/about-us/policy-research/fact-sheets/healthy-schools-and-childhood-obesity/food-marketing-and-advertising-to-children-fact-sheet.pdf?la=en.

Askari, M., Heshmati, J., Shahinfar, H., Tripathi, N., & Daneshzad, E. (2020). Ultra-processed food and the risk of overweight and obesity: A systematic review and meta-analysis of observational studies. *International Journal of Obesity*. https://doi.org/10.1038/s41366-020-00650-z.

Benjamin, E.J.*et al.*, on behalf of the American Heart Association Council on Epidemiology and Prevention Statistics Committee and Stroke Statistics Subcommittee. (2019). Heart Disease and Stroke Statistics—2019 Update: A Report From the American Heart Association. *Circulation*, 139(10). https://doi.org/10.1161/CIR.0000000000000659.

Beydoun, M.A., Fanelli-Kuczmarski, M.T., Allen, A., Beydoun, H.A., Popkin, B.M., Evans, M.K., & Zonderman, A.B. (2015). Monetary Value of Diet Is Associated with Dietary Quality and Nutrient Adequacy among Urban Adults, Differentially by Sex, Race and Poverty Status. *PLOS ONE*, 10(11). https://doi.org/10.1371/journal.pone.0140905.

Bhattarai, N., Prevost, A.T., Wright, A.J., Charlton, J., Rudisill, C., & Gulliford, M.C. (2013). Effectiveness of interventions to promote healthy diet in primary care: Systematic review and meta-analysis of randomised controlled trials. *BMC Public Health*, 13(1), 1203. https://doi.org/10.1186/1471-2458-13-1203.

Bleich, S.N. & Vercammen, K.A. (2018). The negative impact of sugar-sweetened beverages on children's health: An update of the literature. *BMC Obesity*, 5(1), 6. https://doi.org/10.1186/s40608-017-0178-9.

Bower, K.M., Thorpe, R.J., Rohde, C., & Gaskin, D.J. (2014). The intersection of neighborhood racial segregation, poverty, and urbanicity and its impact on food store availability in the United States. *Preventive Medicine*, 58, 33–39. https://doi.org/10.1016/j.ypmed.2013.10.010.

Brones, A. (2018). Food apartheid: The root of the problem with America's groceries. *The Guardian*, May 15. Available at: www.theguardian.com/society/2018/may/15/

food-apartheid-food-deserts-racism-inequality-america-karen-washington-interview#
:~:text=America's%20sustainable%20food%20movement,our%20approaches%20to%
20food%20policy.

Bryce, R., Guajardo, C., Ilarraza, D., Milgrom, N., Pike, D., Savoie, K., Valbuena, F.,
& Miller-Matero, L.R. (2017). Participation in a farmers' market fruit and vegetable
prescription program at a federally qualified health center improves hemoglobin A1C
in low income uncontrolled diabetics. *Preventive Medicine Reports*, 7, 176–179. https://
doi.org/10.1016/j.pmedr.2017.06.006.

Bureau of Labor Statistics. (2020). Household spending for July 2018 through June 2019
up 2.7 percent, compared with prior 12 months. *The Economics Daily,* May 7, 2019.
Available at: www.bls.gov/opub/ted/2020/household-spending-for-july-2018-through-
june-2019-up-2-point-7-percent-compared-with-prior-12-months.htm.

Carlson, A. & Frazao, E. (2012). Are Healthy Foods Really More Expensive? It
Depends on How You Measure the Price. *SSRN Electronic Journal.* https://doi.org/
10.2139/ssrn.2199553.

Carlson, A., Lino, M., Juan, W., Hanson, K., & Basiotis, P.P. (2007). Thrifty Food Plan,
2006 (CNPP-19). US Department of Agriculture, Center for Nutrition Policy and
Promotion. Available at: https://fns-prod.azureedge.net/sites/default/files/usda_
food_plans_cost_of_food/TFP2006Report.pdf.

Cavanagh, M., Jurkowski, J., Bozlak, C., Hastings, J., & Klein, A. (2017). Veggie Rx:
An outcome evaluation of a healthy food incentive programme. *Public Health Nutrition*, 20(14), 2636–2641. https://doi.org/10.1017/S1368980016002081.

Center for Responsive Politics. (n.d.). Food and Beverage: Lobbying, 2019. Open-
secrets.Org. Available at: www.opensecrets.org/industries/lobbying.php?cycle=
2020&ind=n01.

Centers for Disease Control and Prevention. (2016). *National Health and Nutrition
Examination Survey Data.* US Department of Health and Human Services, Centers for
Disease Control and Prevention. Available at www.ars.usda.gov/ARSUserFiles/
80400530/pdf/1516/tables_1–36_2015–2016.pdf and www.ars.usda.gov/ARSU
serFiles/80400530/pdf/FPED/ tables_1–4_FPED_1516.pdf.

Centers for Medicare and Medicaid Services. 2018. National Health Expenditures Fact
Sheet. Available at: www.cms.gov/Research-Statistics-Data-and-Systems/Statistics-
Trends-and-Reports/NationalHealthExpendData/NHE-Fact-Sheet#:~:text=NHE%
20grew%204.6%25%20to%20%243.6,16%20percent%20of%20total%20NHE.

Chaloupka, F.J., Powell, L.M., & Warner, K.E. (2019). The Use of Excise Taxes to
Reduce Tobacco, Alcohol, and Sugary Beverage Consumption. *Annual Review of Public
Health*, 40(1), 187–201. https://doi.org/10.1146/annurev-publhealth-040218-043816.

Cohen, J.F.W., Richardson, S., Parker, E., Catalano, P.J., & Rimm, E.B. (2014). Impact
of the New U.S. Department of Agriculture School Meal Standards on Food Selection,
Consumption, and Waste. *American Journal of Preventive Medicine*, 46(4), 388–394. https://
doi.org/10.1016/j.amepre.2013.11.013.

Conrad, Z. (2020). Daily cost of consumer food wasted, inedible, and consumed in the
United States, 2001–2016. *Nutrition Journal*, 19(1), 35. https://doi.org/10.1186/
s12937-020-00552-w.

Cooksey-Stowers, K., Schwartz, M., & Brownell, K. (2017). Food Swamps Predict
Obesity Rates Better Than Food Deserts in the United States. *International Journal of
Environmental Research and Public Health*, 14(11), 1366. https://doi.org/10.3390/ijerp
h14111366.

Cotti, C., Nesson, E., & Tefft, N. (2019). Impacts of the ACA Medicaid expansion on health behaviors: Evidence from household panel data. *Health Economics*, 28(2), 219–244. https://doi.org/10.1002/hec.3838.

Cunningham, P.J. (2008). Trade-Offs Getting Tougher: Problems Paying Medical Bills Increase for U.S. Families, 2003–2007 (Tracking Report No. 21). Center for Studying Health Systems Change. Available at: www.hschange.org/CONTENT/1017/index.html.

Davis, G.C. & You, W. (2010). The Thrifty Food Plan Is Not Thrifty When Labor Cost Is Considered. *The Journal of Nutrition*, 140(4), 854–857. https://doi.org/10.3945/jn.109.119594.

Ebi, K., Hess, J., & Watkiss, P. (Eds.). (2017). Chapter 8: Health Risks and Costs of Climate Variability and Change. Injury Prevention and Environmental Health (3rd Edition). The International Bank for Reconstruction and Development. Available at: www.ncbi.nlm.nih.gov/books/NBK525226/.

Economic Research Service. (2019a). Expenditures for USDA's food assistance programs fall in 2018. US Department of Agriculture. Available at: www.ers.usda.gov/data-products/chart-gallery/gallery/chart-detail/?chartId=58388.

Economic Research Service. (2019b). Local and Regional Foods: Title I (Commodities), Title II (Conservation), Title IV (Nutrition), Title VI (Rural Development), Title VII (Research, Extension, and Related Matters), Title X (Horticulture), Title XII (Miscellaneous). US Department of Agriculture. Available at: www.ers.usda.gov/agriculture-improvement-act-of-2018-highlights-and-implications/local-and-regional-foods.

Economic Research Service. (2020). Food Prices and Spending. US Department of Agriculture. Available at: www.ers.usda.gov/data-products/ag-and-food-statistics-charting-the-essentials/food-prices-and-spending.

Elsheikh, E. & Barhoum, N. (2013). Structural Racialization and Food Insecurity in the United States. Haas Institute for a Fair and Inclusive Society. Available at: https://haasinstitute.berkeley.edu/sites/default/files/Structural%20Racialization%20%20%26%20Food%20Insecurity%20in%20the%20US-%28Final%29.pdf.

Environmental Protection Agency. (2016). Climate Change and the Health of People with Existing Medical Conditions (EPA 430-F-16–059). Available at: www.cmu.edu/steinbrenner/EPA%20Factsheets/existing-conditions-health-climate-change.pdf.

Food Chain Workers Alliance and Solidarity Research Cooperative. (2016). *No Piece of the Pie: US Food Workers in 2016*. Los Angeles, CA: Food Chain Workers Alliance.

Food Research and Action Center. (2012). Replacing the Thrifty Food Plan in Order to Provide Adequate Allotments for SNAP Beneficiaries. Available at: https://frac.org/wp-content/uploads/thrifty_food_plan_2012.pdf.

Fox, M.K. & Gearan, E. (2019). School Nutrition and Meal Cost Study: Summary of Findings. Mathematica Policy Research. Available at: https://fns-prod.azureedge.net/sites/default/files/resource-files/SNMCS_Summary-Findings.pdf.

French, S.A., Tangney, C.C., Crane, M.M, Wang, Y., & Appelhans, B.M. (2019). Nutritional quality of food purchases varies by household income: The SHoPPER study. *BMC Public Health* 19(231). https://doi.org/10.1186/s12889-019-6546-2.

Fulgoni, V. & Drewnowski, A. (2019). An Economic Gap Between the Recommended Healthy Food Patterns and Existing Diets of Minority Groups in the US National Health and Nutrition Examination Survey 2013–14. *Frontiers in Nutrition*, 6 (37). https://doi.org/10.3389/fnut.2019.00037.

Gleason, S., Wolford, B., Wilkin, M., et al. (2018). Analysis of Supplemental Nutrition Assistance Program Education (SNAP-Ed) Data for All States Study. Altarum

Institute and Gabor and Associates Consulting, Inc. for the US Department of Agriculture, Food and Nutrition Service. Available at: https://fns-prod.azureedge.net/sites/default/files/ops/SNAPED-Data-AllStates-Summary.pdf.

Graff, S., Kunkel, D., & Mermin, S.E. (2012). Government Can Regulate Food Advertising To Children Because Cognitive Research Shows That It Is Inherently Misleading. *Health Affairs*, 31(2), 392–398. Available at: https://doi.org/10.1377/hlthaff.2011.0609.

Harris, J.L., Fleming-Milici, F., Kibwana-Jaff, A., & Phaneuf, L. (2020). Sugary Drink F. A.C.T.S. 2020. Rudd Center for Food Policy and Obesity, Council on Black Health, and Salud America! Available at: http://uconnruddcenter.org/files/Pdfs/Sugary_Drink_FACTS_Full%20Report.pdf.

Harris, J.L., Frazier III, W., Kumanyika, S., & Ramirez, A.G. (2019). Increasing disparities in unhealthy food advertising targeted to Hispanic and Black youth. Rudd Center for Food Policy and Obesity, Council on Black Health, and Salud America! Available at: http://uconnruddcenter.org/files/Pdfs/TargetedMarketingReport2019.pdf.

Haynes-Maslow, L. & Stillerman, K. P. (2016). Fixing Food: Fresh Solutions from Five US Cities. Union of Concerned Scientists. Available at: www.ucsusa.org/sites/defa ult/files/attach/2016/01/ucs-fixing-food-report-jan-2016.pdf.

He, X., Boehm, R., & Lopez, R.A. (2020). Medicaid expansion and non-alcoholic beverage choices by low-income households. *Health Economics*. Available at: https://doi.org/forthcoming.

Hiza, H.A.B., Casavale, K.O., Guenther, P.M., & Davis, C.A. (2013). Diet Quality of Americans Differs by Age, Sex, Race/Ethnicity, Income, and Education Level. *Journal of the Academy of Nutrition and Dietetics*, 113(2), 297–306. https://doi.org/10.1016/j.ja nd.2012.08.011.

Honeycutt, S., Leeman, J., McCarthy, W.J., Bastani, R., Carter-Edwards, L., Clark, H., Garney, W., Gustat, J., Hites, L., Nothwehr, F., & Kegler, M. (2015). Evaluating Policy, Systems, and Environmental Change Interventions: Lessons Learned From CDC's Prevention Research Centers. *Preventing Chronic Disease*, 12, 150281. https://doi.org/10.5888/pcd12.150281.

Hu, L., Kaestner, R., Mazumder, B., Miller, S., & Wong, A. (2016). *The Effect of the Patient Protection and Affordable Care Act Medicaid Expansions on Financial Wellbeing* (No. w22170; p. w22170). National Bureau of Economic Research. https://doi.org/10.3386/w22170.

Imamura, F., O'Connor, L., Ye, Z., Mursu, J., Hayashino, Y., Bhupathiraju, S.N., & Forouhi, N.G. (2015). Consumption of sugar sweetened beverages, artificially sweetened beverages, and fruit juice and incidence of type 2 diabetes: Systematic review, meta-analysis, and estimation of population attributable fraction. *BMJ*, h3576. https://doi.org/10.1136/bmj.h3576.

Jardim, T.V., Mozaffarian, D., Abrahams-Gessel, S., Sy, S., Lee, Y., Liu, J., Huang, Y., Rehm, C., Wilde, P., Micha, R., & Gaziano, T.A. (2019). Cardiometabolic disease costs associated with suboptimal diet in the United States: A cost analysis based on a microsimulation model. *PLOS Medicine*, 16(12), e1002981. https://doi.org/10.1371/journal.pmed.1002981.

John Hancock Insurance. (2020). Vitality Program. Available at: www.johnhancock insurance.com/vitality-program.html.

Joy, A.B., Pradhan, V., & Goldman, G. (2006). Cost-benefit analysis conducted for nutrition education in California. *California Agriculture*, 60(4), 185–191. https://doi.org/10.3733/ca.v060n04p185.

Kraak, V.I., Vandevijvere, S., Sacks, G., Brinsden, H., Hawkes, C., Barquera, S., Lobstein, T., & Swinburn, B.A. (2016). Progress achieved in restricting the marketing of high-fat, sugary and salty food and beverage products to children. *Bulletin of the World Health Organization*, 94(7), 540–548. https://doi.org/10.2471/BLT.15.158667.

Lee, M.M., Falbe, J., Schillinger, D., Basu, S., McCulloch, C.E., & Madsen, K.A. (2019). Sugar-Sweetened Beverage Consumption 3 Years After the Berkeley, California, Sugar-Sweetened Beverage Tax. *American Journal of Public Health*, 109(4), 637–639. https://doi.org/10.2105/AJPH.2019.304971.

Lee, Y., Mozaffarian, D., Sy, S., Huang, Y., Liu, J., Wilde, P.E., Abrahams-Gessel, S., Jardim, T. de S.V., Gaziano, T.A., & Micha, R. (2019). Cost-effectiveness of financial incentives for improving diet and health through Medicare and Medicaid: A microsimulation study. *PLOS Medicine*, 16(3), e1002761. https://doi.org/10.1371/journal.pmed.1002761.

Lee, Y., Mozaffarian, D., Sy, S., Liu, J., Wilde, P.E., Marklund, M., Abrahams-Gessel, S., Gaziano, T.A., & Micha, R. (2020). Health Impact and Cost-Effectiveness of Volume, Tiered, and Absolute Sugar Content SugarSweetened Beverage Tax Policies in the United States: A Microsimulation Study. *Circulation*, CIRCULATIONAHA.119.042956. https://doi.org/10.1161/CIRCULATIONAHA.119.042956.

Liu, J., Rehm, C.D., Onopa, J., & Mozaffarian, D. (2020a). Trends in Diet Quality Among Youth in the United States, 1999–2016. *JAMA*, 323(12), 1161. https://doi.org/10.1001/jama.2020.0878.

Liu, J., Steele, E., Karageorgou, D., Micha, R., Monteiro, C., & Mozaffarian, D. (2020b). Consumption of Ultra-Processed Foods and Diet Quality Among U.S. Adults and Children. *Current Developments in Nutrition*, 4(Supplement_2), 543–543. https://doi.org/10.1093/cdn/nzaa046_043.

Long, M.W., Gortmaker, S.L., Ward, Z.J., Resch, S.C., Moodie, M.L., Sacks, G., Swinburn, B.A., Carter, R.C., & Claire Wang, Y. (2015). Cost Effectiveness of a Sugar-Sweetened Beverage Excise Tax in the U.S. *American Journal of Preventive Medicine*, 49(1), 112–123. https://doi.org/10.1016/j.amepre.2015.03.004.

Malik, V.S. & Hu, F.B. (2019). Sugar-Sweetened Beverages and Cardiometabolic Health: An Update of the Evidence. *Nutrients*, 11(8), 1840. https://doi.org/10.3390/nu11081840.

Micha, R., Peñalvo, J.L., Cudhea, F., Imamura, F., Rehm, C.D., & Mozaffarian, D. (2017). Association Between Dietary Factors and Mortality From Heart Disease, Stroke, and Type 2 Diabetes in the United States. *JAMA*, 317(9), 912. https://doi.org/10.1001/jama.2017.0947.

Mitchell, B. & Franco, J. (2018). *HOLC "Redlining" Maps: The persistent structure of segregation and economic inequality.* National Community Reinvestment Coalition. Available at: https://ncrc.org/wp-content/uploads/dlm_uploads/2018/02/NCRC-Research-HOLC-10.pdf.

Monsivais, P., Aggarwal, A., & Drewnowski, A. (2014). Time spent on home food preparation and indicators of healthy eating. *American Journal of Preventive Medicine* 47(6):796–802. https://doi.org/10.1016/j.amepre.2014.07.033.

Mozaffarian, D., Liu, J., Sy, S., Huang, Y., Rehm, C., Lee, Y., Wilde, P., Abrahams-Gessel, S., de Souza Veiga Jardim, T., Gaziano, T., & Micha, R. (2018). Cost-effectiveness of financial incentives and disincentives for improving food purchases and health through the US Supplemental Nutrition Assistance Program (SNAP): A microsimulation study. *PLOS Medicine*, 15(10), e1002661. https://doi.org/10.1371/journal.pmed.1002661.

Mulik, K. & Haynes-Maslow, L. (2017). The Affordability of MyPlate: An Analysis of SNAP Benefits and the Actual Cost of Eating According to the Dietary Guidelines. *Journal of Nutrition Education and Behavior*, 49(8), 623–631.e1. https://doi.org/10.1016/j.jneb.2017.06.005.

Murphy, K. & Topel, R. (2005). *The Value of Health and Longevity* (No. w11405; p. w11405). National Bureau of Economic Research. https://doi.org/10.3386/w11405.

Naja, F. & Hamadeh, R. (2020). Nutrition amid the COVID-19 pandemic: A multilevel framework for action. *European Journal of Clinical Nutrition*. https://doi.org/10.1038/s41430-020-0634-3.

Naja-Riese, A., Keller, K.J.M., Bruno, P., Foerster, S.B., Puma, J., Whetstone, L., McNelly, B., Cullinen, K., Jacobs, L., & Sugerman, S. (2019). The SNAP-Ed Evaluation Framework: Demonstrating the impact of a national framework for obesity prevention in low-income populations. *Translational Behavioral Medicine*, 9(5), 970–979. https://doi.org/10.1093/tbm/ibz115.

National Nutrition Monitoring and Related Research Act of 1990. (1990). No. H.R. 1608, House—Agriculture. Available at: www.congress.gov/bill/101st-congress/house-bill/1608.

Nikpay, S., Buchmueller, T., & Levy, H.G. (2016). Affordable Care Act Medicaid Expansion Reduced Uninsured Hospital Stays In 2014. *Health Affairs*, 35(1), 6–110. https://doi.org/10.1377/hlthaff.2015.1144.

Ohri-Vachaspati, P., Isgor, Z., Rimkus, L., Powell, L.M., Barker, D.C., & Chaloupka, F.J. (2015). Child-Directed Marketing Inside and on the Exterior of Fast Food Restaurants. *American Journal of Preventive Medicine*, 48(1), 22–30. https://doi.org/10.1016/j.amepre.2014.08.011.

Olsho, L.E., Klerman, J.A., Wilde, P.E., & Bartlett, S. (2016). Financial incentives increase fruit and vegetable intake among Supplemental Nutrition Assistance Program participants: A randomized controlled trial of the USDA Healthy Incentives Pilot. *The American Journal of Clinical Nutrition*, 104(2), 423–435. https://doi.org/10.3945/ajcn.115.129320.

Parks, C.A., Stern, K.L., Fricke, H.E., Clausen, W., Fox, T.A., & Yaroch, A.L. (2019). Food Insecurity Nutrition Incentive Grant Program: Implications for the 2018 Farm Bill and Future Directions. *Journal of the Academy of Nutrition and Dietetics*, 119(3), 395–399. https://doi.org/10.1016/j.jand.2018.12.005.

Rajgopal, R., Cox, R.H., Lambur, M., & Lewis, E.C. (2002). Cost-Benefit Analysis Indicates the Positive Economic Benefits of the Expanded Food and Nutrition Education Program Related to Chronic Disease Prevention. *Journal of Nutrition Education and Behavior*, 34(1), 26–37. https://doi.org/10.1016/S1499-4046(06)60225-X

Rehm, C.D., Monsivais, P., & Drewnowski, A. (2015). Relation between diet cost and Healthy Eating Index 2010 scores among adults in the United States 2007–2010. *Preventive Medicine*, 73, 70–75. https://doi.org/10.1016/j.ypmed.2015.01.019.

Reinhardt, S. (2019). *Delivering on the Dietary Guidelines: How Stronger Nutrition Policy Can Cut Healthcare Costs and Save Lives*. Union of Concerned Scientists. Available at: www.ucsusa.org/resources/delivering-dietary-guidelines.

Richardson, S. *et al.* (2020). Presenting Characteristics, Comorbidities, and Outcomes Among 5700 Patients Hospitalized With COVID-19 in the New York City Area. *JAMA*, 323(20), 2052. https://doi.org/10.1001/jama.2020.6775.

Rico-Campà, A., Martínez-González, M.A., Alvarez-Alvarez, I., de Deus Mendonça, R., de la Fuente-Arrillaga, C., Gómez-Donoso, C., & Bes-Rastrollo, M. (2019) Association between consumption of ultra-processed foods and all cause mortality:

SUN prospective cohort study. *BMJ, 365*(11949). https://doi.org/10.1136/bmj. l1949.

Ridberg, R.A., Bell, J.F., Merritt, K.E., Harris, D.M., Young, H.M., & Tancredi, D.J. (2019). Effect of a Fruit and Vegetable Prescription Program on Children's Fruit and Vegetable Consumption. *Preventing Chronic Disease*, 16, 180555. https://doi.org/10. 5888/pcd16.180555.

Riddle, M.C., Buse, J.B., Franks, P.W., Knowler, W.C., Ratner, R.E., Selvin, E., Wexler, D.J., & Kahn, S.E. (2020). COVID-19 in People With Diabetes: Urgently Needed Lessons From Early Reports. *Diabetes Care*, 43(7), 1378–1381. https://doi. org/10.2337/dci20-0024.

Rose, D. (2007). Food Stamps, the Thrifty Food Plan, and Meal Preparation: The Importance of the Time Dimension for US Nutrition Policy. *Journal of Nutrition Education and Behavior*, 39(4), 226–232. https://doi.org/10.1016/j.jneb.2007.04.180.

Sadeghirad, B., Duhaney, T., Motaghipisheh, S., Campbell, N.R.C., & Johnston, B.C. (2016). Influence of unhealthy food and beverage marketing on children's dietary intake and preference: A systematic review and meta-analysis of randomized trials: Meta-analysis of unhealthy food and beverage marketing. *Obesity Reviews*, 17(10), 945–959. https://doi.org/10.1111/obr.12445.

Schwartz, M.B. (2017). Moving Beyond the Debate Over Restricting Sugary Drinks in the Supplemental Nutrition Assistance Program. *American Journal of Preventive Medicine*, 52(2), S199–S205. https://doi.org/10.1016/j.amepre.2016.09.022.

Seguin, R., Connor, L., Nelson, M., LaCroix, A., & Eldridge, G. (2014). Understanding barriers and facilitators to healthy eating and active living in rural communities. *Journal of Nutrition and Metabolism 2014*:146502. https://dx.doi.org/10.1155% 2F2014%2F146502.

Seiler, S., Tuchman, A., & Yao, S. (2018). The Impact of Soda Taxes: Pass-through, Tax Avoidance, and Nutritional Effects. *SSRN Electronic Journal*. https://doi.org/10. 2139/ssrn.3302335.

Simon, K., Soni, A., & Cawley, J. (2017). The Impact of Health Insurance on Preventive Care and Health Behaviors: Evidence from the First Two Years of the ACA Medicaid Expansions: Impact of Health Insurance on Preventive Care and Health Behaviors. *Journal of Policy Analysis and Management*, 36(2), 390–417. https://doi.org/ 10.1002/pam.21972.

Sommers, B.D., Blendon, R.J., Orav, E.J., & Epstein, A.M. (2016). Changes in Utilization and Health Among Low-Income Adults After Medicaid Expansion or Expanded Private Insurance. *JAMA Internal Medicine*, 176(10), 1501. https://doi.org/ 10.1001/jamainternmed.2016.4419.

Story, M., Kaphingst, K.M., Robinson-O'Brien, R., & Glanz, K. (2008). Creating Healthy Food and Eating Environments: Policy and Environmental Approaches. *Annual Review of Public Health*, 29(1), 253–272. https://doi.org/10.1146/annurev. publhealth.29.020907.090926.

The U.S. Burden of Disease Collaborators (2018). The State of US Health, 1990–2016: Burden of Diseases, Injuries, and Risk Factors Among US States. *JAMA*, 319(14), 1444. https://doi.org/10.1001/jama.2018.0158.

US Department of Agriculture. (n.d.-a). Farmers Market Promotion Program. Agricultural Marketing Service. Available at: www.ams.usda.gov/services/grants/fmpp.

US Department of Agriculture. (n.d.-b). Food and Nutrition Service. Available at: www.fns.usda.gov/.

US Department of Agriculture. (n.d.-c). Local Food Promotion Program. Agricultural Marketing Service. Available at: www.ams.usda.gov/services/grants/lfpp.

US Department of Agriculture. (2020). Official USDA Food Plans: Cost of Food at Home at Four Levels, US Average, May 2020. Available at: https://fns-prod.azur eedge.net/sites/default/files/media/file/CostofFoodMay2020.pdf.

US Department of Agriculture and US Department of Health and Human Services. (2015). 2015–2020 Dietary Guidelines for Americans (Figure 3–1. A Social-Ecological Model for Food and Physical Activity Decisions). US Department of Agriculture and US Department of Health and Human Services. Available at: https://health.gov/our-work/food-nutrition/2015-2020-dietary-guidelines/guidelines/infographic/3-1/.

US Department of Health and Human Services. (2019). CED Healthy Food Financing Initiative. Office of Community Services. Available at: www.acf.hhs.gov/ocs/programs/community-economic-development/healthy-food-financing.

Ver Ploeg, M., Larimore, E., & Wilde, P. (2017) The Influence of Foodstore Access on Grocery Shopping and Food Spending. Economic Research Service. Economic Information Bulletin Number 180. Available at: www.ers.usda.gov/webdocs/publications/85442/eib-180.pdf?v=4138.3.

Wang, D.D., Leung, C.W., Li, Y., Ding, D.L. Chiuve, S.E., Hu, F.B., & Willett, W.C. (2014). Trends in Dietary Quality Among Adults in the United States, 1999 through 2010. *JAMA Internal Medicine*, 174(10), 1587–1595. https://doi.org/10.1001/jamainternmed.2014.3422.

Wang, D.D., Li, Y., Afshin, A., Springmann, M., Mozaffarian, D., Stampfer, M.J., Hu, F. B., Murray, C.J.L., & Willett, W.C. (2019). Global Improvement in Dietary Quality Could Lead to Substantial Reduction in Premature Death. *The Journal of Nutrition*, 149(6), 1065–1074. https://doi.org/10.1093/jn/nxz010.

World Health Organization. (n.d.). Metrics: Disability-Adjusted Life Year (DALY). Health Statistics and Information Systems. Available at: www.who.int/healthinfo/global_burden_disease/metrics_daly/en/.

World Health Organization. (2017). Taxes on sugary drinks: Why do it? (WHO/NMH/PND/16.5 Rev.1). World Health Organization. Available at: https://apps.who.int/iris/handle/10665/260253.

Zhang, F.F., Liu, J., Rehm, C.D., Wilde, P., Mande, J.R., & Mozaffarian, D. (2018). Trends and Disparities in Diet Quality Among US Adults by Supplemental Nutrition Assistance Program Participation Status. *JAMA Network Open*, 1(2),e180237. https://doi.org/10.1001/jamanetworkopen.2018.0237.

11 True Cost Principles in Public Policy

How Schools and Local Government Bring Value to Procurement

Paula A. Daniels

Food Production and Social Values: A Mid-Century Disconnect

It has been a half century since the 1968 speech that US Senator Robert Kennedy gave in which he spoke of the intellectual fallacy of measuring a nation's success by the economic yardstick of the Gross National Product. "Too much and for too long," he said, "we seemed to have surrendered personal excellence and community values in the mere accumulation of material things" (Kennedy, 1968). He went on to list the limitations of measuring success by those prevalent economic indicators:

> Our Gross National Product…counts air pollution and cigarette advertising, and ambulances to clear our highways of carnage. It counts special locks for our doors and the jails for the people who break them. It counts the destruction of the redwood and the loss of our natural wonder in chaotic sprawl. It counts napalm and counts nuclear warheads and armored cars for the police to fight the riots in our cities.…Yet the gross national product does not allow for the health of our children, the quality of their education or the joy of their play. It does not include the beauty of our poetry or the strength of our marriages, the intelligence of our public debate or the integrity of our public officials. It measures neither our wit nor our courage, neither our wisdom nor our learning, neither our compassion nor our devotion to our country, *it measures everything in short, except that which makes life worthwhile.*
>
> (emphasis added)

His challenge to the nation arose from the urgent core of his galvanizing role as a leading progressive voice in a society experienceing a then unprecedented level of cultural and political upheaval (McLaughlin, 2014).

Senator Kennedy spoke of other issues during his landmark speech, including the value of protest. The power of protest has carried through to 2020: recent events in the USA relating to racial inequities have seen large-scale protest participation, including the presence of mayors and other civic or political leaders. However, Kennedy's caution about the limited utility of society's heavy reliance on the Gross Domestic Product (GDP) has not experienced the same traction.

Or has it? If not through an explicit rejection of the GDP as a singular measure, a more subtle but effective version of success measurement in the food system might be underway: in the US school food system.

The Past is Prologue

As with our prevailing commercialized food system, the entrenchment of Gross National Product as a primary measure of progress is considered to be an out-growth of World War II political culture (Debroy and Kapoor, 2019; Costanza *et al.*, 2014). The current unhealthy state of the US food system is often attributed to mid-Century economics and the Cold War era of American economic expansion, a political layer built on the post-World War II use of military chemicals for farmland fertilizer, ushering in the age of agricultural industrialization (Pollan, 2006). The US National School Lunch Program (NSLP)—the second largest nutrition assistance program in the United States (United States Department of Agriculture Economic Research Service, 2019)—was also an outgrowth of World War II. Signed into law by President Harry Truman in 1946, it was intended to increase demand for agricultural commodities and to provide nutrition to lower income school children. Today, it is in 94% of US schools, spends about $13 billion annually, and serves around 30 million children (United States Department of Agriculture Economic Research Service, 2019) in nearly 100,000 public and private K-12 schools across the USA (United States Department of Agriculture Economic Research Service, 2019). The volume of food purchased makes the school system a large component of the $120 billion annual institutional food service market (United States Department of Agriculture Economic Research Service, 2017).

But the NSLP operates on a cost reimbursement basis, and in 2019 the average reimbursement rate was set at between $3.11 and $3.51 per meal, for students at or below established poverty levels (School Nutrition Association, 2020). At those prices, school districts have tended to source highly commoditized, price subsidized food products. One could argue that in the early decades of the program, school food contributed to the exacerbation of the singular commercial value of economic scale, carrying out the Cold War era imperative.

The era which fostered the rise of the US school food program was a turning point in the American food system, ushering in a precipitous decline in farm populations ("a 'free fall' situation leading to "trauma," according to a former US Department of Agriculture [USDA] demographer, Calvin Beale) (United States Department of Agriculture, 1981) as farms consolidated toward large scale operations. With the shift toward highly consolidated, vertically integrated and industrially efficient agriculture came a rise in obesity (Centers for Disease Control and Prevention, 2020), a loss of agricultural biodiversity, and a rise in nitrate pollution and greenhouse gas emissions owing to concentrated methods of farming and animal rearing (Lilliston, 2019).

Before that point, the national obesity rate was about 12%. (Ogden *et al.*, 2010). There was more diversification of farm ownership and type: around 40%

of the US workforce was in agriculture, and there were over six million farms (Dimitri *et al.*, 2005). By contrast, in the second half of the twentieth century, the obesity rate climbed to 60%, and agriculture became consolidated: there is now less than 2% of the workforce in agriculture, and fewer than two million farms, while average farm size has increased by over 60%, and agricultural output has tripled and become increasingly specialized (Dimitri *et al.*, 2005). In the meat sector over that same time frame, meat supply consolidated into just four companies (Ostland, 2011). Ten multinational companies now control most of the world's food system (Taylor, 2016).

For too long the yardstick of success, particularly in terms of public invest-ment, was measured in terms of more volume production, more dollars in returns, and more delivery of calories, without regard to the quality of nutrition or the quality of the relationships along the food supply chain.

As pointed out by Robert Gottlieb and Anupama Joshi in their book *Food Justice*, by being bound to the agriculture commodity program, the school food program was part of the problem (Gottlieb and Joshi, 2017). Despite an interest in nutritional goals in the 1970s, the politics of the 1980s led to a stigmatization of poverty assis-tance programs, and the school food program was no exception. The business-orientated ethos of President Reagan and his Administration also led to infamous characterizations of ketchup as a vegetable; it was a regulatory move to ensure that the commercial product fit within the short-sighted administrative dietary guidelines of the time and could be sold widely throughout the massive federal school food program. The era also saw a rise in vending machines in schools, stuffed with junk food and sugary sodas, followed by corporate sponsorships from food companies such as the American beverage giants Coca-Cola and PepsiCo (Gottlieb and Joshi, 2017; Levine, 2008), succumbed to by financially desperate administrators who were operating under the thumb of federal reimbursement guidelines. The school food program wielded considerable influence on food economics.

Not surprisingly, advocacy for food system reform, which has been on the rise in the past few decades, has often included school food as a key policy area. First Lady Michelle Obama elevated the issue area as a priority with a particular focus on nutrition; her initiatives included the Healthy Hunger Free Kids Act (HHFKA) signed by President Obama in 2010. HHFKA authorized the USDA to make significant reforms to the nutrition guidelines for the school lunch program, for the first time in decades, and included funding for local farm to school and garden programs. This effort became politicized, however, and the subsequent federal administration rolled back a number of the key changes of the HHFKA (Green and Piccoli, 2019).

Embedding Community Values in Institutional Food at a Municipal Level

The federal politics of food, and the regulatory seesaw of the USDA under different administrations, has led many food system advocates and political lea-ders toward recognition of the role of municipal governments as a key area for creating change in the status quo. Indeed, it is a unit of governance that seems increasingly more effective at responding to the needs of modern populations, organized as they are in intensified urban centers. As Benjamin Barber observed in his book *If Mayors Ruled the World: Dysfunctional Nations, Rising Cities*: "We have come full circle in the city's epic history (Barber, 2013). Humankind began its march to politics and civilization in the polis—the township. It was democracy's original incubator." It might again be its best hope, particularly in times of crisis.

In the food system, the Milan Urban Food Policy Pact was launched in 2014, driven by the fact that "[m]ore than 50 percent of the world's population currently lives in urban areas – a proportion that is projected to increase to almost 70 percent by 2050 – and ensuring the right to food for all citizens, especially the urban poor, is key to promoting sustainable and equitable development." The Pact acknowledges that:

> …current food systems are being challenged to provide permanent and reliable access to adequate, safe, local, diversified, fair, healthy and nutrient rich food for all; and that the task of feeding cities will face multiple constraints posed by inter alia, unbalanced distribution and access, environmental degradation, resource scarcity and climate change, unsustainable production and consumption patterns, and food loss and waste.
>
> ("Milan Urban Food Policy Pact," 2015)

The signatory cities to the Pact commit to, among many other things:

- Develop sustainable food systems that are inclusive, resilient, safe and diverse, that provide healthy and affordable food to all people in a human rights-based framework, that minimize waste and conserve biodiversity while adapting to and mitigating impacts of climate change; and
- Encourage interdepartmental and cross-sector coordination at municipal and community levels, working to integrate urban food policy considerations into social, economic and environment policies, programs and initiatives, such as, inter alia, food supply and distribution, social protection, nutrition, equity, food production, education, food safety and waste reduction.

Over 200 mayors around the world have signed the Milan Urban Food Policy Pact (www.milanurbanfoodpolicypact.org/text) since its launch, but as of 2020 only nine of them were US city mayors. However, increasingly, many US cities are seeing the establishment of food policy councils, which in many regions have served the role of food system advocacy as well as government accountability. The Johns Hopkins Center for a Livable Future describes food policy councils as "networks that represent multiple stakeholders and that are either sanctioned by a government body or exist independently of government, and address food-related issues and needs within a city, county, state, tribal, multi-county or other designated region." (Food Policy Networks, n.d.-a) Their database of food policy councils shows them at over 300 in North America (Food Policy Networks, n.d.-b). For many of the food policy councils, school food is a top priority (Bassarab, 2019).

Among the more prominent food policy councils is the Los Angeles Food Policy Council (LAFPC), launched in 2011 as an initiative of Mayor Villaraigosa of Los Angeles, in the second largest city in the USA and the 23rd largest world city. It was launched with a mandate developed by a task force, to advance 55 action steps in six priority areas, directed toward the goal of building a more sustainable and equitable regional food system in the Los Angeles region of southern California (Los Angeles Food Policy Task Force, 2010).

The well-staffed, municipal government supported council gave rise to the Good Food Purchasing Program (the Program), adopted by the City of Los Angeles and Los Angeles Unified School District (LAUSD) in 2012. It was developed through an extensive multi-sector, interdisciplinary, multi-stakeholder collaboration and review process within the LAFPC. The Program provides a metric based, flexible framework that is the basis for a feedback and rating tool for its enrolled institutions, and it embeds five core community values in its Program design.

The Good Food Purchasing Program as a True Cost Influencer

As mentioned, each year food service institutions—from hospitals to jails to school districts—spend nearly $120 billion on food. Yet institutions rarely have the information, resources, or expertise that they need to align purchasing with community values, and local communities lack the data or tools that they need to build public support for aligning purchasing with their values and achieving policy change. For example, the adoption of the Good Food Purchasing Program has served as a basis for municipal council resolutions which explicitly recognized The Good Food Purchasing Program as a lever to address the issues identified in its five core values, along with other issues such as food justice and equity (Board of Commissioners of Cook County, 2018; City of Boston in City Council, 2019).

The Good Food Purchasing Program provides the information and resources that institutions and communities need to break through the murkiness of informational opacity. Similar to how the Leadership in Energy and Environmental Design (LEED) (United States Green Building Council, n.d.) certification works in rating energy efficiency and environmental design in buildings, the Program combines a unique, flexible framework for values-based purchasing with metric-based targets; the staff of the Center provide analysis of the purchasing data against that rubric, and review, verify and rate the compliance levels of participating institutions.

Key aspects of the Program include:

- Program enrollment through organizational collaboration toward shared goals in partnership with municipal leaders and food service providers;

- A rating system for institutions which provides feedback and account-ability, helping institutions redirect food budgets towards more sustainable suppliers and those that adhere to fair labor practices, and providing a basis for transparency and accountability;
- A point system which offers institutions a flexible roadmap toward values based purchasing, through a metric based, tiered structure driving institutional purchasing toward five food system values around which the Program is deeply architected and designed;
- The rating system includes collection of purchasing data from an enrolled institution, for evaluation against the five value categories; and
- The five core value categories are: local economies, environmental sustainability, fair labor, animal welfare, and nutritional health.

As an outcome, the Program promotes the purchase of more sustainably produced food from local economies, especially small, mid-sized, and historically disadvantaged farms and food processing operations, which results in production returns at a more regional and local level, ensures that suppliers' workers are offered safe and healthy working conditions and fair compensation, that livestock receives healthy and humane care, and that consumers—foremost school children, patients, the elderly—enjoy better health and well-being, thanks to higher quality nutritious meals.

Within one year of the Program adoption at LAUSD in 2012, the institution, with its $120 million per year food budget, achieved the success that its design was intended to promote: local sourcing of produce rose from an average of 10% per year to an average of 60% per year, redirecting $12 million to the local food economy. As a result, some 150 new well-paid food chain jobs

Figure 11.1 Good Food Purchasing Program values.

were created in Los Angeles County, including food processing, manufacturing, and distribution. In the ensuing years, 160 truck drivers in LAUSD's supply chain received higher wages and improved working conditions.

Owing to the immediate success of the Program at LAUSD, interest in adoption by other cities was piqued, initially through the Food Policy Task Force of the US Conference of Mayors (convened by Mayor Villaraigosa), and eventually to greater effect through alliances with labor, environmental, and food system advocacy organizations. In 2015 the Program was spun off from the LAFPC and became the program of the Center for Good Food Purchasing, established to advance the national expansion of the Program. As at July 2020, over 50 municipal institutions in 20 major cities across the USA were enrolled in the Program, over 2.5 million students were being served under the scheme, and over $1 billion in institutional purchasing was being analyzed and rated by the Center.

The Program is the first procurement model designed to elevate food service (whether in municipal agencies/departments, universities, schools and/or hospitals) as a transformative tool, using the purchasing power of these institutions to support the five food system values of local economies, environmental sustainability, valued workforce, animal welfare, and nutrition in equal measure. Those five core value categories, together with the public accountability structure of the Program itself, are reflective of the ten Elements of Agroecology as defined by the United Nations (Food and Agriculture Organization, n.d.-a; n.d.-b) and align with at least 12 of the 17 of the United Nations Sustainable Development Goals. The systemically holistic Good Food Purchasing Program was favorably recognized in 2018 (FuturePolicy.org, n.d.) by the World Future Council, the Food and Agriculture Organization of the United Nations, and IFOAM Organics International as a Future Policy Scaling Up Agroecology.

The Good Food Purchasing Program, in measuring how enrolled institutions direct their purchasing toward its five value categories, also aligns with the TEEBAgrifood framework (as described elsewhere in this book). In short, the Program can be characterized as True Cost Accounting (TCA) at work in enrolled municipal institutions.

The Power of Procurement

In his Briefing Note 8 (April 2014), *The Power of Procurement: Public Purchasing in Realizing the Right to Food* (De Schutter, 2014), UN Special Rapporteur De Schutter recognized that:

> Governments have few sources of leverage over increasingly globalized food systems – but public procurement is one of them. When sourcing food for schools, hospitals and public administrations, Governments have a rare opportunity to support more nutritious diets and more sustainable food systems in one fell swoop.

Procurement is also one of the recommended actions of category five of the Milan Urban Food Policy Pact, which calls for a review of:

> ...public procurement and trade policy aimed at facilitating food supply from short chains linking cities to secure a supply of healthy food, while also facilitating job access, fair production conditions and sustainable production for the most vulnerable producers and consumers, thereby using the potential of public procurement to help realize the right to food for all.
> (Milan Urban Food Policy Pact, n.d.)

As pointed out by the Union of Concerned Scientists in their 2017 report on the impacts of the Good Food Purchasing Program in Los Angeles, the "benefits of a better supply chain are amplified across institutions and regions" (Mulik and Reinhardt, 2018). The incremental shifts created by the institutions enrolled in the Program show combined totals across institutions of over $56 million in supporting local economies, over $32 million in supporting fair labor, over $20 million toward meat raised without routine use of antibiotics, and an additional $10 million supporting environmental sustainability. Some other key achievements of the Program:

- Increase in fruits and vegetables served in enrolled institutions
- Increase in worker wages, including a 40% wage increase for 320 warehouse workers and truck drivers along LAUSD's supply chain
- In 2017, LAUSD awarded $70 million in contracts for chicken produced without routine antibiotics.
- In the Austin, Texas school district, expenditures on organic products tripled over the first two years in the Program.
- Over one school year, Oakland Unified School District reduced its carbon and water footprint by roughly 20% through sourcing less meat and doubled its purchase of sustainable and humane food products, without increasing costs.

Notably, the ability to increase purchases even more in key value categories has been hampered by the cost of food products that fall within the identified categories of environmental sustainability, fair trade, or humane. Those food products which are certified as, for example, USDA Organic, Fair Trade certified, or Animal Welfare Approved, are sold at a price premium, which reflects the more positive relationship to the environment, labor, or animal welfare than industrially produced foods; in other words, those positively certified food products are closer to "true cost." However, that price premium is difficult for a school district to bear within budgetary constraints.

Incentives for True Cost Food

As mentioned, school districts have very tight budgets for their food purchases, and the reimbursement amount typically also pays for labor (LAUSD Food

Service Director, personal interview by Paula Daniels, 2020). The procurement bid process of these large institutions does allow them to obtain reasonable percentages of the value-based food within their budgets, as conveyed to the food service bidders through Requests for Proposals. The value certified food products can be incorporated into vendor bids at a price break made possible owing to the volume of demand, but not yet in quantities sufficient to make larger-scale shifts.

A key recommendation to address this gap is the creation of a financial incentive fund to support the purchase of the more "true cost" foods by municipal food service providers. The fund could be created in a number of ways, including through public-private-philanthropic partnership arrangements that incorporate fund criteria and oversight. Local and state governments could lead the way in developing and directing these financial incentives to the anchor institutions to enable purchasing support for fair wage and climate friendly food production practices, such as soil health.

Based on conversations with food service providers at school districts and other municipal institutions, an additional $0.15 to $0.25 per meal would provide the ability to purchase food that reflects the true cost of its production in one or more of the value categories (School Food Service Directors of Austin, Minneapolis and San Francisco, personal conversation by Alexa Delwiche, n.d.).

The incentive concept builds on the pioneering local food incentive models already established in Michigan's "10 Cents a Meal for Michigan's Kids & Farms" (www.tencentsmichigan.org/), New Mexico (New Mexico Department of Health, 2019), Oregon (Kane *et al.*, 2011), and New York (New York State Department of Agriculture and Markets, n.d.; New York State Health Foundation, 2018; New York State, 2019). For example, in Michigan, the state legislature created a 10 Cents Per Meal program in 2017 which provides 10 cents per meal to provide 57 school districts with those extra funds to buy local fruits, vegetables, and dry beans. Since its inception, 121 school districts throughout the state have applied. According to the 10 Cents Per Meal 2018–2019 legislative report, school food service directors reported an increase in the variety of produce served to students in school meals. The return to the local food economy supported 143 Michigan farms and 20 supply chain business, which also benefited from the advance planning the school food service directors could undertake (Michigan Department of Education, 2019). New Mexico operates its program as a grant program with an annual award of, for example, $85,000 for Albuquerque public schools. In Oregon, the state legislature accessed economic development funds for a 7 cents per meal incentive for local food purchases in the school lunch and breakfast programs. A report on two school districts which participated in the program (Ecotrust, n.d.) found that the school districts indeed used the extra amount to leverage the investment to purchase local items "that cost slightly more than items they had previously been purchasing non-locally" and also that "new vendor relationships were formed at both schools", bolstering the local food economy. The report observed:

Schools can easily funnel the money through a mainline food service distributor, and the more that these companies experience requests for local products, the more likely they are to expand their local purchases and product offerings, with direct implications for the scale and effects of farm to school programming nationwide.

The most generous reimbursement rate, in a program with the most ambitious targets, is in the state of New York, which offers "25 cents per meal for any district that purchases at least 30 percent ingredients for their school lunch program from New York farms" (New York State Department of Agriculture and Markets, n.d.).

An important next step is expanding this model to other valued attributes of an agroecological food system, including in the incentive criteria that there be proportional purchasing of food that supports environmental sustainability (including climate friendly and humane production practices) and fair labor.

This undertaking could serve as a proof point for the field in a few ways:

- Quantify how much money school districts need to increase Good Food (or True Cost) procurement.
- Demonstrate feasibility of the model to inform future school meal reimbursement policy initiatives at a federal level.
- Evaluate broad-based health outcomes and project changes in long-term health costs.

A Twenty-First-Century Path

The inclusion of values other than monetary as a measure of success has been taking place over the past decade or so in intricate ways, organization by organization. In our example, school districts and other municipal food service institutions in the US are measuring their success by the point system of the Good Food Purchasing Program and how well they support the five community-based values of the Program. They are measuring what matters, as Senator Kennedy had urged in 1968.

Will this TCA framework for food have the same ripple effect? Will it set the US on a new trajectory toward an agroecological food system? It depends. Whether formally adopted or infused in the decision-making of even more food service entities, measuring progress toward more of what matters will make a difference; indeed, it already has.

References

Barber, B. (2013). *If Mayors Ruled the World: Dysfunctional Nations, Rising Cities*, (pp. 3–28). New Haven, CT: Yale University Press. Available at: www.jstor.org/stable/j.ctt5vksfr.4.

Bassarab, K.*et al.* (2019, April 1). Food Policy Council Report 2018. Johns Hopkins Center for a Livable Future, Food Policy Networks. Available at: https://assets.jhsph.edu/clf/mod_clfResource/doc/FPC%20Report%202018-FINAL-4-1-19.pdf.

Board of Commissioners of Cook County. (2018). To Adopt the Good Food Purchasing Program. Available at: https://cook-county.legistar.com/LegislationDetail.aspx?ID=3309826&GUID=ED1C9BDF-90BD-4355-AB3C-AE1738EC6A38&Options=&Search=.

Centers for Disease Control and Prevention. (2020). Adult Obesity Facts. *Overweight & Obesity*. Centers for Disease Control and Prevention. Available at: www.cdc.gov/obesity/data/adult.html.

City of Boston in City Council. (2019). An Ordinance Regarding Good Food Purchasing Standards in the City of Boston. Available at: www.boston.gov/sites/default/files/file/document_files/2019/02/0139.pdf.

Costanza, R.*et al.* (2014). A Short History of GFP: Moving Towards Better Measures of Human Well-being. *The Solutions Journal, (5)*1, (pp. 91–97). Available at: www.thesolutionsjournal.com/article/a-short-history-of-gdp-moving-towards-better-measures-of-human-well-being.

De Schutter, O. (2014). The Power of Procurement: Public Purchasing in the Service of Realizing the Right to Food. (Briefing Note 08), United Nations. Available at: www.srfood.org/images/stories/pdf/otherdocuments/20140514_procurement_en.pdf.

Debroy, B. and Kapoor, A. (2019). GDP Is Not a Measure of Human Well-Being. *CNN*. Available at: https://hbr.org/2019/10/gdp-is-not-a-measure-of-human-well-being.

Dimitri, C., Effland, A., & Conklin, N. (2005). The 20th Century Transformation of U. S. Agriculture and Farm Policy. Economic Information Bulletin, Number 3. United States Department of Agriculture, Economic Research Service. Available at: www.ers.usda.gov/webdocs/publications/44197/13566_eib3_1_.pdf?v=7007.

Ecotrust. (n.d.). The Impact of Seven Cents. Available at: https://ecotrust.org/media/7-Cents-Report_FINAL_110630.pdf.

Food and Agriculture Organization of the United Nations. (n.d.-a). 10 Elements of Agroecology. Available at: www.fao.org/agroecology/knowledge/10-elements.

Food and Agriculture Organization of the United Nations. (n.d.-b). Agroecology Knowledge Hub. Available at: www.fao.org/agroecology/home/en.

Food Policy Networks. (n.d.-a). *About Us*. Available at: www.foodpolicynetworks.org/about/.

Food Policy Networks. (n.d.-b). *Food Policy Council Map*. Available at: www.foodpolicynetworks.org/councils/fpc-map.

FuturePolicy.org. (n.d.). *Good Food Purchasing Program*. Available at: www.futurepolicy.org/healthy-ecosystems/los-angeles-good-food-purchasing-program.

Gottlieb, R. & Joshi, A. (2017). *Food Justice*. (p. 88). Cambridge, MA: The MIT Press.

Green, E. & Piccoli, S. (2019). Trump Administration Sued Over Rollback of School Lunch Standards. *The New York Times*, April 3. Available at: www.nytimes.com/2019/04/03/us/politics/trump-school-lunch-standards.html.

Kane, D., Kruse, S., Ratcliffe, M.M., Sobell, S.A., & Tessman, N. (2011). The Impact of Seven Cents. Ecotrust. Available at: https://ecotrust.org/media/7-Cents-Report_FINAL_110630.pdf.

Kennedy, R. (1968). Remarks at the University of Kansas. John F. Kennedy Presidential Library and Museum. Available at: www.jfklibrary.org/learn/about-jfk/the-kennedy-family/robert-f-kennedy/robert-f-kennedy-speeches/remarks-at-the-university-of-kansas-march-18-1968.

Levine, S. (2008). *School Lunch Politics*. Princeton, NJ: Princeton University Press.

Lilliston, B. (2019). Latest agriculture emissions data show rise of factory farms. Institute for Agriculture and Trade Policy. Available at: www.iatp.org/blog/201904/latest-agriculture-emissions-data-show-rise-factory-farms.

Los Angeles Food Policy Task Force. (2010). Good Food For All Agenda. Available at: https://goodfoodlosangeles.files.wordpress.com/2010/07/good-food-full_report_single_072010.pdf

McLaughlin, K. (2014). Eight unforgettable ways 1968 made history. CNN, July 31. Available at: www.cnn.com/2014/07/31/us/1968-important-events/index.html.

Michigan Department of Education. (2019). *10 Cents a Meal for School Kids & Farms: State Pilot Project Overview (2018/2019 Legislative Report)*. Available at: https://d3n8a8pro7vhmx.cloudfront.net/tencentsmichigan/pages/26/attachments/original/1553192603/10_Cents_a_Meal_2018-2019_Legislative_Report.pdf?1553192603.

Milan Urban Food Policy Pact. (n.d.). Food supply and distribution. MUFPP Recommended actions. Available at: www.milanurbanfoodpolicypact.org/mufpp_food-supply-and-distribution/.

Milan Urban Food Policy Pact. (2015). *Milan Urban Food Policy Pact*. Available at: www.milanurbanfoodpolicypact.org/wp-content/uploads/2016/06/Milan-Urban-Food-Policy-Pact-EN.pdf.

Mulik, K. & Reinhardt, S. (2018). Purchasing Power: How Institutional "Good Food" Procurement Policies Can Shape a Food System That's Better for People and Planet. Union of Concerned Scientists. Available at: www.ucsusa.org/sites/default/files/attach/2017/11/purchasing-power-report-ucs-2017.pdf.

New Mexico Department of Health. (2019). Students to Celebrate Farm to School Programs for New Mexico Grow Week. Available at: www.nmhealth.org/news/healthy/2019/9/?view=800.

New York State. (2019). Governor Cuomo Announces $1.5 Million Available to School Districts Through State's Farm-To-School Program. Available at: www.governor.ny.gov/news/governor-cuomo-announces-15-million-available-school-districts-through-states-farm-school.

New York State Department of Agriculture and Markets. (n.d.). Farm to School. Available at: https://agriculture.ny.gov/farming/farm-school.

New York State Health Foundation. (2018). Issue Brief: Local Lunches: A "Win-Win" for New York State's Farmers and Students. Available at: https://nyshealthfoundation.org/wp-content/uploads/2018/07/issue-brief-local-lunches-farm-to-school.pdf.

Ogden, C.*et al.* (2010). Prevalence of Overweight, Obesity, and Extreme Obesity Among Adults: United States, Trends 1960–1962 Through 2007–2009. Centers for Disease Control and Prevention. Available at: www.cdc.gov/nchs/data/hestat/obesity_adult_07_08/obesity_adult_07_08.pdf.

Ostland, E. (2011). The Big Four Meat Packers. *High Country News*, March 21, 2011. Available at: www.hcn.org/issues/43.5/cattlemen-struggle-against-giant-meatpackers-and-economic-squeezes/the-big-four-meatpackers-1.

Pollan, M. (2006). What's Eating America. Smithsonian. Available at: https://michaelpollan.com/articles-archive/whats-eating-america/.

School Nutrition Association. (2020). School Meal Trends & Stats. Available at: https://schoolnutrition.org/aboutschoolmeals/schoolmealtrendsstats/.

Taylor, K. (2016). These Ten Companies Control Everything You Buy. *Business Insider*, September 28, 2016. Available at: www.businessinsider.com/10-companies-control-the-food-industry-2016-9?op=1.

United States Department of Agriculture. (1981). *A Time to Choose: Summary Report on the Structure of Agriculture*. Available at: https://archive.org/stream/timetochoosesumm 00unit/timetochoosesumm00unit_djvu.txt.

United States Department of Agriculture. (2019). National School Lunch Program Fact Sheet. Available at: www.fns.usda.gov/nslp/nslp-fact-sheet.

United States Department of Agriculture, Economic Research Service. (2017). *Market Segments*.

United States Departments of Agriculture Economic Research Service. (2019). National School Lunch Program. Available at: www.ers.usda.gov/topics/food-nutrition-assista nce/child-nutrition-programs/national-school-lunch-program.

United States Green Building Council. (n.d.). LEED Rating System. Available at: www. usgbc.org/leed.

12 Embedding TCA Within US Regulatory Decision-Making

Kathleen A. Merrigan

True Cost Accounting (TCA) is not a brand-new concept or approach but rather an evolved, modern, and hopefully improved variant of Cost Benefit Analysis (CBA). For this reason, it is important to study CBA, understand how it has been used historically, and glean lessons learned for TCA designers, advocates, and practitioners.

CBA is deeply embedded in US policymaking at the federal level. The goal of CBA is to facilitate rational decision-making by adding up and comparing the negative and positive consequences of an action to determine whether the action will lead to a gain or a loss. CBA has been particularly applicable to appraising the desirability of proposed policy. Given the similarity between TCA and CBA, it is useful to study how CBA has been used in US federal rulemaking. This chapter provides a brief overview of that history and extracts relevant lessons learned from CBA application, with particular attention to CBA shortcomings. The purpose in doing so is twofold. First, it is important to alert TCA designers of potential methodological and implementation challenges that have hindered CBA, with the hope that such knowledge will facilitate creative TCA design to circumvent such challenges. Second, it is common for new US presidents to review rulemaking processes and issue directives for improved rulemaking aligned with their philosophical leanings. It is possible that TCA could be substituted for CBA during any forthcoming administration changeover; thus, an understanding of relevant US regulatory processes and laws could provide insight into how to elevate TCA as the best choice analytical tool for those newly in charge of the executive branch.

TCA, A Variant of CBA

The Politics of TCA Versus CBA

There is more similarity between TCA and CBA than there is difference. Both are methodologies intended to fully describe and make transparent the costs and benefits associated with proposed actions and, in doing so, help decision-makers understand the implications of contemplated actions. To the extent that there are differences, they mainly relate to the scope of analysis. The degree to which TCA extends beyond the typical scope of CBA is significant, as the goal is to encompass and evaluate a wide range of externalities, both positive and negative, including in-depth attention to aspects related to social and human capital. While there is nothing inherent in CBA that narrows the analysis from the ambitious scope of TCA, in practice it has not covered the breadth of issues evaluated in TCA.

From the beginning of CBA, there have been strenuous objections to placing a value—particularly a monetary value—on certain things based on ethical and/or moral grounds. For example, putting a price tag on a life or a limb (although insurance does this) or monetizing the intrinsic value of a forest. Martha Nussbaum writes that while CBA is helpful in answering obvious questions, it does not help with confronting tragic questions—those in which addressing the question surfaces unpalatable choices and, perhaps, obscures the presence of moral dilemmas (Nussbaum, 2000). It would be tragic to use CBA to choose between a proposal that allows children to spray pesticides and one that allows children to drive tractors, when both are too dangerous for children to undertake; in each case, costs are not disadvantageous, but flat out wrong. Although objections to monetizing nonmarket benefits, like happiness, have largely come from liberal critics, the irony could be that finding successful ways to monetize social and human capital could be what ultimately justifies stronger regulations (Sunstein, 2018). It has yet to be determined how TCA methodology will address social and human capital externalities, as these methodological issues are now being refined.

Much of the foundational work behind TCA has been generated and funded by left-leaning people and organizations, some of which have given up hope that CBA can deliver the full transparency across the four capitals that is foundational to TCA philosophy and, eventually, practice. Does that make TCA a liberal thing? Is CBA a conservative thing? As will be discussed, Democrats and Republicans alike seem to be generally supportive of CBA in the USA, although both parties are also looking to improve upon current practice. Revesz and Livermore issue a challenge: "It is time for progressive groups, as well as ordinary citizens, to retake the high ground by embracing and reforming CBA" (Revesz and Livermore, 2018). Perhaps taking the high ground means touting TCA as the cure for historic CBA failures. If so, it is critical to understand how CBA has worked, and not worked, over time, so that TCA can be shaped and promoted as a desirable alternative.

The History of CBA

Philosophical Roots

CBA has long been viewed as important to rational decision-making (Katzen, 2006). The basic idea that no action can be taken unless the benefits justify the costs has deep philosophical roots, with threads of its origin seen in the writings of Alexander Hamilton, Adam Smith, Jeremy Bentham, John Stuart Mill, Friedrich Hayek, Walter Lippmann, and Amartya Sen, among others (Sunstein, 2019). It has been a cornerstone of welfare economics, even with the inherent difficulties of defining and measuring welfare (Zerbe et al., 2010). And it has been a mainstay of US federal policymaking, particularly since the Reagan era (1981–89), when it was enshrined as a tool for producing objective analysis to aid regulatory decision-making (Bipartisan Policy Center, 2020). While it is interesting to understand the philosophical roots of CBA, in practice, it is untethered to a particular philosophical camp. Rather, it is a widely used technocratic tool that aids decision-making, no matter the direction a particular decision might take.

International Organization, Journal, Recognition

Although this chapter is focused on use of CBA by the US federal government, the science and practice of CBA extends well beyond US borders. For example, in 2014 the European Commission published a how-to-guide to CBA to appraise investment projects (European Commission, 2014). The Society for Benefit-Cost Analysis, established in 2007, had members from 35 countries. The Society launched the open-source *Journal of Benefit-Cost Analysis*, published by Cambridge University Press, and which is international in scope; a 2019 special issue volume was titled, "Benefit Cost Analysis in Low- and Medium-Income Countries: Methods and Case Studies" (2019). The point here being that CBA has been recognized as a valid tool to advance rational decision-making across the globe; this conceivably opens a space for TCA to expand the scope of the decision-making process.

CBA in US Federal Policymaking

At this writing, CBA must be undertaken for most major and significant US proposed rules, including those likely to result in an annual impact on the economy of $100 million or more, rules considered to be novel, and rules that materially alter entitlements. The CBA is available for review by government officials and the public, and ideally it identifies all costs and benefits of a proposed rule with attendant dollar values and is intended to inform decision-makers and the public of the potential impacts of proposals and thereby empower them to provide substantive feedback.

The historical roots of CBA are found in the Administrative Procedures Act of 1946, which established notice and comment rulemaking and required agencies to provide explanations for their actions. Beginning with the Reagan Administration, a series of executive orders (E.O.s) have been issued that relate to rulemaking and CBA (Note that an E.O. is not statutory law, but it does bind executive agencies and require their adherence until such time as it is repealed by a later president). In 1981, President Reagan issued E.O. 12291, which stated that regulatory action should not be undertaken unless the potential benefits to society outweighed the potential costs. The E.O. mandated CBAs for major rules and established the Office of Information and Regulatory Analysis within the Office of Management and Budget (OMB) at the White House which oversees all federal rulemaking to this day. In 1993 President Clinton issued E.O. 12866, affirming the earlier Reagan E.O. and the role of CBA. President Obama issued three E.O.s on rulemaking, with the most significant being E.O. 13563, which, among other things, recognized inherent difficulties in CBA, stating that "each agency may consider (and discuss qualitatively) values that are difficult or impossible to quantify, including equity, human dignity, fairness, and distributive impact" (E.O. 13563). The E. O. also required retrospective analysis of existing rules (E.O. 13563). President Trump issued two E.O.s related to CBA, E.O. 13771 that established new ways to cut costs of regulations and E.O. 13777, which required all agencies to appoint a regulatory reform officer and a regulatory reform task force to identify regulations ripe for repeal, replacement, or modification (E.O. 13777).

Box 12.1 US Regulatory Instruments Referenced in this Chapter

Year	Instrument	Title
1946	Statute	Administrative Procedures Act
1981	Reagan E.O. 12291	Federal Regulation
1993	Clinton E.O. 12866	Regulatory Planning & Review
1997	Appropriations Directive	Section 645 of PL 104–208 (Treasury Approps. Act)
2001	Statute	Regulatory Right to Know
2011	Obama E.O. 13514	Leadership in Environmental, Energy, & Economic Performance
2011	Obama E.O. 13563	Improving Regulation & Regulatory Review
2012	Obama E.O. 13610	Identifying & Reducing Regulatory Burdens
2017	Trump E.O. 13771	Reducing Reg. & Controlling Regulatory Costs
2017	Trump E.O. 13777	Enforcing the Regulatory Reform Agenda

Lessons Learned from CBA for TCA Consideration

Many of the CBA flaws that have been identified over time are unlikely to be solved by simply substituting a TCA approach. Scholars over the decades have documented and debated CBA shortcomings, and advocates of TCA would be well advised to absorb the lessons learned and, to the extent possible, design TCA methodologies in recognition of the four greatest CBA challenges: lack of sufficient knowledge; complexity of analysis; implementation realities; and regulatory budgeting.

Knowledge Gaps

A CBA is only as good as the data used to produce it, and the kinds of hard data needed to produce a successful analysis are sometimes unavailable or unreliable. One important aspect of this is that regulated entities are often the primary source for data on potential costs of policy proposals. Going directly to the source is a good instinct in many ways, and requesting data from regulated entities is sometimes the only way to get necessary information, but it does have the potential to produce compromised data. Regulated entities have an incentive to exaggerate the costs of proposed regulations in order to avoid or soften them (Katzen, 2006). Moreover, it might be difficult to separate out the costs of specific actions. For example, a company might implement worker protections on its own, absent regulation, and in contemplating the costs of proposed worker protection regulations, it might be difficult to accurately calculate the differential between what the company intended to do on its own and the what the proposed regulation would cost (Congressional Research Service, 2014; United States Government Accountability Office, 1995).

The articulation of certain benefits, particularly in dollar values, is challenging. A 2014 US Government Accountability Study of 203 federal rules emphasized that agency officials found monetizing benefits more difficult than monetizing costs (Government Accountability Office, 2014). In particular, monetizing welfare benefits, such as quality of life, freedom, relationships, and happiness is difficult for many reasons, and because these values are not traded in the marketplace, analysts struggle to give them a price tag within a CBA. But absent a benefit attribution, these values become invisible. Among the examples cited in CBA literature of non-traded benefits that often go unmeasured is biodiversity (Congressional Research Service, 2014), which of course was the concern that led to the launch of TEEBAgriFood, the TCA effort initiated by the United Nations (UN) Environment Programme in 2015. Techniques such as "willingness to pay" and "statistical life" calculations are used to approximate hard to measure benefits, but few are satisfied by the state of the art. There is almost universal agreement that measurement of welfare benefits and other nontraded goods is a longstanding and significant weakness of CBA.

Absent full information in the face of risk, some argue that "The Precautionary Principle" should prevail. The Precautionary Principle holds that when conclusive evidence is not available related to an action, especially actions that could have irreversible consequences, the government should not proceed—or proceed with extreme caution—until extensive scientific knowledge is obtained. This principle has been applied particularly in issues surrounding the environment, and is the bedrock of many international agreements (e.g., the Rio Declaration, Kyoto Protocol). CBA and TCA could push up against the precautionary principle, as neither methodology can promise complete information necessary to understand risks.

Overwhelming Complexity

Some federal rules are hundreds of pages long, without counting the hundreds of additional pages in accompanying analyses, including the CBA. Consider the proposed rule introduced in 2020 on continuous improvement in the National Organic Program, which covers a range of activities such as labeling of nonretail containers, training of inspectors, supply chain traceability, and grower group certifications, with each of these issues and others contained in the proposal requiring consideration of the relevant costs (approximately $7.4 million) and benefits (approximately $87 million) (Agriculture Marketing Service, 2020). The complexity of CBAs that accompany significant rules is often overwhelming, and, in many cases, that complexity renders the documents useless. Many CBAs are left unread by busy policymakers and inaccessible to average citizens. For these reasons, some CBA scholars have urged greater simplicity in CBAs, brought about by the construction of system boundaries, aggregation techniques, and, perhaps most importantly, elimination of the expectation that the CBA must be definitive and complete (Zerbe et al., 2010). Reducing the complexity might provide greater clarity and allow citizens to reflect on CBA findings in their comments submitted for consideration in the rulemaking process (Sunstein, 2018).

Implementation

The quality and success of CBA, as implemented by the US federal government, has been compromised in three major ways. First, as we have learned through the TCA efforts described in this volume, in depth analysis of externalities takes time and resources. Yet Congress does not appropriate a separate budget to conduct CBA associated with rulemaking. Rather, each individual agency must carve out funding from existing program budgets to conduct the CBA. In other words, money spent outsourcing a CBA contract and/or internal staff time devoted to CBA are resources no longer available for programmatic priorities (e.g., CBA costs subtracted from cost-share incentives for soil enhancing practices in the case of rulemaking undertaken by the United States Department of Agriculture [USDA] Natural Resources Conservation Service). This leads to agency staff wanting to "cut corners" (Katzen, 2006) and conduct CBA "on the cheap" in order to protect programmatic budgets.

Second, producing a quality CBA requires skill and experience. Not all agencies have economists or otherwise qualified analysts on staff capable of executing a CBA, and even for those that do, the staff available to undertake CBA work do so infrequently. Other than at the OMB, government lacks a trained and dedicated CBA workforce. As a result, the quality of CBA work is uneven and, in the worst cases, inconsistent.

Third, and most significantly, it has become standard practice for program staff to first draft a proposed rule and then pass it on to their colleagues who subsequently develop a corresponding CBA. This linear progression of work reduces the CBA to a paper exercise that is designed to justify the proposal rather than shape it (Katzen, 2006). Rather, CBA should be undertaken in parallel with development of the proposal, so that the analysis can inform the rulemaking docket in progress and allow policymakers to make informed adjustments along the way.

Pitfalls of Regulatory Budgeting

Politicians are always promising to reduce regulations and historically this has been attempted through "regulatory budgeting." In 1997 a directive was inserted into an appropriations bill that required OMB to estimate the aggregate costs and benefits of federal regulations as a way to keep track of what was going on and reign in the regulatory state; this provision became permanent law in 2001 (Congressional Research Service, 2014). Execution proved difficult, however, and eventually OMB concluded that coming up with aggregate estimates for all federal rules on an annual basis was infeasible and, instead, began providing Congress with a ten-year rolling summary of costs and benefits for major rules only. It was with great fanfare, then, when the Trump Administration announced E.O. 13771 in 2017, directing agencies to cut costs of regulations by 1) repealing two existing regulations for every new regulation issued; and 2) capping the total incremental costs imposed by all new regulations in a given year to zero. In 2018 the Trump Administration announced that its regulatory reforms eliminated $50.9 billion in overall regulatory costs across the federal government in just one year, and it placed more onerous regulatory budget caps on certain agencies (e.g., Environmental Protection Agency) that required costs of regulatory actions to be less than zero. The aggregation of CBAs—or TCAs if TCA becomes the methodology of choice—facilitates politicians' temptation to put the federal government on a regulatory "diet" and limit the total volume of rules and overall compliance costs imposed on the economy.

Opportunities to Substitute TCA for CBA in US Policymaking

Retrospective Review

Given the elaborate work that goes into public notice and comment rulemaking, it is not surprising that, over time, studies have been conducted to evaluate the accuracy of CBAs *ex post*. How well did the CBA identify and predict costs and benefits? Generally, these studies have revealed significant errors but no systematic bias (Sunstein, 2018). While too few in number, such studies have led to great interest among CBA scholars in retrospective review—formalized, systemic empirical research of *ex post* costs and benefits (Katzen, 2006). The dedicated regulatory person and task force required for every agency, as established by E.O. 1377, could take on the responsibility of retrospective review and even go so far as making it a public process with an opportunity for the public to comment.

The emerging interest in, and enthusiasm for, retrospective review can be deployed in support of TCA. To help policymakers to consider the potential value of substituting TCA for CBA, it might be useful to select several CBAs over the past several years and, in a retrospective way, apply TCA in those contexts to determine whether, and how, TCA methodology might have delivered different results. For example, would a TCA have resulted in different information and valuation of 2019 USDA rulemaking related to line speeds for poultry plant workers, perhaps because TCA might ascribe higher value to workers' well-being than a traditional CBA? If so, might TCA have changed the course of rulemaking at the time, and if so, would the meat processing industry have fared better during the 2020 coronavirus disease (COVID-19) pandemic, where many workers became ill and some died owing to working conditions? While that is speculation on top of speculation, the example is given to suggest the potential of TCA retrospective review.

The TCA Revolution

Cass Sunstein, the former Director of the Office of Information and Regulatory Analysis during President Obama's first term and well-known CBA scholar, says that he would not be surprised to see Congress enact CBA into law in the coming years. In his 2018 book, *The Cost-Benefit Revolution*, he writes: "Democrats and Republicans alike embrace CBA and many of them support legislation that requires it" (Sunstein, 2018, pp. 4). Sunstein recommends that Congress enact existing executive orders related to CBA (see chart) into law to provide affected people the opportunity to go to court to require that regulations be issued when benefits unambiguously justify the costs (Sunstein, 2018). Furthermore, he recommends establishment of an Office of Regulatory Accountability separate from the White House and empowered to investigate inaction, as well as actions of federal agencies; in 2019 five Republican Senators introduced the "Independent Agency Regulatory Analysis Act" to do just that, although it was not enacted into law.

Sunstein is right—there is great interest in CBA, and there are demands from Republicans and Democrats to improve upon its current administration. For example, in 2018 the Environmental Protection Agency (EPA) published a list of questions in the Federal Register in which it sought comment on how to achieve consistent and transparent interpretations of costs and benefits in rulemaking (Environmental Protection Agency, 2018); in 2020 the agency published a proposed rule related to the use of CBA in Clean Air Act rulemaking that builds on what was learned through the public response to those 2018 questions as well as in multiple public hearings on CBA (Environmental Protection Agency, 2020). The "CFTC [Commodity Futures Trading Commission] Cost-Benefit Analysis Improvement Act" was introduced in 2020 by a Republican member to strengthen CBA in commodity futures trading. In the same Congress, a Democratic member introduced the "Smart Building Acceleration Act" to require CBA for federal investments in smart buildings. In 2020 the Bipartisan Policy Center, a think tank in Washington, DC known to host conversations and produce reports on cutting edge issues, held an event on the evolution of CBA (Bipartisan Policy Center, 2020). Clearly, interest in the design and value of CBA is evident across the policymaking space at the federal level.

History has shown that new Administrations seek to put a mark on executive branch regulatory proceedings. The Trump Administration's roll-back of regulations and the E.O.s enacted by the administration, particularly the requirement that before any new regulation is issued, two existing regulations must be removed, are difficult to implement and are viewed by many Democrats as an attack on the regulatory state. A new Administration is likely to introduce new

E.O.s to guide rulemaking and this would provide an opportune time to introduce the concept and practice of TCA into those new E.O.s.

The time is right to advance TCA, as the "revolution" that Sunstein and others are calling for in the practice of CBA. Between interest in the US Congress, calls from scholars and think tanks, and an eventual change of leadership in the executive branch, there is substantial appetite to improve upon current CBA practice. TCA should be held up as the solution that US policymakers seek.

Conclusion

TCA is not the radical departure from CBA portrayed by some TCA advocates. If TCA designers ignore the lessons of CBA shortcomings, there is no reason to believe that TCA will overcome the difficulties that have plagued CBA practice. It is the intention of this chapter to alert those interested in advancing TCA of the potential challenges ahead and to encourage creative design to address and counter CBA shortcoming briefly described herein. A final thought—if the degree of difference between TCA and CBA is greatly exaggerated, it will be very difficult politically to substitute TCA for CBA in upcoming US federal reforms of regulatory processes. Rather than promoting TCA as something brand new, the best course of action, for adoption in US rulemaking, is to harness the power of familiarity and argue that TCA is simply an improved form of CBA. Indeed, it is.

References

Agricultural Marketing Service. (2020). National Organic Program: Strengthening Organic Enforcement (AMS-NOP-17-0065-0001). Regulations.Gov. Available at: https://beta.regulations.gov/document/AMS-NOP-17-0065-0001.

Bipartisan Policy Center. (2020). EPA's Cost-Benefit Analysis Rule: Too Prescriptive? [Webinar]. Available at: https://bipartisanpolicy.org/event/epas-cost-benefit-analysis-rule-too-prescriptive/.

Congressional Research Service. (2014). Cost-Benefit and Other Analysis Requirements in the Rulemaking Process (No. R41974, p. 33). Congressional Research Service. Available at: https://crsreports.congress.gov/product/pdf/R/R41974#page=24&zoom=100,116,690.

Environmental Protection Agency. (2018). Increasing Consistency and Transparency in Considering Costs and Benefits in the Rulemaking Process. *Federal Register*, June 13, 83(114), 27524–27528. Available at: www.epa.gov/sites/production/files/2018-06/documents/epa_frdoc_0001-22431.pdf.

Environmental Protection Agency. (2020). Increasing Consistency and Transparency in Considering Benefits and Costs in the Clean Air Act Rulemaking Process. *Federal Register*, June 11, 85(113), 35612–35627. Available at: www.govinfo.gov/content/pkg/FR-2020-06-11/pdf/2020-12535.pdf.

European Commission, Directorate-General for Regional and Urban Policy. (2014). *Guide to cost-benefit analysis of investment projects: Economic appraisal tool for cohesion policy*

2014–2020. Publications Office of the European Union. Available at: https://ec. europa.eu/regional_policy/sources/docgener/studies/pdf/cba_guide.pdf.

Executive Order No. 13563. (2011). Improving Regulation and Regulatory Review. *Federal Register*, January 21, 76(14), 3821–3823. Available at: www.reginfo.gov/ public/jsp/Utilities/EO_13563.pdf.

Executive Order No. 13771. (2017). Reducing Regulation and Controlling Regulatory Costs. *Federal Register*, January 30, 82(22), 8339–8341. Available at: www.govinfo. gov/content/pkg/FR-2017-02-03/pdf/2017-02451.pdf.

Executive Order No. 13777. (2017). Enforcing the Regulatory Reform Agenda. *Federal Register*, March 1, 82(39), 12285–12287. Available at: www.govinfo.gov/content/ pkg/FR-2017-03-01/pdf/2017-04107.pdf.

Katzen, S. (2006). Cost-Benefit Analysis: Where Should We Go From Here? *Fordham Urban Law Journal*, 33(4), 1313. Available at: https://ir.lawnet.fordham.edu/ulj/ vol33/iss4/12.

Nussbaum, M.C. (2000). The Costs of Tragedy: Some Moral Limits of Cost-Benefit Analysis. *The Journal of Legal Studies*, 29(S2), 1005–1036. JSTOR. https://doi.org/10. 1086/468103.

Revesz, R. & Livermore, M. (2008). *Retaking Rationality: How Cost Benefit Analysis Can Better Protect the Environment and Our Health*. Oxford: Oxford University Press.

Robinson, L.A., Hammitt, J.K., Jamison, D.T., & Walker, D.G. (2019). Conducting Benefit-Cost Analysis in Low- and Middle-Income Countries: Introduction to the Special Issue. *Journal of Benefit-Cost Analysis*, 10(S1), 1–14. https://doi.org/10.1017/ bca.2019.4.

Sunstein, C.R. (2018). *The Cost-Benefit Revolution*. Cambridge, MA: MIT Press.

United States Government Accountability Office. (1995). Regulatory Reform: Information on Costs, Cost-Effectiveness, and Mandated Deadlines for Regulations (Briefing Report to the Ranking Minority Member, Committee on Governmental Affairs, US Senate PEMD-95–18BR). Available at: http://archive.gao.gov/t2pbat1/ 153774.pdf.

United States Government Accountability Office. (2014). Federal Rulemaking: Agencies Included Key Elements of Cost-Benefit Analysis, but Explanations of Regulations' Significance Could Be More Transparent [Reissued on September 12, 2014] (GAO-14–714). Available at: www.gao.gov/products/gao-14-714.

Zerbe, R.O., Davis, T.B., Garland, N., & Scott, T. (2010). *Toward Principles and Standards in The Use of Benefit-Cost Analysis: A Summary of Work*. Available at: https://evans.uw.edu/ wp-content/uploads/files//public/Final-Principles-and%20Standards-Report.pdf.

13 International Policy Opportunities for True Cost Accounting in Food and Agriculture

Barbara Gemmill-Herren, Zoltán Kálmán and Alexander Müller

Why the International Level?

Food and agriculture policy might often be seen as being formulated, adopted, and implemented on the level of the nation state. But national policies in turn are embedded within a set of international agreements ranging from World Trade Organization (WTO) rules to those regional bodies seeking to harmonize agricultural policies throughout a continent such as the European Union in Europe and Mercosur in Latin America. There are a number of compelling reasons for considering international entry points for the True Cost Accounting (TCA) for externalities in food and agriculture. In at least three respects (at a minimum) the international agenda is pivotal to the adoption of TCA in the food and agriculture sector: 1) addressing the seemingly intractable paradigm of "feeding the world" and its implicit coda, "with cheap food"; 2) facing the reality that—through international trade—the burden of externalities is so readily shifted across borders and continents, often from the Global North to the Global South; and 3) recognizing that externalities of the food system have impacts on many levels, from local to global.

Feeding the World with Cheap Food

With great regularity, alarms have been set off as to how the world will feed a growing population, using a common reference point of the year 2050. Also, almost uniformly, the question is met with estimates of how much food production needs to be increased. Estimates have varied from a need to double food production (often repeated but not actually backed up by analysis), to a 70% increase (Food and Agriculture Organization of the United Nation, 2009), and then a 60% recalculation (Alexandratos and Bruinsma, 2012). Nonetheless, the need for vast increases in food production has been the dominant narrative that continues to find its way into the popular media[1] and drives the productionist agendas of international agricultural research, development aid, and philanthropy (Wise, 2020; Pimbert and Moeller, 2018; Biovision and IPES-Food, 2020). This reference point and narrative is the rationale behind government support for high-input intensive industrial farming and farm consolidation, in developed and developing countries alike.

The weaknesses in these estimates and their underlying models have been enumerated by many scholars; including:

- The fact that the root causes of hunger—extreme poverty and gaping inequalities—persist stand as a damning indictment of the global food system. However, several estimates indicate that enough food is produced today to feed from 9 billion (IPES-Food, 2016; High Level Panel of Experts on Food Security and Nutrition, 2014, 2017; Chappell, 2018) to almost 10 billion people (Berners-Lee et al., 2018). At the same time, around one-third of this is lost or wasted (High Level Panel of Experts on Food Security and Nutrition, 2014). Access, equity, distribution, and addressing food waste remain key problems.

- Global business models that bank on the expectation that over the next 30 years, the global community will adopt Western, grain-fed meat-centered diets, despite growing environmental and health concerns. This leads to policy and practice that continues to divert food grains to livestock feed and biofuel production, despite considerable criticism of biofuel policy (Wise, 2013). Nor is food loss and waste being reduced to the greatest extent possible (High Level Panel of Experts on Food Security and Nutrition, 2014), and as committed to within the Sustainable Development Goals ("By 2030, halve per capita global food waste at the retail and consumer levels and reduce food losses along production and supply chains, including post-harvest losses"). Berners-Lee *et al.* (2018) make intricate global quantifications of the extent to which reductions in the amount of human-edible crops fed to animals and, less importantly, reductions in waste, could increase food supply. They find that no nutritional case can be made for feeding human-edible crops to animals, which reduces calorie and protein supplies for global food security.

- The assumption that food and nutritional security can be resolved through increased production, "intensification" and technical change, when hunger and malnutrition has been thoroughly documented to be first and foremost an issue of different entitlements (Sen, 1981; Smith and Haddad, 2015; High Level Panel of Experts on Food Security and Nutrition, 2017). Smith and Haddad (2015), for instance, reviewed studies to address child malnutrition carried out over a 42-year period, from 1970 to 2012, spanning 116 countries. They found that the predominant and strongest contributors to reducing hunger were not related to agricultural production but were social measures and issues of entitlement: access to safe water and female education (Figure 13.1). Contributors related to production figure next, although access to dietary energy from non-staples (thus, primarily legumes, fruits, and vegetables) are almost as important as dietary energy from staples (rice, maize, wheat, root crops, etc.). Two of the other predominant determinants to reducing child nutrition are also social measures: access to sanitation and the ratio of female to male life expectancy.

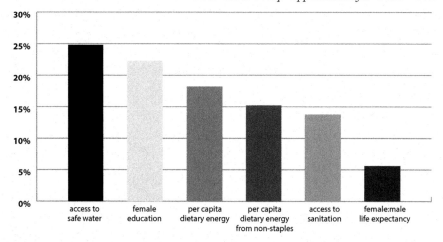

Figure 13.1 Contributions of underlying determinants to reducing hunger.
Source: adapted from Smith & Haddad, 2015.

- Inadequate attention in policy and practice to nutrition and equitable prices for food. The most recent "State of World Food Security and Nutrition in the World" finds that hunger has increased globally since 2015. This same report finally considered healthy diets to be integral to food security, not just limited to the consumption of staples. The authors also explored the fact that costs of healthy diets at current prices are 60% higher than the cost of a diet that might not be healthy, but does provide adequate nutrition, and five times the cost of an energy sufficient diet. While most of the poor around the world can afford an energy-sufficient diet, as defined in the report, they cannot afford either a nutrient-adequate or a healthy diet (Food and Agriculture Organization of the United Nations *et al.*, 2020). As the report notes, "the unaffordability of healthy diets is due to their high costs relative to people's incomes, a problem likely to be exacerbated by COVID-19". As observed by the new Special Rapporteur on the Right to Food, Michael Fakhri, government policies always directly or indirectly influence food prices, and the overwhelming trend has been to drive down food prices for merely energy efficient (calories only) diets (Fakhri and Tzouvala, 2020). Countries need guidance to genuinely address hunger, nutrition, and healthy diets. Cheap food currently replaces social safety nets, although inadequately; the paradigm of cheap food is responsible for poor and often hungry famers and causes high health risks for poor people. The recognition of nutritious food as a human right, together with incomes to sustain healthy diets among farming communities and food chain workers, would ensure food and nutrition security as envisioned in the United Nations' Sustainable Development Goals.
- The inherent inaccuracy of predictions of world food production, most often based on global estimates of supply and demand, yet, as eloquently phrased by Wise (2013):

".... ecosystems and agricultural production occur at local and regional scales. So too does hunger. Thus, global estimates of "our" ability to feed "the world" immediately break down, begging the more important questions of how these systems develop across widely differing landscapes, societies, and levels of economic development, and how equitably the food is then distributed. In the end, "the world" is not fed, in aggregate, and there is no collective "we" doing the feeding."

There is no question that yield potential varies considerably by crop and region, and that many agricultural lands with significant yield gaps are in rainfed zones in developing countries. However, the specific pathways for increasing such production must recognize that food and farming systems are complex socio-ecological systems, each with distinct needs to be addressed in terms of resources, power imbalances, and ecological constraints. Here again, governments can learn from each other and benefit from guidance in navigating such complex landscapes, rather than implementing programs based on the belief that technological fixes to production are the solution. The ongoing discussions of the Committee on World Food Security's (CFS) Voluntary Guidelines on Food Systems and Nutrition (based on the related High Level Panel of Experts on Food Security and Nutrition [HLPE] Report) will also provide appropriate guidance in this regard.

International Trade

The global food system is, in the words of *The Economist*, "the unsung star of 21st century logistics," making up an $8 trillion global supply chain that accounts for about a tenth of global Gross Domestic Product (GDP) (*The Economist*, 2020) and employing one out of every three economically active workers (Food and Agriculture Organization of the United Nations, 2014). Although farming is inherently local, the food industry is increasingly global. Food exports have grown sixfold over the past 30 years. The companies that dominate this trade operate on a global basis to source and ship agricultural commodities to food processing facilities and then to consumers. As *The Economist* notes, their size and global reach permit them to generate substantial profits on narrow margins, by quickly swapping one source or one market for another to accommodate changes in supply or demand (The Economist, 2020).

At the same time, this global food system is responsible for widespread degradation of land, water, and ecosystems; high greenhouse gas emissions; biodiversity losses; chronic over- and undernutrition and diet-related diseases; and livelihood stresses for farmers (Pengue *et al.*, 2018).

Industrial, input-intensive food systems have been found globally responsible for 19% to 29% of greenhouse gas emissions (Andrieu and Kebede, 2020), 61% of fish population decline (United Nations Environment Programme, 2016) and the use of 20% of aquifers (United Nations Environment Programme, 2016). Agriculture is the main driver of land degradation (Intergovernmental Panel on Climate Change, 2019). Estimated costs of inaction continue to mount; for example, it is estimated that the world lost between $6 trillion and $11 trillion in ecosystem services between 1997 and 2011, owing to land degradation (Organisation for Economic Co-operation and Development, 2019). Under current food consumption patterns, diet-related health costs are projected to reach $1.3 trillion per year in 2030, and diet-related social costs of greenhouse gas emissions associated with the current food system are projected to exceed $1.7 trillion per year (Food and Agricultural Organisation of the United Nations *et al.*, 2019)

An emerging feature of global food systems is the existence of multiple, insidious forms of visible and invisible flows of natural resources and externalities, across borders and continents. For each shipment of food being transported beyond national borders, the natural resources used in the production of such shipments are also, in a sense, being "virtually" transported to the recipient country. This was highlighted in Pengue *et al.* (2018), considering the growing quantities of trade in biomass, nutrients and "embedded" water over time. Overall, about 15 per cent of all biomass materials globally extracted are redistributed from one country to another through trade (www.materialflows.net/home; United Nations Environment Programme, 2015). As food and other products are traded internationally, water and nutrient resources in one country are used to support consumption in another country (Figure 13.2). Argentina, for example, as a large food and biomass supplier to the world, is equally a main extractor of water and nutrients, largely consumed in Asia (Pengue *et al.*, 2018).

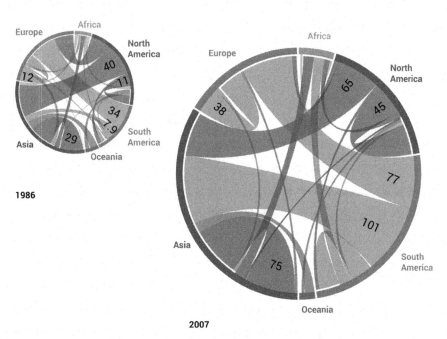

Figure 13.2 Virtual water flows between the six world regions, 1986 and 2007.
Source: TEEB, 2018; adapted from Dalin *et al.* 2012.

As an example of how local practices add up to major global trends, biogeo-chemical flows throughout the world have been profoundly transformed as farming systems have turned from traditional means of maintaining soil fertility (through fallowing, integrating livestock with crops, applying crop rotation, intercropping, cover crops, use of pulses and legumes, reduced tillage, and use of composted material) to the increased use of fossil-fuel-based and mined fertilizers. Agricultural intensification, carried out without restorative practices, ultimately leads to soils unable to sustain their fertility. To compensate, modern conventional farming enterprises have increased the use of NPK (nitrogen, phosphorus and potassium) fertilizers (Figure 13.3). Nitrogen and phosphorous both are notoriously "leaky" nutrients that end up in waterways when applied in excess. As an outcome of agriculture's increased biomass production, "cascades" of nitrogen and phosphorus are causing serious pollution of water bodies, leading to nitrate contamination of drinking water and "dead zones" in oceans and other water bodies (Lassaletta *et al.*, 2016; Ribaudo *et al.*, 2011; Cox and Schechinger, 2018; Townsend and Howarth, 2010).

Thus, accounting for negative externalities such as soil depletion, nutrient pollution, and overextraction of water along the value chain is a transboundary issue. Global trade leads to significant incomes for exporting countries, but TCA questions would be: Are the negative externalities accounted for? And do the export earnings contribute to restoring the natural resources that a country exports?

While globally traded foodstuffs, together with inputs such as pesticides and fertilizers flow fairly abundantly between borders, the human workforces that sustain agriculture and food systems are subject to a different dynamic. The labor force in agriculture and food systems is among the least valued of all sectors, in all countries of the world, with many areas of concern around the lack of protections for human rights. In the USA, for example, farmworkers have been marginalized by laws that exempt agriculture from many labor

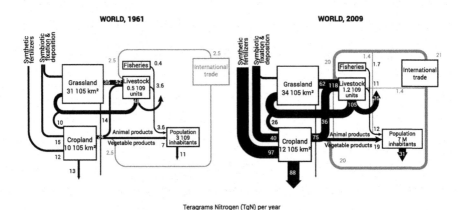

Figure 13.3 Generalized representation of Nitrogen transfers through the world agro-food system, 1961 and 2009.
Source: TEEB, 2018; adapted from Lassaletta *et al.* 2016.

protections and by many policies that make them vulnerable to exploitation (Wilde, 2018). International borders are permeable for people, but the fate of undocumented immigrants who enter North America or Europe in search of work puts them at particular risk; the threat of deportation robs them of what small opportunities they have to organize and collectively bargain.

In the pursuit of the lowest cost of production for globally traded commodities, global food companies tend to source from countries where the cost of labor is least expensive, all other things being equal. The resulting terms of employment of the agricultural workforce, under such conditions, is hardly acceptable, with respect to wages, hours of work, and health and safety. Yet such terms of employment are essentially a "necessary ingredient" of cheap food. A caveat that will need to be addressed, in exploring how measures to introduce TCA might capture the costs of externalities, is that the international nature of the global food system allows burdens to be shifted almost imperceptibly along the food value chain. Within food value chains, assigning costs for negative externalities could fall heavily on farming communities, the agricultural labor force, and low-income consumers. Equally, attributing benefits for positive externalities might rarely accrue to the less powerful actors in food value chains—again, farming communities and the agricultural labor force, unless policy exists to assign equitable allocations. The challenge to do so is even greater with the international trade in foodstuffs and thus the need for international policy development. It is unfortunate, but the reality is that trade rules negotiated internationally strongly lack in negative externalities. WTO trade rules favor the lowest-cost producers and refuse to consider how such costs are reduced. In critical rulings, national or local governments have been prevented from taking measures that internalize external costs or restrict trade when imported goods fail to internalize costs. This feeds a race to the bottom instead of the desired "harmonization upward" of environmental standards and practices (Wise, 2019).

As nation-states position themselves with respect to international markets, all countries must decide what they want to import and what they want to produce domestically. To decide to commit to domestic food production inevitably affects international markets, as does the decision to import affects domestic markets. The application of TCA to food policy could be a helpful way to define these "virtual borders" and to understand what the actual costs and trade-offs are that they are dealing with.

Box 13.1 Beyond GDP: Multidimensional Indicators of Well-Being

Amanda Jekums

Gross Domestic Product (GDP) has long been the standard metric for assessing the national economic prosperity and societal progress of countries around the globe. However, its basis in extractive and damaging practices, coupled with increasing rates of pervasive social injustice and income inequality, demonstrates that GDP is an inadequate and inaccurate measure of individual living standards and collective well-being. Over the past 25 years, income inequality among Organisation for Economic Co-operation and Development countries has increased seven times over. The average income of the richest 10% of the population is now nine times that of the poorest 10% (2019). Clearly the benefits of GDP growth are not reaching everyone.

Given its limitations, countries and citizens around the world are rejecting GDP as the sole measure of success. Similar to the aims of True Cost Accounting to identify metrics that go well beyond single and linear measures of success, the examples provided below illustrate creative examples of how countries are moving beyond GDP towards multidimensional indicators of well-being.

In 2008 Bhutan formally adopted a new policy principle to promote conditions that will enable the pursuit of Gross National Happiness (GNH) (Kingdom of Bhutan, 2008). The multidimensional concept of GNH takes a systems approach, which measures nine domains of GNH: psychological well-being, health, time use, education, cultural diversity and resilience, good governance, community vitality, ecological diversity, resilience, and living standards. The practice allows governments to incorporate this information in decisions on policies and projects and enables targeted responses to specific situations or causes of unhappiness. The process has also encouraged public citizens and private entities alike to think more holistically (Ura *et al.*, 2012).

Vanuatu is a small island country in the southwestern Pacific Ocean. As the world's most at-risk country for natural disasters (Bündnis Entwicklung Hilft, 2019), it is not surprising that their highest level policy framework is composed of indicators directly linked to the United Nations' Sustainable Development Goal. Launched in 2016, the Vanuatu 2030 People's Plan focuses on 15 priority policy objectives: happy and healthy people, an inclusive and equitable society, sustainable land management and food production, conservation and biodiversity, climate resilience, supportive infrastructures, and strong economic and employment opportunities (Department of Strategic Policy, Planning and Aid Coordination, 2017). Collecting data on novel social indicators present a challenge, but progress has been positive, particularly in connecting their national vision for well-being and sustainability to action on the ground in villages and urban centres across the country (Government of The Republic of Vanuatu, 2018).

Most recently, New Zealand introduced its first well-being focused budget in 2019 (New Zealand Treasury, 2019a). The framework measures similar domains as Bhutan's GHN and the Vanuatu 2030 People's Plan, which are categorized under four capitals: financial and physical, human, natural, and social. The data is collected in an online Living Standard Framework (LSF) Dashboard (New Zealand Treasury, 2019b), which informs Ministers on priorities for improving well-being. It is also open to the public in an effort to promote transparency and civic engagement. One of the LSF indicators measures trust in government institutions. Recent research has attributed New Zealand's success in eliminating the coronavirus to a high trust in authorities (Wilson, 2020), demonstrating the importance of measuring alternative indicators of societal progress and using the data to improve well-being.

Economic wealth and social well-being are both derived from capital stocks—natural, social, human, and produced—and these capitals must be used and managed in ways that ensure that they maintain their value over time. Single measures of success like GDP (and yield per acre in agriculture, for example) promote growth at all costs, ignoring the diversity of inputs and compounding negative impacts. Despite challenges related to defining appropriate indicators, collecting data, and reporting on these holistic well-being frameworks, they illustrate—in a profoundly hopeful way—the opportunity to move beyond GDP as the dominant economic measure. By focusing more broadly, these enterprising countries are using alternative indicators of success to reveal transformational pathways towards sustainable and equitable societal progress.

Referenced works

Bündnis Entwicklung Hilft. (2020). The WorldRiskReport. Available at: www.worldriskreport.org

Department of Strategic Policy, Planning and Aid Coordination. (2017). Vanuatu 2030 | The People's Plan: National Sustainable Development Plan 2016–2030 Monitoring and Evaluation Framework. Available at: www.gov.vu/images/publications/NSDP%20M&E%20Framework.pdf.

Government of The Republic of Vanuatu. (2018). Annual Development Report 2018. Available at: www.gov.vu/images/publications/Annual_Development_Report_2018.pdf.

New Zealand Treasury. (2019a). The Wellbeing Budget. Available at: https://treasury.govt.nz/sites/default/files/2019-06/b19-wellbeing-budget.pdf.

New Zealand Treasury. (2019b). Measuring Wellbeing: The LSF Dashboard. Available at: https://treasury.govt.nz/information-and-services/nz-economy/higher-living-standards/measuring-wellbeing-lsf-dashboard.

Organisation for Economic Co-operation and Development. (2019). Inequality. Available at: www.oecd.org/social/inequality.htm.

Kingdom of Bhutan. (2008). The Constitution of The Kingdom of Bhutan. Available at: www.nationalcouncil.bt/assets/uploads/files/Constitution%20%20of%20Bhutan%20English.pdf.

Ura, K., Alkire, S., Zangmo, T., & Wangdi, K. (2012). An Extensive Analysis of GNH Index. Centre for Bhutan Studies. Available at: www.grossnationalhappiness.com/wp-content/uploads/2012/10/An%20Extensive%20Analysis%20of%20GNH%20Index.pdf.

Wilson, S. (2020). Pandemic Leadership: Lessons from New Zealand's Approach to COVID-19. Leadership, 16(3). https://doi.org/10.1177/1742715020929151.

International Entry Points

Building on the rationales identified above for an international agenda on TCA in food and agriculture, mention has been made in a number of policy venues and documents negotiated and adopted on an international level.

The first of these is the CFS's High Level Panel of Experts report (High Level Panel of Experts on Food Security and Nutrition, 2019), on "Agroecological and other innovative approaches for sustainable agriculture and food systems that enhance food security and nutrition," as adopted by the Steering Committee of the HLPE and presented at the October 2019 meeting of the intergovernmental body. The report, in its summary, made the following points:

> It is clear that market forces, left to themselves, are unlikely to result in transitions towards [sustainable food systems] SFSs. This is because there are **many externalities associated with production, processing and distribution of food that are not priced and because the power exerted from the increasingly concentrated agri-food input and retail sector often works against addressing these externalities** (para 29).
>
> A considerable inertia, manifest in public policies, corporate structures, education systems, consumer habits and investment in research, favors the currently dominant model of agriculture and food systems, representing a series of lock-ins. **In the dominant model, environmental and social externalities are not properly considered and, therefore, not appropriately factored into decisions influencing the development of food systems. To overcome this inertia and challenge the status quo....** (para 30).
>
> Key changes in agriculture and food policies that could contribute to transitions towards SFSs for FSN include: putting greater emphasis on health and nutritional benefits; **implementation of true cost accounting**; [inter alia]. (para 32).

In its recommendations the report urged that:

> States and IGOs, in collaboration with academic institutions, civil society and the private sector, should: (inter alia) recognize **the importance of true cost accounting for negative as well as positive externalities in food systems and take steps to effectively implement it where appropriate**; (Recommendation 5)

Secondly, the 194 member nations of the FAO, two associate members, and the European Union adopted a strategy on biodiversity mainstreaming across agricultural sectors in December 2019 (Food and Agriculture Organization of the United Nations, 2020). The logic behind this strategy has been a recognition of the spiraling declines of biodiversity for agricultural reasons, on the one hand (Díaz *et al.*, 2020), and of the critical dependence of sustainable agriculture on biodiversity and ecosystem services on the other (Food and Agriculture Organization of the United Nations, 2020). The strategy, as adopted, calls for:

> Support provided to Members, at their request, to enhance their capacity to mainstream biodiversity (Outcome 1), [specifically to]
> **Provide advice on options to internalize the positive and negative economic, environmental and social impacts (externalities) of different agriculture and food systems** (Activity 1.10); and
> Advocate the recognition of the role of biodiversity for food security and nutrition (Outcome 3), [specifically to]
> **Raise awareness of stakeholders along the value chain of the positive and negative environmental and social impacts (externalities) of the different agriculture and food systems** (Activity 3.2).

Other entry points looming on the horizon are the development of "Systems of Environmental-Economic Accounting (SEEA)"; is a framework for national accounting to go beyond GDP by integrating economic and environmental data (https://seea.un.org), together with other work on developing global TCA standards for the private sector, as described in this volume. The Sustainable Development Goals (www.un.org/sustainabledevelopment/sustainable-development-goals/) speak directly to the need to bring a far broader perspective than GDP, along the lines of TCA, into statistics, planning, and development, through at least two targets:

> 15.9 By 2020, integrate ecosystem and biodiversity values into national and local planning, development processes, poverty reduction strategies and accounts.
> 17.19 By 2030, build on existing initiatives to develop measurements of progress on sustainable development that complement gross domestic product, and support statistical capacity-building in developing countries.

We are woefully behind on the first target and have much work to do on both. The UN Food Systems Summit (www.un.org/sustainabledevelopment/food-systems-summit-2021) to be convened by the Secretary-General in autumn 2021 (www.un.org/sustainabledevelopment/food-systems-summit-2021/) is a historic opportunity for food system transformation. However, it would only be able to meet its goal to drive this transformation if it genuinely embraces TCA in food and agriculture.

National governments have many multilateral venues available to explore and develop true cost policies in food and agriculture. Significantly, both the UN Convention on Biological Diversity and the UN Framework Convention on Climate Change (UNFCCC), in recent decisions (such as mainstreaming biodiversity into sectors including agriculture, CBD/COP/DEC/XIII/3; UNFCCC's decision 4/CP.23 on the "Koronivia joint work on agriculture") have turned increasingly to focus on the role of food and agriculture in both biodiversity loss and climate change. Their related bodies have issued recent reports and assessments underscoring the dependences and linkages between ecosystem services, biodiversity, climate change adaptation and mitigation and productive lands (Intergovernmental Panel on Climate Change, 2019; Intergovernmental Science-Policy Platform on Biodiversity and Ecosystem Services, 2019). All of these reports seek policies that can stem the tide of ecosystem degradation and build regenerative systems, for which TCA holds great potential as a mechanism to change the dynamic. National commitments under these conventions (National Biodiversity Strategies and Action Plans for the Convention on Biological Diversity and the National Determined Contributions of the UNFCCC) are key areas where national polices can be presented. However, as the *Global Biodiversity Outlook 5* has summarized after a ten-year period of implementation of an agreed global goal: "None of the Aichi Biodiversity Targets will be fully met, in turn threatening the achievement of the Sustainable Development Goals and undermining efforts to address climate change." (Global Diversity Outlook 5, 2020, pp. 2). This is the second time that a set of ten-year global biodiversity goals have failed to have been met. The question needs to be addressed as to what lessons can and must be learned from the implementation failure of global targets; simply agreeing on general global targets without a clear implementation strategy and a sound monitoring mechanism will not solve the problem.

True cost policies in food and agriculture—if implemented at the level of companies and national governments and used for the monitoring of the flows of values of the different capitals—have the potential to shed light on progress or failure of implementation of such agreements. This requires the engagement and buy-in of multiple stakeholders through inclusive processes and the development of an agreed system of standard reporting beyond productive capital. TCA should not be seen as another attempt to hide the real costs of our lifestyle—"greenwashing" unsustainable production—but to display all positive and negative externalities of production and consumption. So far, the aforementioned processes have not adopted a rigorous TCA but have continued to work on new global goals. Experience made so far with two decades of global biodiversity goals without an appropriate monitoring and reporting framework provides a clear message: There is no real progress without changing the economic drivers of unsustainable production and consumption.

One of the most respected governance structures is the Committee on World Food Security (CFS), a foremost inclusive, international, and intergovernmental platform. In addition to government representatives, all stakeholders from civil society, academia, and the private sector can channel their inputs and are actively engaged in the discussions. CFS is widely recognized also among the UN organizations and could be followed by national governments (and by the Food Systems Summit) as an inclusive model. The high-quality, neutral, science-based CFS HLPE Reports and the CFS "products" (adopted by consensus, after a multi-stakeholder policy convergence process) could be excellent tools for governments for the elaboration and design of their integrated, systemic food policies. In particular, negotiations on Voluntary Guidelines on Food Systems and Nutrition and on Agroecology Policy Recommendations are ongoing and highly relevant to TCA in agriculture and food; TCA should be both guiding and driving principles of CFS discussions and Summit preparations as well.

Concrete steps on national governance levels that can realize the reforms needed would include:

- Trade reform that allows environmental and other true-cost considerations to inform and shape trade agreements;
- Elimination of policies that promote forms of agriculture and food production with high negative externalities; and
- Recognition of healthy and nutritious food as a human right, secured through income equity; and
- Based on TCA, elaboration of policy incentives (positive and negative) to orient all stakeholders (including smallholder farmers and private multinationals) to opt for the appropriate decisions

Many key actors in the intergovernmental processes, national governments, and the private sector can promote, incorporate, and respect new investment guidelines that account for positive and negative externalities in Food and Agriculture.

Thus, the door is open on both international and national levels, for advancing on the concept and application of TCA in food and agriculture, reinforced by the work of the UNEP's The Economics of Biodiversity and Ecosystem Services to develop and refine approaches, frameworks and tools for the agrifood sector (http://teebweb.org/agrifood/). Projects are also currently underway to fully integrate TCA in the standard accounts of private sector to ensure that all capitals involved in the food systems can be reported and assessed (https://tca2f.org/reports/) and (https://futureoffood.org/wp-content/uploads/2020/07/TCA-Inventory-Report.pdf).

Note

1 For example, Bayer states that "By the middle of the century, the demand for agricultural products will be 50 percent higher on average than in 2013. An increase of 112 percent is forecast for the Sub-Saharan Africa and South Asia regions" (Bayer, 2017). Available at: www.bayer.com/en/the-future-of-agriculture-and-food.aspx.

References

Alexandratos, N. & Bruinsma, J. (2012). World agriculture towards 2030/2050: the 2012 revision. ESA Working paper. Rome: Food and Agriculture Organization of the United Nations.

Andrieu N. & Kebede Y. (2020). Climate Change and Agroecology and case study of CCAFS. CCAFS Working Paper no. 313. Wageningen: CGIAR Research Program on Climate Change, Agriculture and Food Security.

Berners-Lee, M., Kennelly, C., Watson, R., & Hewitt, C.N. (2018). Current global food production is sufficient to meet human nutritional needs in 2050 provided there is radical societal adaptation. *Elementa: Science of the Anthropocene*, 6(1).

Biovision & IPES-Food. (2020). Money Flows: What is holding back investment in agroecological research for Africa? Biovision Foundation for Ecological Development & International Panel of Experts on Sustainable Food Systems.

Cox, C. & Schechinger, A.W. (2018). America's nitrate habit is costly and dangerous: Prevention is the solution, but voluntary actions fall short. EWG, October 2, 2018. Available at: www.ewg.org/research/nitratecost.

Dalin, C., Konar, M., Hanasaki, N., Rinaldo, A., & Rodriguez-Iturbe, I. (2012). *Evolution of the global virtual water trade network*. Proceedings of the National Academy of Sciences, 109(16), 5989–5994. doi:10.1073/pnas.1203176109.

Díaz, S., Settele, J., Brondízio, E., Ngo, H., Guèze, M., Agard, J., ... & Chan, K. (2020). *Summary for policymakers of the global assessment report on biodiversity and ecosystem services of the Intergovernmental Science-Policy Platform on Biodiversity and Ecosystem Services*.

Fakhri, M. & Tzouvala, N. (2020). *To reduce hunger governments need to think beyond making food cheap. The Conversation*, July 17. Available at: https://theconversation.com/to-reduce-world-hunger-governments-need-to-think-beyond-making-food-cheap-142361.

Food and Agriculture Organization of the United Nations. (2009). How to Feed the World in 2050. In *Executive Summary-Proceedings of the Expert Meeting on How to Feed the World in 2050*. Rome: Food and Agriculture Organization of the United Nations.

Food and Agriculture Organization of the United Nations. (2014). *The state of food and agriculture 2014: Innovation in family farming*. Rome: Food and Agriculture Organization of the United Nations.

Food and Agriculture Organization of the United Nations. (2019). *The State of the World's Biodiversity for Food and Agriculture*. Food and Agriculture Organization Commission on Genetic Resources for Food and Agriculture Assessments, Rome.

Food and Agriculture Organization of the United Nations. (2020). *FAO Strategy on Mainstreaming Biodiversity across Agricultural Sectors*. Rome: Food and Agriculture Organization of the United Nations. doi:10.4060/ca7722en.

Food and Agriculture Organization of the United Nations, International Fund for Agricultural Development, UNICEF, World Food Programme & World Health Organization. (2020). *The State of Food Security and Nutrition in the World 2019. Safeguarding against economic slowdowns and downturns*. Rome: Food and Agriculture Organization of the United Nations.

Global Diversity Outlook 5. (2020). *Global Biodiversity Outlook 5–Summary for Policy Makers*. Montréal, QC: Secretariat of the Convention on Biological Diversity.

High Level Panel of Experts on Food Security and Nutrition. (2014). Food losses and waste in the context of sustainable food systems. A report by the High Level Panel of Experts on Food Security and Nutrition of the Committee on World Food Security, Rome. Available at: www.fao.org/3/a-i3901e.pdf.

High Level Panel of Experts on Food Security and Nutrition. (2017). Nutrition and food systems. A report by the High Level Panel of Experts on Food Security and Nutrition of the Committee on World Food Security. Rome. (No. 8). CFS Committee on World Food Security HLPE. Available at: www.fao.org/3/a-i7846e.pdf.

High Level Panel of Experts on Food Security and Nutrition. (2019). Agroecological and other innovative approaches for sustainable agriculture and food systems that enhance food security and nutrition. A report by the High Level Panel of Experts on Food Security and Nutrition of the Committee on World Food Security. Rome. (No. 14). CFS Committee on World Food Security HLPE. Available at: www.fao.org/3/ca5602en/ca5602en.pdf.

Intergovernmental Science-Policy Platform on Biodiversity and Ecosystem Services. (2019). *Global assessment report on biodiversity and ecosystem services of the Intergovernmental Science-Policy Platform on Biodiversity and Ecosystem Services.* Bonn: IPBES Secretariat.

Intergovernmental Panel on Climate Change. (2019). Chapter 4: Land degradation. In: *Special Report on Climate Change and Land.*

Lassaletta, L., Billen, G., Garnier, J., Bouwman, L., Velazquez, E., Mueller, N.D., & Gerber J.S. (2016). Nitrogen use in the global food system: past trends and future trajectories of agronomic performance, pollution, trade, and dietary demand. *Environmental Research Letters,* 11(9), 095007.

Organisation for Economic Co-operation and Development. (2019). Biodiversity: Finance and the Economic and Business Case for Action. Report prepared for the G7 Environment Ministers' Meeting, 5–6 May 2019. Organisation for Economic Co-operation and Development. Available at: www.oecd.org/environment/resources/biodiversity/Executive-Summary-and-Synthesis-Biodiversity-Finance-and-the-Economic-and-Business-Case-for-Action.pdf.

Pengue, W., Gemmill-Herren, B., Balázs, B., Ortega, E., Viglizzo, E., Acevedo, F., Diaz, D. N., Díaz de Astarloa, D., Fernandez, R., Garibaldi, L.A., Giampetro, M., Goldberg, A., Khosla, A., & Westhoek, H. (2018). *'Eco-agri-food systems': today's realities and tomorrow's challenges. In TEEB for Agriculture & Food: Scientific and Economic Foundations* (pp. 57–109). Geneva: UN Environment.

Pimbert, M. & Moeller, N. (2018). Absent agroecology aid: on UK agricultural development assistance since 2010. *Sustainability,* 10(2): 505. doi:10.3390/su10020505.

Ribaudo, M., Delgado, J., Hansen, L., Livingston, M., Mosheim, R., & Williamson, J. (2011). Nitrogen in agricultural systems: Implications for conservation policy. *USDA-ERS Economic Research Report,* 127.

Sen, A. (1981). *Poverty and famines: an essay on entitlement and deprivation.* Oxford: Oxford University Press.

Smith, L. C. & Haddad, L. (2015). Reducing child undernutrition: past drivers and priorities for the post-MDG era. *World Development,* 68, 180–204.

The Economics of Biodiversity and Ecosystem Services. (2018). *TEEB for Agriculture & Food: Scientific and Economic Foundations.* Geneva: UN Environment.

The Economist. (2020). The global food supply chain is passing a severe test; keeping the world fed, May 9, 2020.

Townsend, R. A. & Howarth, R. (2010). Fixing the Global Nitrogen Problem. *Scientific American,* 302(2), 64–71.

United Nations Environment Programme. (2015). *International Trade in Resources: A Biophysical Assessment,* Report of the International Resource Panel, United Nations Environment Programme

United Nations Environment Programme. (2016). *Food Systems and Natural Resources. A Report of the Working Group on Food Systems of the International Resource Panel* (pp. 1210–1226).Wilde, P. (2018). *Food Policy in the United States: An Introduction.* Abingdon: Routledge.

Wise, T.A. (2013). *Can we feed the world in 2050? A scoping paper to assess the evidence.* Global Development and Environment Institute Working Paper No. 13–04. Medford, MA: Tufts University.

Wise, T.A. (2019). *Eating Tomorrow: Agribusiness, Family Farmers, and the Battle for the Future of Food.* New York, NY: The New Press.

Wise, T.A. (2020). *Failing Africa's Farmers: An Impact Assessment of the Alliance for a Green Revolution in Africa.* Medford, MA: Tufts University. Available at: https://sites.tufts.edu/gdae/files/2020/10/20-01_Wise_FailureToYield.pdf.

Section 5

Through the Value Chain

The 21st Century has seen the rise of B Corp corporations (an incorporation status created in 2006 by B Lab), through which a company legally commits to a requirement that it take into account the impact of its decisions on workers, customers, suppliers, community, and the environment. B Corp companies further commit to standards of social and environmental performance, transparency, and accountability. In this regard, it is a significant departure from the prevailing for-profit model of incorporation, which has only one leg on its stool, in that it solely requires a return of shareholder value in monetary terms. The B Corp represents a fundamental shift in business orientation that embraces responsibility toward people and planet (in addition to profit) that True Cost Accounting in food seeks to have permeate throughout society, thus creating a more balanced, three-legged stool. There are now over 3,500 B Corps globally (https://bcorporation.net), including the well-known brands Patagonia, Ben & Jerry's, and 17 of Danone's subsidiaries (Danone, 2020).

In addition to B Corp proliferation, the concept of sustainability appears to be, thankfully, writ large across the global business landscape. A number of large brands in the food business sector have set carbon emissions targets, including Horizon Organic (which aims to be the first "carbon positive" dairy company) and Danone, which set a carbon emissions target about a decade ago on its entire value chain, approximately 60% of which is related to agriculture. In 2019 at the United Nations Climate Action Week 87 multinational companies (with a combined market capitalization of over $2.3 trillion and annual direct emissions equivalent to 73 coal-fired power plants) set targets to align their value chains to limit global temperature rise to 1.5°C and reaching net-zero by no later than 2050.

However, Hank Cardello, Director of the Food Policy Center at the Hudson Institute, sees the food industry sustainability pledges in this way:

> While the food industry claims to be attuned to the consumer, its risk aversion means it makes major changes only when it's forced, as the pandemic has shown. When food companies don't see a crisis, 'innovation' amounts to line extensions and retro, iconic boxes...Simply being slavish

to CSR [Corporate Social Responsibility] reporting no longer goes far enough. Last year, 90% of companies on the S&P 500 Index published sustainability reports. CSR reporting is no longer a differentiator; it is a minimum ante to be relevant in today's consumer and business climate.

(Cardello, 2020)

There might be more health- and sustainability-driven innovations in the food sector than immediately meets most eyes. The outdoor clothing company Patagonia has launched a food line called Patagonia Provisions which produces and sells packaged food produced in a farming method characterized as "regenerative agriculture," and has partnered with seven other companies in creating the Regenerative Organic Certification, which will certify farming practices that produce "healthier soils, higher animal welfare, and fairness for farmers and workers." The certification is designed to boost market share, such as the brand benefit provided by the United States Department of Agriculture Organic seal. As we saw in Chapter 7, the market share for organic is growing, aided by its accessible branding and growing consumer awareness of its better true cost ratio, in that it is of more benefit to the environment.

The two chapters that follow here are but a few highlights of the evolution of the food business sector in the direction of incorporating sustainability and equity into corporate bottom lines. In the more specific framing of emergent True Cost Accounting (TCA) in financial and investment decision-making in food, this upcoming section offers a valuable case study focus on a fundamental lever of business decision-making: financial risk.

In Chapter 14, the authors offer four case studies reflecting, first, a corporate perspective on assessing the True Cost of various regional and global supply chains; second, a bank's experience with TCA; third, insights into True Cost considerations from an insurance sector view; and finally, the experience of a financial auditor. These cases illustrate "proof of concept" for TCA in the private sector, leading companies willing to differentiate themselves through their interest in understanding and addressing their externalities

Chapter 15 addresses the question of how investment and creative finance can support healthy and equitable food systems shifts, pointing out the need for a common metrics framework and aligned accounting standards to measure impact, so that investors and companies can "de-risk" the financial viability of their investments and scale with a meaningful range of sustainability metrics in their balance sheets. The authors provide three case studies to demonstrate how values-conscious investment companies have balanced risk and impact in their portfolios.

These case studies are valuable as illustrative insights into a larger picture of how multiple actors in private enterprise are acting with the sense of purpose needed to influence the large gears of commerce on this fragile planet.

References

Cardello, H. (2020). Food Industry is a No-Show in New Sustainability Study. *Forbes*. Retrieved September 22, 2020. Available at: www.forbes.com/sites/hankcardello/2020/10/20/food-industry-is-a-no-show-in-new-sustainability-study/#7388d6591c57.

Danone. (n.d.). B Corp. Retrieved September 22, 2020. Available at: www.danone.com/about-danone/sustainable-value-creation/BCorpAmbition.html#AMBITION.

14 The Business of TCA

Assessing Risks and Dependencies Along the Supply Chain

Tobias Bandel, Jan Köpper, Laura Mervelskemper,
Christopher Bonnet and Arno Scheepens

Introduction

Climate change, resource scarcity, consumer awareness, and new regulations trigger practice changes in global supply chains regarding environmental and social aspects. These better practices go along with additional costs, which, based on current accounting schemes, could negatively impact the economic performance of companies. This causes a dilemma for the private sector: while trying to comply with these new requirements, the companies get financially punished as the higher costs for sustainable measures reduce their profits. True Cost Accounting (TCA) can be used to show the benefits of better practices at the company or supply chain levels, not only using sustainability language but in tangible financial terms. This chapter presents the experience of different actors from the corporate and financial sectors in applying TCA. The first case study offers a corporate perspective on assessing the True Cost of various regional and global supply chains, the second case study discusses a bank's experience with TCA, the third case study provides insights into the True Cost considerations from an insurance sector view, and the fourth case study shares the experience of a financial auditor.

A key finding from all case studies is that a true cost assessment across entire supply chains is possible, allowing for an assessment that crosses private and financial sector initiatives, integrating sustainable performance into financial market requirements. However, although data and models to assess the true cost of ecological or natural capital aspects already exist, there is still a substantial need for further research regarding social and human capital aspects such as health. The following four case studies demonstrate how TCA is a valuable tool for agri-food companies, banks, insurances, and financial auditors.

Case Study 1: Assessing the True Cost of Various Regional and Global Supply Chains

What are the true costs of food production, and what can be done to reduce these externalities to society? How can we quantify and monetize better farming practices to show that sustainably produced food costs society and taxpayers less? These questions arose in 2014 when various companies had identified financial and reputational exposure and started to assess their true cost profile. The initial motivation to conduct true cost assessments was based purely on pioneering entrepreneurial spirit, trying to secure and further develop their future business cases by minimizing current and future risks.

In November 2019 Boston Consulting Group (BCG) published a report (Boston Consulting Group, 2019) about how to secure the future of German agriculture. The key finding was that today's German agricultural system causes externalities—that is, costs to the society and the environment, amounting to €90 billion. This is in addition to another €10 billion of subsidies and other direct payments, which are currently borne by society, in the form of taxpayers. This €100 billion only covers externalities related to climate, air, water, soil, livestock, and ecosystem services from the German agricultural sector. Social aspects are not covered. The study assumed that more sustainable production would reduce the costs to society. At the same time, Christian Heller, CEO of the Value Balancing Alliance presented to the European Business and Nature Summit in Madrid on how today's costs to society will become costs to businesses over time (Heller, 2019).

The cumulative experience of conducting true cost assessments with the following companies are included in this case study: Alnatura, Bauck, Demeter, Eosta, GLS Bank, Haciendas Bio, Hipp, Lebensbaum, Martin Bauer, Rapunzel, Tradin, Triodos, and Weleda. The assessments analyzed products and supply chains covering a variety of agricultural products from different origins worldwide and were conducted by Soil & More Impacts (SMI), in some cases in collaboration with EY. The focus was to assess the impact on natural capital aspects (biodiversity, climate, soil, and water) (Natural Capital Coalition, 2016). Selected social and human capital aspects were analyzed as well. The intention of these pilot assessments was not only to generate true cost value but also to test the model for its applicability and scalability to global complex supply chains.

Priority was given to primary data available through existing audits such as organic, Fairtrade, Rainforest/UTZ, or financial accounts. To maximize the comparability and acceptance in the food and agricultural market, commonly used impact assessment models, reference values and monetization factors were used such as the Cool Farm Tool (Hillier et al., 2011), the RUSLE (Revised Universal Soil Loss Equation) (Renard, 1997), Aqueduct maps (Gassert *et al.*, 2014), ClimWat (Muñoz and Grieser, 2006), CropWat (Smith, 1992), the DALY (Disability-adjusted life year) concept (Homedes, 1996) and EcoMatters (van Maurik *et al.*, 2016). In most cases, the assessed supply chains were benchmarked against the common practice in the region, a baseline, or an improved scenario.

The overall finding was that despite the fact that TCA is a rather young and developing science, the most commonly used approaches, assumptions, and models seem to be good and detailed enough to generate meaningful results, identifying and highlighting strengths and weaknesses, costs and benefits of the different products and supply chains.

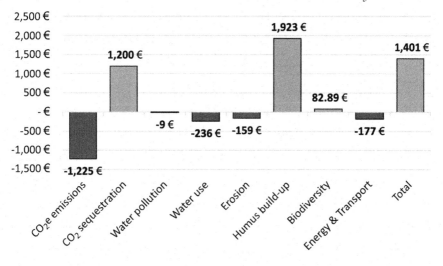

Figure 14.1 Calculated external costs in €/hectare for an organic farm in Germany.

The following figures show selected results from true cost assessments of the participating companies.

Figure 14.1 shows the true cost assessment of a cereal- and vegetable-producing German organic farm. The external cost due to CO_2 emissions was nearly offset by the amount of CO_2 sequestered. The major benefit of this farm was generated due to humus (soil) build-up. Overall, that farm created an external benefit of €1,401/hectare. This is a weighted average across the entire crop rotation which could be broken down to external costs and benefits per kilogram of product, factoring in the yield. From a scientific and modelling perspective, one of the key learnings was that the entire crop rotation of a farm needs to be assessed in order to identify the real external costs or benefits of a farming system.

Figure 14.2 shows the true cost result in €/hectare of an intensively managed vegetable farm which generates external costs of €702/hectare. Figure 14.3 illustrates the same farm after implementing some better practices such as intercropping and improved compost management, resulting in a reduction of the external costs to €106/hectare.

As the currently prevailing standard accounting and economic valuation systems do not consider these positive or negative externalities, there is no direct financial incentive for better practices, which leads to distorted markets and false accounting. Therefore, apart from the necessity of political interventions, it is required that the financial market institutions start considering these externalities by including them in credit ratings, insurance policies, annual accounts, and company valuations. In order to foster this process, Soil & More Impacts and TMG Thinktank for Sustainability started an initiative together

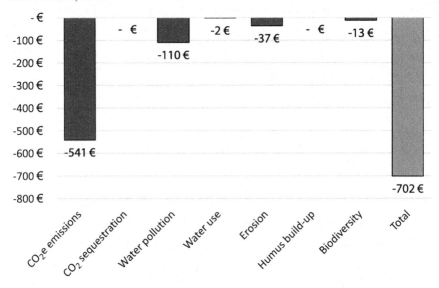

Figure 14.2 Calculated external baseline costs in €/hectare for an intensively managed vegetable farm in Germany.

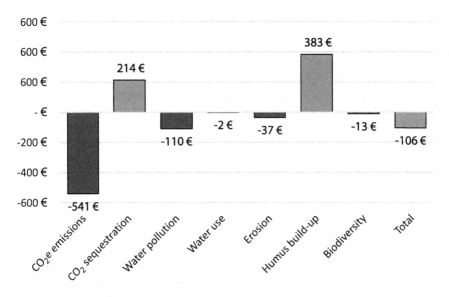

Figure 14.3 Calculated external scenario costs in €/hectare for an intensively managed vegetable farm in Germany.

with EY and some market-leading companies to develop guidelines on how to include both positive and negative externalities in annual reports as a basis to generate financial incentives for better farming practices.

Case Study 2: True Cost Accounting at GLS Bank

Founded in 1974 in Bochum Germany, GLS Bank is the first social-environmental bank globally, with a specific focus on financing the basic needs of people in line with regenerative environmental practices. Taking these two focal points as the core of all business activities of GLS Bank, economic sustainability is the logical consequence instead of the key imperative of doing business. To date, GLS Bank has a staff of 700 employees and a balance sheet total of around €7 billion.

The need to drastically rethink the current patterns of economic systems along the lines of planetary boundaries, common goods, and social justice finally seems to be a common understanding by an increasing number of market participants, supervisory authorities, and citizens. In line with this, the European Union, Central Banks, and supervisors have been calling for a more proactive integration of sustainability-related risks and opportunities into business management and target setting. The predominant focus of these initiatives currently lies on climate-related issues, as the short-, medium-, and long-term impacts of this challenge are more tangible and have a (better) data basis. However, this is just the starting point for a wide-ranging revision of how sustainable business models need to be framed. The interplay between the buildup, use, and degradation of values, as well as their long-term relevance for business performance and socio-environmental sustainability needs to be put in focus.

With a view to understanding, translating, and managing sustainability-related risks and opportunities, economic actors in general and financial institutions in particular need to (re)define the parameters that (will) affect economic value.

As this viewpoint is accompanied by a great opportunity for greater consideration of sustainability aspects, GLS Bank has been engaging in a profound rethink of risk management and accounting. As the first German socio-environmental bank, its mission is to redefine capital as a means for positive societal change and using money to finance a variety of exclusively sustainable projects and businesses.

Accordingly, the bank defines the value of an economic activity or organization to lie far beyond financial capital as the core driver of short- to long-term value creation. Rather, value is created, sustained, and strengthened by mutual impacts on and across multiple capitals: human capital, social capital, natural capital, financial capital, intellectual capital, and production capital. These capitals and their interactions represent the true values that determine an organization's holistic value creation or degradation and therefore its future viability and competitiveness.

In this context, TCA represents a concrete methodology for a far more holistic view on value drivers by integrating quantified sustainability aspects into the well-known logic of financial accounting, following a similar understanding of the dependency of capitals: in times of globally scarce raw materials such as soil and water, it is of strategic importance not only for the agricultural sector, but also for national economies as a whole, to take a close look at the availability and use of vital resources such as soil, water, and energy and, if necessary, to intervene with effective measures to secure these resources. Whether and at what price raw materials can be processed and traded in the future is determined based on the agricultural practices applied today. Those who take appropriate care of, for example, soil and water today will be able to offer agricultural products competitively and in line with planetary boundaries in the future. In turn, it can be argued that sustainable investments in multiple capitals lower economic risks.

GLS sees its mission to strive for a (more) sustainable future and to implement a more holistic view of sustainability-related risks and therefore engages in TCA. Considering the first aspect, the market-based approach of TCA monetizes harmful activity and financially rewards sustainable activity, thus making the conservation of resources financially attractive and leveraging sustainable behavior. Regarding the second aspect, the approach of TCA provides an opportunity to improve risk and opportunity management in the lending process. Former intangible or invisible risks and return potentials are given a monetary value and, as a result, can be considered when assessing the creditworthiness and credit default risk of a project or organization.

As a first pilot, GLS Bank and GLS Treuhand have applied the method of TCA together with Soil & More Impacts for three organic partner farms. The results show that the current agriculture practices generate high costs: while organic farms generate an average profit of around €720/hectare, the conventional comparable farms cause net costs averaging €3,670/hectare. These costs have so far been paid by society—either directly, for example through higher water treatment costs, or indirectly in terms of environmental damage. In the medium term at the latest, these costs will also return to the farmers and their land when assets like soil fertility are destroyed or political countermeasures are taken that will affect farmers. In the ongoing criticism of agriculture and the debate about the need for agricultural transformation, TCA reveals that organic agriculture provides valuable socio-environmental services and makes a beneficial contribution to society.

The application of TCA might not lead to a fundamental change in the granting of loans by GLS Bank. The bank instead aims to create a leverage effect that can be achieved when other banks without a normative view on sustainability act in the same way, realizing the financial risks of sustainability aspects and thus considering them when granting loans. In return, this can help to steer capital towards sustainable agriculture.

Although not all ecosystem services or capitals can nor should be (fully) monetized, the view of manifold impacts opens the playground for business decisions that are multidimensional with a high probability of identifying current and future risks and opportunities. Hence, TCA paves the way to understand and disclose social and ecosystem services that have tangible impacts on the viability of business models.

Case Study 3: The Research of Allianz in Assessing Natural Capital for Risk Management Solutions in the Insurance Sector

Allianz Global Corporate & Specialty (AGCS) is the Allianz Group's dedicated carrier for corporate and specialty insurance business. AGCS provides an insurance and risk consultancy across the whole spectrum of specialty, alternative risk transfer and corporate business. Their role as the leading corporate insurance company demands an in-depth awareness and understanding of the emerging sustainability-related trends that impact their clients and their operations. To do this, AGCS has built a dedicated team of experts in sustainability risks from an industrial insurance perspective.

AGCS supports its clients to identify and assess material risks along their value chain and identify and design risk management solutions in a collaborative manner. Environmental, social, and governance (ESG) factors are increasingly relevant in risk management, and the sustainable use of natural capital is one important element. By many scientific and macro-economic indicators, it is becoming increasingly evident that natural capital is being depleted at a far faster rate than the planet can replenish it, and with consequences that extend well beyond the direct effects on the environment. Consequently, businesses face new risks from the ongoing depletion of natural capital.

In 2018 AGCS published an exploratory report "Measuring And Managing Environmental Exposure: A Business Sector Analysis Of Natural Capital Risk" (Allianz, 2018b) outlining potential exposure to natural capital risks, based on an analysis of 2,500 companies across 12 industry sectors. The report compares and analyzes selected sectors and assigns each of them to one of three risk categories: danger zone (sectors where risks are generally greater than mitigation), middle zone (sectors where risks are roughly matched to mitigation), or safe haven (sectors that generally do not seem to face high risks and/or are reasonably well prepared for risk). According to the study, the following sectors have been assigned to the following risk categories:

- Danger zone: Oil and gas; mining; food and beverage; transportation
- Middle zone: Automotive, chemical, clothing, construction, manufacturing, pharmaceutical, and utilities
- Safe haven: Telecommunications

Natural capital risk assessment is expected to become increasingly important for corporates as numerous liability and business interruption cases have been revealed around the globe. These types of losses are expected to increase unless these risks are mitigated.

A significant number of companies have started to address natural capital risk in their enterprise risk management. Factoring natural capital costs into business decision-making can help companies to anticipate potential threats. For example, when opening a new plant, factors such as future water availability and the emerging emissions regime should be considered. Natural capital risk exposure will become increasingly important, as it is expected that companies will have to actively disclose these risks to governmental agencies and investors as both risks and related management expectations evolve.

"With threats to the environment coming from many different areas, there will be no such thing as business as usual in future," says Chris Bonnet, Head of ESG Business Services from Allianz and co-author of the report. "Companies need to understand, quantify and even monetize their dependence on natural capital and the impacts their operations have on it to ensure their organizations are resilient and future-proof." More information about natural capital risk and the report insights can be found in Allianz (2018a).

Case Study 4: Natural Capital Inclusion for Sustainable Innovation and Risk Management: The Perspective of a Sustainable Industrial Design Engineer from EY Climate Change and Sustainability Services

Back in the 1930s the Hawthorne Works in Chicago had commissioned a study to look into worker productivity in the factory under varying conditions. Researchers saw that productivity increased with changes in light intensity. However, the workers fell back into lower productivity as soon as the study ended. The conclusion was drawn that the light intensity was not the cause for the increase in productivity, but rather the increased attention on individual workers and their performance.

Traditionally the attention of the financial sector with regards to the performance of companies has been on their financial/economic performance. In recent years, there has been a steady increase in attention to non-financial information, also in the financial sector. The realization that non-financial information is just as important, or perhaps even more important than financial information to evaluate the potential for long-term value creation of companies has spurred the disclosure of all kinds of different non-financial metrics and other performance indicators in sustainability reports, integrated annual reports, and sometimes even in financial statements.

According to the Global Investor survey conducted by EY in 2018, nearly all investors who responded to the survey (97%) say that they conduct an evaluation of non-financial disclosures; just 3% of respondents say they conduct little or no review. At the same time, investors' clients are increasingly asking about non-financial information and expecting it to be integrated into mandates. Furthermore, non-financial information plays an increasingly important role in the investment decision-making process, and nearly all respondents (96%) say that such information has played a pivotal role. In interviews, investors stressed the importance that sustainability disclosures play in determining appropriate market valuations. Therefore, companies should focus on ensuring that their non-financial information has the same level of scrutiny as financial information.

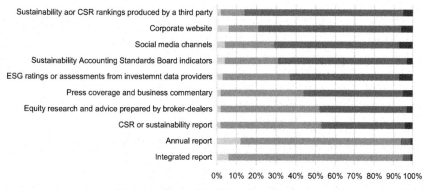

Figure 14.4 Survey results: How useful do you find the following sources of non-financial information when making an investment decision? (adapted from the Global Investor Survey, EY, 2018).

Investors are requesting broader and higher-quality non-financial information from companies, and seeking consistent, investment-grade information to support their decision-making. For investors, the most useful non-financial reports come from companies that understand material non-financial risks and opportunities which are most important to their industry and business model. Investors report that, governance aspects aside, the main non-financial factors in investment decision-making are related to supply chain, human rights, and climate change risks.

Respondents also say that non-financial information must be standardized to create a useful basis of comparison, to establish benchmarks and to mark trends. Investors say that national regulators are best suited (70%) to lead efforts to close the gap between investors' need for non-financial information. In

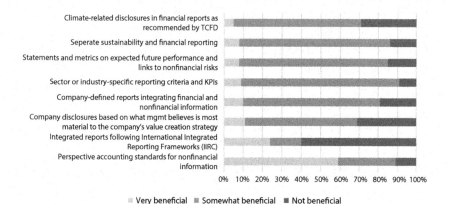

Figure 14.5 Survey results: How beneficial would each of the following reports or disclosures be to your investment decision-making? (adapted from the Global Investor Survey, EY, 2018).

addition, investors are looking for intelligent collaboration among themselves, regulators, and organizations such as trade groups and non-governmental organizations to establish appropriate and effective reporting standards.

In the agri-food sector, it is known that at least some material environmental (and social) impacts and dependencies (risks) occur at the farm level. Large national and multinational food and beverage companies rely on vast amounts of natural (and social) capital, such as agricultural land, biodiversity, healthy soils, etc. mainly through purchasing agricultural products from a large number of agricultural suppliers.

In order to identify, quantify, and eventually mitigate the associated impacts and risks associated with the environmental impacts and dependencies, large food and beverage product companies will need to obtain data on the non-financial performance of their supply chain in order to report reliably on their own non-financial performance. But most importantly, it is essential that this information is then also utilized to improve the non-financial performance, similar to what we are used to with financial performance information. Both for non-financial performance reporting as well as strategic decision-making, it is essential that the data that companies collect to use for these purposes is reliable. Obtaining assurance can provide the increased credibility and reliability of non-financial information, similar to financial information.

Business activities can lead to multiple different environmental impacts that can occur locally and/or globally and measuring these impacts is always complex. Scientific research and development have led to standardized methods for assessing impacts, but the way that they are applied often leaves room for "manipulation," which can have a large effect on the identified non-financial risks and opportunities portrayed in the reporting of companies.

Given the previously discussed trends and developments, there is an emerging need for standardized TCA, which brings together the different environmental (and social) impacts into a single monetary unit, allowing for full integration with annual reports, integrated reports, as well as strategic decision-making for companies and investors to better balance their financial performance with their non-financial performance. Therefore the main need for the coming decade is to develop and align a sector-specific, highly automated, standardized method, approach and guidelines in order to eventually come to sector-specific reporting standards for non-financial information similar to the standards already available for financial reporting.

The real benefit of TCA is in "turning on the light" with regards to the required transition towards a sustainable society. The attention that the financial sector is giving to non-financial performance of assets spurs companies to think about their non-financial performance. The pilots that EY involved in the *True Cost: from Cost to Benefit* project confirmed that farms are open to supplying non-financial information to their clients if they are able to. By "turning on the light", movement towards a more sustainable way of doing business is already visible. If we can manage to turn on the light on a larger scale, where standardization plays an essential role, we should be able to see a bigger movement towards more sustainable production and consumption.

Conclusion

These case studies showcase the versatility in application and use of TCA across different business players and emphasize the potential TCA has in becoming a relevant tool for assessing impacts and dependencies in the financial sector.

By using TCA for analyzing and evaluating the environmental impact of different agricultural management practices, agri-food companies can base their supply chain decisions on comparable and transparent results. Value-driven corporations like the GLS Bank can substantiate their mission and corresponding decisions with monetary figures of their impact. Insurance providers have realized that capital dependencies and impacts are highly interconnected, leading to immense natural capital risks that are barely considered in existing tools used by the financial and insurance industry. In addition, financial auditors like EY are increasingly acknowledging the need for a standardized way of assessing the long-term value and impact of companies to create a comprehensible basis for investors and other readers of annual reports. Even though TCA is a young field, it is built on existing scientific knowledge and can be further developed, standardized, and integrated into practical tools. With this, it can be a powerful lever for transformative change towards a new definition of value—based on capitals thinking—in the business world.

References

Allianz. (2018a). Allianz report: Failure to manage natural resources brings increasing interruption and liability risks for businesses. Available at:www.agcs.allianz.com/news-and-insights/news/natural-capital-risks.html.

Allianz. (2018b). Measuring and Managing Environmental Exposure: A Business Sector Analysis Of Natural Capital Risk. Allianz Global Corporate & Specialty SE. Available at: www.agcs.allianz.com/content/dam/onemarketing/agcs/agcs/reports/AGCS-Natural-Capital-Risk-Report.pdf.

Boston Consulting Group. (2019). Die Zukunft der deutschen Landwirtschaft nachhaltig sichern).

Gassert, F., Luck, M., Landis, M., Reig, P., & Shiao, T. (2014). Aqueduct global maps 2.1: Constructing decision-relevant global water risk indicators. World Resources Institute.

Heller, C. (2019). Presentation at the European Business & Nature Summit 2019 in Madrid. Value Balancing Alliance. Available at: https://naturalcapitalsummit.com/wp-content/uploads/2019/11/EBNS-PL3_Christian-Heller_BASF-and-value-balancing-alliance.pdf.

Hillier, J., Walter, C., Malin, D., Garcia-Suarez, T., Mila-i-Canals, L., & Smith, P. (2011). A farm-focused calculator for emissions from crop and livestock production. *Environmental Modelling & Software*, 26(9), 1070–1078.

Homedes, N. (1996). The disability-adjusted life year (DALY) definition, measurement and potential use. (*Human Capital Development papers*, No. 16128, p. 1). The World Bank.

Muñoz, G. & Grieser, J. (2006). *Climwat 2.0 for CROPWAT*. Water Resources and Development and Management Service of Food and Agriculture Organization of the United Nations, pp. 1–5.

Natural Capital Coalition. (2016). Natural capital protocol. Available at: https://naturalcap italcoalition.org/natural-capital-protocol.

Renard, K.G. (1997). *Predicting soil erosion by water: a guide to conservation planning with the Revised Universal Soil Loss Equation (RUSLE)*. US Department of Agriculture. United States Government Printing.

Smith, M. (1992). *CROPWAT: A computer program for irrigation planning and management (No. 46)*. Rome: Food and Agriculture Organization of the United Nations.

van Maurik, J.H., Sonnen, M.J., & Damen, L. (2016). *Natural capital–Ecomatters imple-mentation guide*. Ecomatters. Available at: www.ecomatters.nl/wp-content/uploads/2018/09/Ecomatters-Natural-Capital-Ecomatters-implementation-guide.pdf.

15 Investing in the True Value of Sustainable Food Systems

Tim Crosby, Jennifer Astone and Rex Raimond

How can investment and creative finance support healthy and equitable food systems shifts? More donors, investors, and members of the finance community are seeking integrated, holistic methods to assess investments. Social enterprises are also looking to demonstrate the value add of their impact. Recently, the Transformational Investing in Food Systems Initiative (TIFS Initiative) (www.tifsinitiative.org) adapted The Economics of Ecosystems and Biodiversity for Agriculture and Food's True Cost Accounting Framework (TEEB for Agriculture & Food) (http://teeb-web.org/agrifood/) into a tool for investors and entrepreneurs and have piloted the application of the tool. This chapter examines the strengths and weaknesses of such a tool for impact investing for food systems transformation and offers early case studies of how entrepreneurs and fund managers are looking to support equitable, agroecological food systems.

- Sistema Bio designs, builds, and sells patented biodigesters to small-scale farmers in Mexico, Nicaragua, Colombia, Kenya, and India. The biodigesters convert cow manure into energy and fertilizer, saving farmers money while protecting groundwater and improving soil health.
- Community Markets for Conservation (COMACO) works with 81 community cooperatives in Zambia to provide incentives for biodiversity conservation, support 188,500 small-scale farmers, apply climate smart agriculture, and run a business to manufacture and sell healthy foods.
- Root Capital is a non-profit social investment fund that invests in businesses that collect, aggregate, process, and market crops for rural farmers in Latin America, sub-Saharan Africa, and Southeast Asia.

Each case demonstrates a business model or investment approach that aims to create positive outcomes across Natural, Social, Human, and Produced Capital. They provide insight about how an applied TCA framework can be a helpful tool to shift how investment decisions are made.

A True Cost and Value Approach to Investing

Investing towards personal and ethical values has been promoted since ancient times (CNote, 2019), while values-aligned investing in North America got started in the 1960s social and political movements boycotting companies involved in the Vietnam War. Today, sustainable investing has expanded into Socially Responsible Investing,

Environmental and Social Governance screening, the Global Reporting Initiative, Impact Investing, and most recently Blended and Integrated Capital Investing (RSF Social Finance, n.d.). By 2018 $30 trillion was invested globally with considerations of ethics, social and environmental values (Global Sustainable Investment Alliance, 2019). Data show that funds aligned with sustainable investing better estimate the true value of the underlying assets and their future values and are also outperforming many of their peers (Mooney, 2002).

While the field of sustainable investing has exploded, the ability to qualify and quantify the positive non-financial impact of an investment has not been standardized. However, the hunger for a global set of standards is witnessed by the way investors and fund managers quickly adopted the United Nations (UN) Sustainable Development Goals (SDGs). While the 17 SDGS include a total of 247 indicators, they contain scant guidance on what data sets are valid and standardized. In order to scale sustainable investing, investors and companies need a common approach and shared metrics to measure impact, and have these approaches and metrics align with accounting standards.

Prior innovations in investing have worked within the structure of Generally Accepted Accounting Principles (GAAP) and International Financial Reporting Standards (IFRS). With pressure from investors and associations like the Sustainability Accounting Standards Board (www.sasb.org), GAAP has begun implementing revisions that allow hitherto undervalued accounting items to be included in formal reporting requirements.

Investors, especially asset managers, are demanding more. They want to know that their impact is measurable and that those measurements are linked to over-arching frameworks like the SDGs. By aligning investment criteria with True Cost Accounting (TCA), there is an opportunity to account for underlying material costs not currently captured in GAAP and IFRS as well as align values-based investment interests with needed accounting standard revisions. This should better connect the front-end decision-making of where to invest with the reporting of outcomes from that investment.

Opportunities for True Value Investing in Aligned Food Enterprises

Food systems transformation is central to the agenda for achieving a more just and sustainable world (Global Alliance for the Future of Food, 2020). Given food systems' undeniable links to climate change, migration, zoonotic disease, biodiversity loss, structural inequality, and public health—the myriad global emergencies that people currently face—transformation of these systems has emerged as a global priority.

Despite the rising voice for systems change, institutional investors, governments, philanthropists, and private sector companies seek enterprise or fund level success, not always system success. They generally seek to reduce their exposure to risks that they perceive as significant (e.g., financial loss, climate change, natural resource degradation, social inequality, food insecurity, or rising costs of health-care), and might have divergent mandates (e.g., capital preservation, generating income and profit, and/or creating public goods). The need is to address these issues and mandates together in a more holistic approach to investing.

Investors—individuals, asset managers, and institutions—are being asked at an ever-increasing rate to prove the impact of their investments beyond financial returns. Social entrepreneurs are learning how to demonstrate the social and environmental value of their business to investors and donors. These needs require harmonizing multiple priorities, risk mitigation, and return expectations with the metrics to show positive outcomes for food producers, food workers, natural systems as well as consumer and community health. The biggest hurdles for the needed innovations in investment practices involve redefining risk, reward, efficiency, and scale to become more systems-focused, internalizing those considerations into decision-making structures, and agreeing on missing impact metrics. These innovations must overcome the current biased metrics for food systems investors that primarily reward two dimensions: increasing productivity and profit, a reductionist approach to food production.

Enter the United Nations Environment Program's TEEBAgriFood initiative which delineates costs and values across four types of "Capitals": Natural, Human, Social, and Produced. This holistic framework is being applied to create tools for business to develop comprehensive profit and loss reports that integrate value added or lost (e.g., I360X, True Price/Impact Institute, Harvard Impact Weighted Accounts, Common Land), explore the creation of new accounting standards (e.g., Sustainable Accounting Standards Board), and help agri-food companies to measure and manage their impacts, dependencies, and risks, and unlock new opportunities (e.g., World Business Council on Sustainable Development and Capitals Coalition). In the long run, these efforts can assist investors in identifying companies that are performing financially and creating better environmental and social benefits than their peers. To help diverse investors work together more effectively, an investing approach that incorporates the TEEBAgriFood framework analysis can elucidate and organize the anticipated negative and positive outcomes of a given enterprise's activities before an investment is made, while leveraging the framework's focus on the accounting sector.

Applying TCA Rapid Assessments

The Transformational Investing in Food Systems (TIFS) initiative has identified the need for practical decision-making tools that holistically identify the social and environmental impacts of enterprises. TIFS is piloting two TCA Rapid Assessment tools: one to assess impacts of social enterprises and one for investment funds (prototypes are being tested and will be made available publicly).

The goal of these TCA tools is to help investors and entrepreneurs assess enterprises' impacts on Natural, Human, Social, and Produced Capital stocks of food systems. The TCA tools are a set of questions that make each of the four Capitals visible to an investor and those seeking investment. The TCA tools pose outcome-oriented questions related to the four Capitals for consideration during normal due diligence processes; for example, "Does this investment increase or decrease Pollination? Does this project improve or reduce Community Wellbeing?" TIFS advances these tools as one lever to influence up-front investment decisions and back-end reporting and helps entrepreneurs to make a case in the absence of robust public policy that prices externalities. This converges with ongoing efforts to harmonize impact strategies, metrics, and data, as well as efforts to create new accounting standards and profit and loss reports.

Just as the development of financial analysis methods has enhanced investors' capacity to predict and improve financial returns, systematic impact analysis—both before and after investment—is in high demand and still needs work. The following examples highlight funds and enterprises that aim to achieve system-level impact. These established enterprises and investment funds used our first version of the TCA tools to track changes in Capital from a systems perspective. The following cases agreed to be engaged in this process. They are illustrative examples involving two enterprises and one fund that are leaders in their fields. The analysis and synthesis that follow incorporate feedback from almost 30 interviews with investors, fund managers, enterprises, and other experts, and a comparative review of major impact management and measurement frameworks. The results of the case examples, interviews, and review of the field of impact investing have informed the tool and the ensuing analysis.

Sistema Bio

Sistema Bio (https://sistema.bio) designs, builds, and sells patented biodigesters to small-scale farmers to convert manure into energy and fertilizer. Their low-cost, modular biodigesters save farmers money while protecting groundwater and improving soil health. Since 2010 Sistema Bio has installed over 11,000 units in Mexico, Nicaragua, Colombia, Kenya, and India.

The company measures tons of CO_2e mitigated, tons of treated waste, biogas produced, trees saved, and hectares per year enhanced with biofertilizer. It does not measure the benefits of avoided deforestation, time saved, or money saved.

Beneficial Returns (www.beneficialreturns.com) is an impact investment debt fund that provided a loan for Sistema Bio to purchase trucks to strengthen their infrastructure and follow up on their customers. Beneficial Returns uses its own internal assessment tool to evaluate a business's ability to contribute to the environment and community well-being while running a profitable enterprise. In a financial innovation to recognize impact and financial health, Beneficial Returns waives borrowers' final payment if they exceed a predetermined impact target and make their other payments on time. This innovation incentivizes continued attention to impact over maximizing profits during growth, which is a challenge faced for small and growing social enterprises.

Sistema Bio finances its enterprise with grants, equity, and debt. To reach smallholder farmers outside tight (highly controlled or coordinated) value chains, Sistema Bio must pilot its model to market, sell, and monitor the biodigesters in rural environments with weak infrastructure and limited farmers' awareness of the benefits of biodigesters, yet great opportunity for adding value. While Sistema Bio could be more profitable if it sold only to larger farmers closer to commercial hubs, it decided to run its for-profit enterprise to reach underserved communities. It envisions financial sustainability with high social impact and enhanced soil health.

TCA Rapid Assessment

In each of the four capitals, Sistema Bio scores well on select services; for example, in Natural Capital, provision and regulating services of air and water were key. In Human Capital, improving farmer livelihood is front and center. The tool underscores the value of Sistema Bio's addition to Social Capital by educating farmers about the biodigester technology and income benefits associated with using biofertilizers for soil health, adding new dimensions because the current questions are focused on "workers" and do not explicitly include customers and other community members. In Produced Capital, recognizing that Sistema Bio works in loose (less controlled) value chains reinforces the value-addition of their business model.

Beneficial Returns as a fund manager found the TCA framework useful to compare one borrower with another, and as a common framework for team members to evaluate an investment opportunity.

Community Markets for Conservation (COMACO)

A Zambian public good, non-profit company since 2009, **Community Markets for Conservation (COMACO)** (https://itswild.org) works with 81 community cooperatives in the Luangwa Valley to provide incentives for biodiversity conservation, training, and support services to 188,500 small-scale farmers. This includes agroforestry, organic composting, minimum tillage, crop rotation, and water conservation strategies. They also run a business to manufacture and sell 17 different healthy, pesticide-free, organic, value-added foodstuffs under the brand *It's Wild!* Products including peanut butter, rice, wild honey, wild mushrooms, dried mango, a soy-based high protein snack, and breakfast cereal, among others.

Their landscape-level conservation approach works on four levels to: 1) engage farmers and former wildlife poachers via cooperatives; 2) ensure food security and improve nutrition; 3) increase individual income through processing and marketing of surplus crops and sustainably harvested wild foods; and 4) enhance biodiversity through payments rewarding collective Conservation Pledges, conservation area set-asides, and soil enhancement practices.

COMACO tracks impact through measuring crop productivity, income, participation, and engagement. Maize yields have improved two to three-fold, on average, and annual incomes for farmers have increased 450% from $79 in 2001 to $393 in 2019. Women represent 52% of farmers, and 76 former poachers now guard crops in elephant-friendly ways. They recently placed some 29,800 beehives in the community conservation areas.

Most important is how they achieve these impacts by ensuring food security. COMACO will not buy a farmer's production if she does not produce a surplus above what the family needs for its own consumption, but she will still get a cash payment in recognition of commitment to the community conservation district.

COMACO started with research grants examining the linkages between wildlife poaching, hunger, and food aid to ensure farm families' food security. COMACO asked farmers to sign a Conservation Pledge to reduce wildlife poaching. Today, grants finance training services and expansion strategies, while the food processing business sustains itself from sales in Zambia's major retail stores as well as in schools and hospitals. Carbon sales on international markets contribute to the payments for conservation. COMACO runs two distinct entities side-by-side (non-governmental organization and public good company) in order to separate their respective funding income and expenditures and accomplish their interrelated goals.

TCA Rapid Assessment

COMACO scores well on all four Natural, Human, Social and Produced Capitals. Their emphasis on increasing Human and Social Capital is particularly strong, with Social Capital as their greatest asset. Since they invest heavily in Human and Social Capital and return all profits within loose supply chains connected to farmer cooperatives, they experience capital constraints in terms of sourcing more investment to expand their approach to other regions.

The model is based on a conservation pledge that is only valid if an entire village signs on, increasing impact by its collective design. Increased pay for crop production is linked to: meeting household food security, adoption of organic farming techniques for increased food production, and the absence of poaching and forest threats. By linking adoption of improved farming techniques to enhanced household nutrition and income, using a lead farmer approach with farmer-to-farmer training as well as a cooperative economic model, COMACO puts farmer ownership at the heart of the work, an intentional strategy that reinforces linkages between different kinds of capital.

COMACO found that having the four areas of capital in one analysis was a helpful way to communicate the impact of the entire enterprise. Despite this, they worry that if a prospective investor only had the tool, they would not be able to weave together the complex self-reinforcing work strands into a meaningful story. A key to success for them is to understand what is working for communities. To what degree is young talent retained in rural areas? To what degree do women participate? How is the enterprise stimulating the environment around it in a way that further engages the communities? The feedback from COMACO points to the importance of expanding the tool to help investors consider how the outcomes in the Four Capitals impact the overall food and agricultural system in which it operates.

Root Capital

Founded in 1999, **Root Capital** (www.rootcapital.org) is a non-profit social investment fund growing prosperity for rural farmers in Latin America, Sub-Saharan Africa, and Southeast Asia by investing in the businesses that collect, aggregate, process, and market their crops. These businesses provide farmers with fertilizer, better seed varieties, and training on agricultural methods; connect farmers to international markets where their crops fetch a better price; and help farmers to achieve higher and more stable incomes. Root Capital links farmers to markets for sustainably produced goods (or "green markets"), provides assistance on sustainable farm practices, and promotes climate change mitigation and adaptation. Root Capital focuses on companies working with smallholder farmers in formalized, consolidated markets in tight value chains (Rural and Agricultural Finance Learning Labs, n.d.). with clear standards and specific contractual obligations, mostly involving high-value crops (e.g., coffee and cocoa cooperatives).

Through the first quarter of 2020 Root Capital provided financing to 726 businesses working with over 1.5 million smallholder farmers, including some half-a-million women. The businesses Root Capital reached paid nearly $5 billion directly to producers. Through on-site training, centralized workshops, and remote engagements, Root Capital has also trained 1,517 enterprises on strategic, financial, and operational skills (Root Capital, 2020).

In order to fulfill its high-impact mission, Root relies on a blend of creative investment capital, philanthropy, and guarantees. For its loan portfolio, Root Capital has solicited investment capital from over 200 institutional, public, and private investors, raising both concessional capital with a small return on investment (ranging from 0.5–2.5%) and patient, subordinated debt through a notes program. Root Capital also raises philanthropic equity that stays on the lending balance sheet to cover write-offs, as well as loan guarantees, which enable Root Capital to expand to new geographies and value chains. Finally, Root Capital raises grants for operational and non-lending programs, such as impact measurement, technical assistance, and training to agri-businesses, and building the impact investing field.

Root Capital focuses on increasing access to finance for agricultural enterprises that are locked out of traditional financial markets. Here, Root Capital uses the concept of financial "additionality" or of "investor contribution," which refers to the agri-business's ability to obtain a similar loan—a loan of similar size, for the same purpose, with similar collateral, and for a similar rate and fee—from another source, such as a commercial bank. Investors and enterprises that are working—and sharing power—with people and communities who are systematically shut out of financial systems are more likely to contribute to systems transformation.

TCA Rapid Assessment

Root Capital scores well in all four Capitals of the TEEBAgriFood framework. Root provides loans and training to agribusinesses who in turn promote sustainable farming practices that enhance Natural Capital, including soil and water quality, and improve ecosystem services. Root Capital's loans improve Human Capital by improving smallholder farmers' working conditions and incomes. The companies they invest in improve employee wellbeing, working conditions, skills and training, and provide employment security. Root Capital contributes to increasing Social Capital by strengthening ties between the agribusiness and their farmer suppliers. Root Capital increases Produced Capital by, for instance, strengthening farmer engagement in supply chains, and building financial infrastructure where none exists.

Root Capital seeks to first and foremost improve the prosperity of rural communities by partnering with agricultural enterprises that increase farmer incomes, create jobs, and contribute to ecosystem health. Root Capital applies a negative screen for egregious practices (e.g., child labor) and then applies a positive impact screen that enables Root to invest in enterprises that are higher risk, more challenging and highly additional and balance those investments with more stable and profitable loans.

Balancing risk and impact in a portfolio

Root Capital has been a pioneer in developing quantitative impact due diligence tools and facilitating investors' movement towards integrating impact into financial decision making. As part of its loan due diligence process, Root Capital compares each loan's prospective social and environmental impact to its risk-adjusted, expected financial returns to ensure that the portfolio effectively balances financial return and expected impact. Root Capital uses this method to improve their decisions around capital allocation and portfolio goal setting (McCreless, 2017; Impact Frontiers Collaboration, 2020).

Analysis

We found that the language of the TCA tool does not always resonate with the investors and entrepreneurs whom we interviewed. In particular, many

Table 15.1 Comparison of Four Capitals Across Different Enterprises

	Natural Capital	*Human Capital*	*Social Capital*	*Produced Capital*
Sistema Bio	Greenhouse gas emissions, food waste, biodiversity loss, provisioning (bio-inputs), soil enhancement	Farmer income and knowledge	Community wellbeing	Units produced, repayments, synthetic inputs
COMACO	Biodiversity loss, nutrient cycling/soil enhancement practices	Farmer well-being, farmer health/consumer health	Food security, social cooperation/cooperatives, collective values	Crop productivity, income, marketing
Root Capital	Provisioning (bio-inputs), soil quality, water quality, ecosystem services	Worker well-being, working conditions	Collective knowledge, training, supply chain relationships	Income, market access, financial infrastructure

This table provides a summary of these three entities' impacts in the four Capitals: Natural, Human, Social, and Produced. As the table shows, the use of the four Capitals enables comparison of their relative strengths as an enterprise or fund.

organizations are working to improve the lives of farmers, their families, and their communities, including their health and well-being. The use of "worker" as the sole term to describe impacts within Human Capital limits the way investors and entrepreneurs understand who is engaged in the "work." Also, the limited exploration of health outcomes such as nutrient density as a result of enhanced soil health, biodiverse local diets, or culturally relevant diets that reflect local practices in the Human and Social Capital section misses key outcomes.

Although the three examples—and many social enterprises—score in all four Capitals, the interactions among the four Capitals differ significantly in local, regional, and global food systems contexts. The TCA tools provide a starting point to explore the interrelationships between impacts in the four Capitals, but should be expanded to more deeply consider systems-level outcomes.

Systems transformation requires important—and difficult—work to strengthen Social Capital. We found that the current Rapid Assessment, while it incorporates questions related to social networks, shared norms, and collective knowledge and values, offers a limited view of Social Capital. For instance, it does not capture important and complex power dynamics between people, within and among communities, and between people and institutions. Finding opportunities for transformative change in local economics and social relationships is part of the genius of these enterprises and funds. These opportunities spring from a complex mix of social knowledge, innovative approaches, the creation or enhancement of markets, creative finance, and other factors which result from strong place-based knowledge, mutual respect, and community relationships.

Understanding the impacts and outcomes of enterprises and investment funds on complex systems requires consideration across Capitals, including the inter-relationships, interactions, and trade-offs across the Capitals. Future versions of the Rapid Assessment tools should create opportunities to explore potential and real transformational effects of investments in the food and agricultural sectors.

Synthesis

The TIFS initiative is developing practical tools that inform investment decisions by holistically considering the human, social, and environmental impacts of enterprises. In writing this chapter, we wanted to test how the TCA tool had additive value for investors and entrepreneurs working to transform food systems. As described above, the cases illustrate how they have combined diverse sources of capital to meet their missions and tailored ways of measuring their outcomes that traverse the four capitals in the model. Our early analysis reveals a mixed outcome for the true cost accounting framework underlying our Assessment tools. In order for the TCA tool to be more relevant for those in the investing and social entre-preneur community interested in transforming food systems, we will need to modify the TCA tool beyond its current form. Below, we outline three high-level considerations for continuing our work with the tools to make it more effectively benefit investors and entrepreneurs seeking systems transformation.

Comprehensive and Standard Frameworks Enable Insights: As demon-strated by the cases, the four Capitals framework enables a comprehensive look at how an enterprise or a fund incorporates each element: Natural, Human, Social, and Produced. Often, when enterprises or funds present themselves, they focus on one or two capitals without acknowledging the relevance of or their impact in the other areas. All of the interviewees noted that the four Capitals approach enabled them to be more inclusive in their self-assessment inventory. Several scored well on multiple dimensions of the four Capitals providing yet another point of comparison. For this reason, the TCA framework helps to build the case for both a comprehensive and a standardized framework that incorporates all four Capitals.

Discrete Metrics Downplay Holistic Analysis: Evaluating the four Capitals through a series of discrete metrics, by necessity, requires the simplification of complex relationships and feedback loops between factors that enable a social enterprise or investment to return value to the community and/or farmer. Our Rapid Assessment adaptation did not examine the interrelationships between catalytic elements in the cases, hiding a critical dimension of the analysis. In each of the cases, the enterprise/fund worked hard to incorporate elements within Produced and Social capital that would enable increases in the Natural and Human capital elements. Holistic analysis pays attention to the whole picture including the interrelationships and feedback loops between the four Capitals.

Transformational Nature of Enterprise/Fund: One of the key reasons for the creation of the TCA framework's four Capital approach is to highlight the extractive nature of an economic system that primarily values only Produced Capital. For enterprises and funds, this remains a critical challenge as they are attempting to create increases within the four Capitals while also being financially positive in contexts of historic and ongoing extraction of people, cultures, and nature, in areas of limited infrastructure, and ongoing political and economic power asymmetries. All of the cases add value through training, infrastructure, knowledge exchange, and engaging with farming communities, each of which requires time, new relationships, and investment—yet another hurdle for profitability. The cases remind us that food system transformation requires asking the uncomfortable question of: how should the profits from enterprises be distributed, and to whom? The TCA framework points to these issues and—with necessary improvements— can inform the development of tools that help investors, fund managers, entrepreneurs, and communities to give equal consideration to the four Capitals.

Conclusion

Frameworks like True Cost Accounting start to make concrete the mantra "what gets measured gets managed" to include non-financial attributes that do not yet have standardized and accepted measurements for return. In the future, tools such as integrated profit and loss statements can create standard approaches to measure companies' performance across the four Capitals. For examples of frameworks for integrated profit and loss statements, see Harvard University Impact Weighted Accounts Project (Harvard Business School, n.d.) and the Impact Institute (Impact Institute, n.d.). This is slow and deliberate work that requires different voices and competing interests to work together. However, investors and entrepreneurs need better tools now to make informed decisions and are not waiting for new accounting agreements. They need and want systems-based tools that inform them about how to place investments and demonstrate the value of their businesses that simultaneously address multiple outcomes and drive towards the transformation required for our food system to meet the synergistic needs of humans and the environment.

The TIFS Community understands the urgency of this work and is organizing partners to develop tools, information, and strategies to address the needs of investors and entrepreneurs to track the systemic and transformational outcomes of their work and the types of financing required for such change.

We offer our tools, community, and values in an effort to influence and persuade the broader field of impact investing to envision the real costs of finance and what it will take to change how those decisions are made. By engaging with investors and entrepreneurs who are making hard choices, doing the real work, and being innovative, we believe that our collective actions can influence capital flows toward those that are truly transforming food systems.

References

CNote. (2019). The History of Socially Responsible Investing. Availablae at: www.mycnote.com/blog/the-history-of-socially-responsible-investing/.

Global Alliance for The Future of Food. (2020). Principles and the Future of Food. Available at: https://futureoffood.org/wp-content/uploads/2020/03/Principles-and-the-Future-of-Food-2020-03-10.pdf.

Global Sustainable Investment Alliance. (2019). *2018 Global Sustainable Investment Review.*

Harvard Business School. (n.d.). Impact Weighted Accounts Project. Available at: www.hbs.edu/impact-weighted-acounts/Pages/default.aspx#:~:text=The%20mission%20of%20the%20Impact,%2C%20social%2C%20and%20environmental%20performance.

Impact Institute. (n.d.). Integrated Reporting and Integrated Profit Loss. Available at: www.impactinstitute.com/areas-of-expertise/integrated-reporting-and-ipl.

Mooney, A. (2020). ESG passes the Covid challenge. *Financial Times,,* June 1, 2020. Available at: www.ft.com/content/50eb893d-98ae-4a8f-8fec-75aa1bb98a48.

Root Capital. (2020). *Q1 2020 Quarterly Performance Report.* Available at: https://rootcapital.org/about-us/financial-information/.

RSF Social Finance. (n.d.). About Us. Available at: https://rsfsocialfinance.org/our-story/how-we-work/integrated-capital.

Rural and Agricultural Finance Learning Labs. (n.d.). Value Chain Development. Available at: www.raflearning.org/topics/value-chain-development.

Section 6

To the Table

Everyone eats, and there is tremendous power in the choices made by consumers. The three chapters in this section offer perspectives on mobilizing consumer knowledge and power as an important part of food systems transformation. How can True Cost Accounting (TCA) help consumers to make different choices? The authors in this section address three thorny issues in food systems: comparing foods created through different production practices; giving fair value to labor; and setting true prices.

In Chapter 16, Kathleen Merrigan uses TCA as an assessment framework to ask questions about new ways of producing meat and meat alternatives. If we are to meet our climate targets, then we need to eat less and better meat. But what does that mean in a practical sense, and how do we evaluate alternatives? Merrigan demonstrates how a TCA assessment can help us to ask new questions and balance key considerations. Here, TCA and the four capitals are used as a guide to thinking systemically and in an integrated way. For the issue of meat alternatives, Merrigan raises important questions about the ecological impacts of different meat production practices (not all are the same) and alternative meat products, their ingredients and health impacts, as well as impacts on livelihoods, bringing nuance to these debates.

Except for Chapter 5 (The Hidden Costs of Industrial Food Systems), there are very few contributions to this volume that address food sector labor, despite its global significance; this is a gap that needs filling. The SARS-Coronavirus-2 disease (COVID-19) pandemic has exposed the incredible precariousness of labor across the food system—from field to factory to table—and in Chapter 17 Saru Jayaraman and Julia Sebastian delve into the "true costs" of labor in restaurants in the USA. This chapter connects the historical legacy of slavery to low-wage work service sector work for workers of color and women in the food system. This structural racism and the associated structural inequities have been laid bare by the pandemic, resulting in skyrocketing unemployment rates and food insecurity. Although the authors focus on the true cost of dining out in restaurants, these structural inequities are worth examining across the food system.

Adrian de Groot Ruiz tells the story of the True Price Store in Amsterdam in Chapter 18. In the store, consumers are asked to reflect on the true price of common products like coffee, milk, apples, and chocolate. Are you willing to pay the true price and/or make a more sustainable choice? What happens when this information is laid bare for consumers? De Groot Ruiz leaves us with four calls to action: transparency (about the true price); transformation (changing production practices to reduce negative impacts); transaction (remediate external costs); and taxation (making sustainable products less expensive by taxing negative externalities).

These chapters speak to the potential of mobilizing consumers in a movement for true cost and true price. When considered together in this section, ("To the Table"), we see the potential for engaging consumers in important discussions about the true costs of their food choices, for building power across and between workers and eaters, and increasing awareness of citizens to demand change across our food systems.

16 Trade-Offs

Comparing Meat and the Alternatives

Kathleen A. Merrigan

In 2019 Impossible Foods won a United Nations Global Climate Action Award for production of a plant-based "climate-positive burger" (United Nations Framework Convention on Climate Change, 2019).

The Rise of Faux Meat

The case "against" animal meat is compelling. Overconsumption has been connected to heart disease, cancer, and obesity, among other health concerns. Several countries issue dietary guidance that advises reduced meat consumption, particularly red meat, with the Netherlands going so far as to advise its citizens to limit meat consumption to 500 grams per week, of which no more than 300 grams should be red meat (Kromhout *et al.*, 2016). Just as the human health implications of eating beef, pork, and chicken vary, so too do the implications of various production regimes that produce those foods. For example, a grass-fed beef operation produces, on a per cow basis, far less methane than a grain-fed beef operation; use of antibiotics varies by species (e.g., in the USA in 2018, 42% of medically important antibiotics were used in cattle, 39% in swine, and 4% in chicken) (US Food and Drug Administration Center for Veterinary Medicine, 2019). Despite many real differences that exist, altogether global meat consumption and production add up to alarming impacts on human and planetary health. And as bad as things are, they are projected to get worse. The dietary transition now underway, brought on by rising Gross Domestic Product and urbanization, is expected to result in increased global meat consumption and a corresponding 80% increase of greenhouse gas (GHG) emission from animal agriculture by 2050, among other life threatening impacts (Tilman and Clark, 2014). Given the valid concerns about meat, it is not surprising to see emergence of faux meat which expands choices for vegetarians/vegans and provides options for meat eaters.

What is Faux Meat?

The term "faux meat" describes a category of products that range from plant-based and other non-meat alternatives to laboratory produced cellular meat and is a logical extension of the historical usage of "faux fur" and "faux leather," both terms describing non-animal-based alternatives that came into vogue in response to consumer demand.

Start-up companies are introducing a range of faux meat products that are gaining market traction in the USA, Europe, Australia, New Zealand, Canada, Israel, and increasingly in Russia and China. While this trend has not significantly impacted overall meat consumption in the developed countries where they are sold (e.g., per capita meat consumption in the USA in 2018 was 221.1 lbs.)—an increase for the fifth year in a row—and was projected to rise to 222.6 lbs. in 2020 prior to the coronavirus disease (COVID-19) pandemic (Food and Agricultural Policy Research Institution, 2020)—consumer interest in faux meat is growing. So too is the range of product offerings. To provide a glimpse of this emerging market, this chapter considers the spectrum of faux meat burgers that range from traditional plant-based burgers (e.g., Beyond Meat) to those genetically engineered to mimic the taste of meat (e.g., Impossible Foods) to cellular options not yet commercially available (e.g., Memphis Meats).

We begin with a brief description of faux meat burgers by category. Plant-based faux meat burgers have been in the market for many years, and the ingredients vary. Plant-based burgers have mostly been made of soy, with several companies choosing to use and label products as made from non-genetically modified organism (GMO) soy protein. In addition, wheat, chickpeas, black beans, mushroom, and pea protein are commonly used, as are, to a lesser extent, jackfruit, oats, algae, and seaweed. As a highly processed food, the plant-based Beyond Meat burger has a total of 18 ingredients, including 380 mg of sodium.

A more recent development is using genetic engineering to create a heme-enriched faux meat as represented by the Impossible Burger, a first in class product with significant intellectual property behind its processes. (As mentioned, it was recognized by the United Nations with a Global Climate Action Award). The Impossible Burger is a highly processed food with a total of 17 ingredients, including 370 mg of sodium. The distinguishing attribute in this category of faux meat is the introduction of heme—the molecule that gives blood its red color and which is abundant in animal muscle, creating the taste that we associate with meat. Heme also exists in plants, particularly legumes. Impossible Foods has succeeded in deriving heme from leghemoglobin—the protein found in nodules attached to the roots of soybeans which is extracted through a process using genetically engineered yeast.

Cellular meat (also known as clean meat, in vitro meat, lab-grown meat, or cultured meat) is still under development, although cell cultures have been used for a long time to produce food enzymes (e.g., microbial rennet for cheesemaking), food ingredients (e.g., monosodium glutamate [MSG]), vitamins (e.g., B12) as well as flavors and fermented foods and beverages (Stephens *et al.*, 2018). The first lab-grown burger was developed in the Netherlands in 2013, at a cost of €220,000; in the USA, Memphis Meats, founded in 2015, has attracted significant venture capital and is finding ways to reduce the cost of producing cellular meat, but it is far from commercially viable. The high cost mostly stems from the use of fetal bovine serum (FBS) which is extracted from cow fetuses and then mixed with growth-inducing proteins, with companies keeping the exact composition of their serum processes secret (Reynolds, 2018). Companies are racing to find ways to forgo use of FBS; Just Inc. has developed a method to grow cultured chicken meat without FBS, but there are no announcements yet about red meat. Because it is extremely difficult to develop cellular meat that mimics complex muscle meat, like steak, expectations are that the first commercialized products will be meat fillers/ingredients and likely burgers. An interesting twist may be 3D food printing: cultured cells (along with added flavor, vitamins, and iron) are the ingredients and a 3D printing technology merges fat and tissue to produce a cut of meat that mimics what consumers see in grocery stores. In 2018 Novameat succeeded in printing an entirely plant-based steak through this method (Shieber, 2019).

Together with start-ups, big business is engaged across the spectrum of faux meat innovation. The venture arm of the iconic chicken giant, Tyson Foods, has invested in Beyond Meat, Memphis Meats, and Future Meat Technologies. According to CB Insights, Tyson is looking to pivot from being solely a meat producer to also having plant protein brands (CB Insights, 2020). Cargill has invested in Memphis Meats. Fast food giants Burger King and White Castle are serving the Impossible Burger; McDonald's is testing Beyond Meat burgers in Canada and Nestlé's Awesome Burger in Germany; Unilever has acquired the Vegetarian Butcher; Starbucks has added meat alternatives to its breakfast menu; and Sysco Corp is introducing the soy based Simply Burger.

The US Plant Based Food Association (a trade group representing nearly 200 plant-based food companies) estimates that the US plant-based meat category in 2019 was worth $939+ million, with 2019 sales up by 18% overall, and the driver for this market—refrigerated plant-based meat—was up by 63% (Plant Based Foods Association, 2020). This is compared with a 2.7% growth rate in the conventional meat category during the same period (Plant Based Foods Association, 2020). Overall, the association claims that in 2019 plant-based meat accounts for 2% of retail packaged meat sales in the USA. The faux meat trend is also reaching into Asia. In 2017 China announced a $300 million deal to import cellular lab-grown meat from three Israeli-based companies as part of a larger plan to decrease the country's meat consumption by 50% (CB Insights, 2020).

Exploring the True Cost of Faux Meat

Comparing "real" meat with faux meat is complicated (Santo et al., 2020; Ritchie *et al.*, 2018, Smetana *et al.*, 2015). First, the faux meat industry is nascent, with much room for innovation, so today's analysis of the subsector might not prove true tomorrow. For example, pea protein costs are high because once the protein is extracted, there is significant unused byproduct, but this could change. Second, the wide range of current livestock practices means that there is a corresponding wide range of true costs for the various kinds of operations. For example, we can expect the true cost of a rotational grazing beef operation to vary considerably from a beef confined animal feeding operation (CAFO). Third, while animal agriculture is not new, it too is innovating. The potential to feed cattle seaweed (to reduce methane emissions) and insects (to replace forages grown with pesticides) are just two examples under development. Undertaking a full True Cost Accounting (TCA) of faux meat is beyond the scope of this chapter, as such an effort will require transdisciplinary teams tackling time-consuming and complicated research that pairs and analyzes specific faux meat products and production processes to various kinds of meat products and production regimes. But it should be done, and there are clear advantages to doing so now before the faux meat industry takes hold, for better or worse.

This chapter is intended to illuminate the potential impacts of a shift, of any magnitude, from animal agriculture to faux meat production and consumption. To simplify potential considerations and identify, at a very high level, major issues for consideration by the four capitals, the following discussion focuses on beef and faux meat substitutes for beef. Similarly, comparisons could be made between livestock and faux chicken, faux pork, and faux seafood, all of which are either on the market or expected soon.

Natural Capital

The 2019 International Panel on Climate Change (IPCC) Special Report attributes 21–37% of total GHG emission to the overall food system (International Panel on Climate Change, 2019), with greater precision in the attribution to crop and livestock production, estimated as 9–14% within the overall food system total (Mbow *et al.*, 2019). Methane from ruminants is of greatest concern, and for this reason, beef and dairy CAFO operations have an especially high contribution. As stated above, there is a need to differentiate costs between the dominant production method of feedlots with various kinds of grass-fed operations, as GHG impacts will vary. A report by the Land Stewardship Project finds that shifting 25% of ruminants to well-managed grazing operations and 25% of cropland to perennial cover, diverse rotations, and cover crops could offset US GHG emissions by as much as 9% (Boody, 2020).

There is an absence of serious critique of plant-based burger ingredients. For crops used in plant-based meat as well as cattle feed, common use of pesticides and nitrogen fertilizers, and their resulting pollution, must be factored in. It has been suggested that fertilizer that is used to support grain-fed animal agriculture generates nearly twice as much nitrous oxide for crops destined for direct human consumption owing to the double whammy of crop fertilization for animal feed and disposal of manure in concentrated livestock operations (Davidson, 2009).

Land use is a consideration for both faux meat and beef, and in some cases, the impact could be related, as much of cropland used to produce cattle feed is also used to produce the crops that become core ingredients in plant-based meat. Cellular meat does not require significant land, and as a result, it might be possible to repurpose land that is currently in production or let it lie fallow if cellular products become a significant source of protein. Yet 40% of global terrestrial land, because of lack of moisture, steepness, and/or heat, is best suited for animals that convert plant materials indigestible for humans into meat.

Water use is a factor in faux meat and beef. As with energy, embedded water needs to be assessed for each ingredient in faux meat and compared with embedded energy and water in beef produced in various production systems.

Produced Capital

Soil health promotion to sequester carbon is widely discussed as a potential strategy to combat climate change and reward farmers for environmental stewardship. For example, practices such as crop rotation, cover cropping, and adaptive paddock management contribute to healthier soils. A 2018 National Academy of Sciences report (The National Academies of Sciences Engineering Medicine, 2019) suggests the potential to remove 250 million metric tons of carbon dioxide per year in the USA alone. Researchers are devising methodologies to measure soil carbon while policy designers and farm advocates are debating market mechanisms to reward ecosystem services to financially reward regenerative agriculture. US-based Indigo Ag is signing up farmers to sell carbon credits; The Nature Conservancy raised $20 million to set up a carbon marketplace. The production of soil carbon and related financial rewards may variously apply to plant-based and heme-infused faux meat (depending on cropping practices) as well as animals produced in regenerative systems.

Profits will be had by companies and shareholders of successful faux meat companies. However, faux meat products currently sell at a significantly higher retail price on a per-pound basis than beef and might be out of reach for many consumers in developed countries and out of sight in developing nations. Looking way into the future, if we consider the cost of faux meat made from cereal crops, such comparisons could shift: the IPCC projects that, based on several models, cereal prices could increase from between 1–29% in 2050 owing to climate change, which could greatly impact the cost.

Energy needs to produce cellular meat are huge—well beyond any other faux meats or beef production. This is even true when considering the embedded energy in feedstuffs for cattle, with the typical conversion rate of six pounds of grain to produce one pound of meat gain. Methane digesters are used by some livestock producers to dispose manure, capture methane, and sell energy to the electrical grid and/or power their operations.

Human Capital

Health factors must be evaluated. Consumption of faux "red" meat is generally comparable to beef in terms of calories and saturated fat, although it is somewhat higher in carbohydrates. While faux meat lacks many vitamins and minerals found in beef, because it is processed, many of these missing components can be, and are, added as ingredients. A significant health concern is sodium, with significantly higher sodium levels across the entire faux meat category compared with real meat with no sodium other than what might be added in cooking. As for beef, the biggest health concern comes from eating it raw or undercooked, risking exposure to E. Coli, which makes people sick and, in worst cases, causes death. Of course, overconsumption of both faux meat and beef lead to other health impacts. In terms of production, faux meat does not entail the use of hormones and/or antibiotics (with the possible secondary impact on cow fetuses). Much of faux meat is made from row crops (e. g., soy, wheat), meaning that core ingredients are typically grown with synthetic pesticides and fertilizers (as pointed out in the Natural Capitals subsection above), which can end up in food and water and can have negative health impacts on the farm and beyond. Growing animal feed could cause similar negative impacts (e.g., corn, sorghum, alfalfa). Notably, GMO corn and soybeans are produced using glyphosate, classified by the World Health Organization as "probably carcinogenic to humans" (World Health Organization International Agency of Research on Cancer, 2017) and the subject of legal suits brought by pesticide applicators suffering from Non-Hodgkin Lymphoma disease (Baum Hedlund Aristei Goldman, 2020).

Employment shifts are likely if the faux meat market continues to grow. Smaller livestock operations are vulnerable and likely to shut down in declining markets (as seen in US dairy, which has lost market share to faux milk products (Sitzer, 2019)) and market consolidation is often accompanied by CAFO expansion and decreased competition. We can expect faux meat companies to be geographically concentrated, likely in peri-urban areas. This could contribute to the problem of declining opportunities in rural places and, certainly, a decline in the quality of life among independent farmers and ranchers, even if they are able to find jobs "in town." Faux meat production will likely require employees with different skills (e. g., molecular engineering), and in some cases, these jobs might pay higher wages than farm and ranch work. Finally, and ironically, a shift to faux meat production could be viewed as a shift from farm to factory in the face of social push-back against factory farming.

Social Capital

Animal welfare is dramatically different between meat and faux meat. Although there is a wide range of livestock practices related to animal welfare, with some operations achieving recognition for humane care, some consumers nonetheless reject outright all livestock reared for human consumption. For such consumers, and in the many cases of operations that compromise animal welfare, plant-based and heme-infused faux meat is clearly superior. However, with the use of FBS, cellular meat remains tethered to traditional livestock systems with its reliance on extraction of bovine fetuses and, as such, will not satisfy all consumers.

Quality of religious life could be a factor. Kosher and Halal dietary law dictate animal slaughter requirements together with other rules related to what animals can be consumed and when. Most plant-based meat is consistent with Kosher and Halal rules, and the Impossible Burger has been certified as Kosher and Halal (Impossible Foods, n.d.). However, faux meat, particularly cellular forms, is challenging old frameworks, leaving religious leaders pondering how these new technologies fit. Could there be a future for Kosher pork and shrimp? There is a possibility that faux meats could enhance food options for certain religious groups.

Governance of faux meat is murky. In the USA, for example, the meat industry is lobbying state legislatures to secure laws that prevent the term "meat" from appearing on faux meat labels. At the federal level, oversight of faux meat is shared by the US Food and Drug Administration and United States Department of Agriculture, with leaders of both institutions side-stepping contentious debates, creating regulatory uncertainty. To the extent that faux meat involves genetic manipulation of any sort, it is unclear whether and how such products would fit into regulatory frameworks for biotechnology across the globe. As the faux meat industry grows, it is reasonable to expect friction over governance within countries and between countries, with potential trade conflict emerging from different approaches to these novel products. This is not to suggest that the livestock sector is exempt from governance concerns (e.g., the years of wrangling in the CODEX Alimentarius Commission over use of ractopamine hydrochloride as an animal growth promoter (Farm and Dairy, 2012)), but the pathways for resolving issues is far clearer.

There are many associations—local, national, and international—that engage farmers and ranchers, providing them support and a sense of community and identity. Farm and ranch life can be isolating, particularly in remote areas, and associations and networks, together with related social activities (e.g., rodeos, livestock auctions) are critical to wellbeing and, often, business success. It is unlikely that such networks and associations will play a similarly significant role in the emerging faux meat industry, which will likely be concentrated in business centers and not contribute substantially to the quality of rural life.

Conclusion

While faux meat is an interesting development and holds promise, it is premature for faux meat champions to declare a sustainability victory. While not possible in this brief chapter, to ultimately declare which product(s) or category of "meat" is best, the goal has been to surface the kinds of questions that TCA would necessarily address and to build support for the kinds of analysis that provide data-infused insights currently absent from decision-making processes. Furthermore, this discussion has been focused on the developed world. A parallel discussion is necessary to consider the two-thirds of rural households globally, many of them poor and food insecure, whose well-being relies on livestock. It is time to apply TCA methodology to faux meat innovations to determine their true cost at a global scale across the four capitals, as identified by the TEEBAgriFood framework.

References

Baum Hedlund Aristei Goldman. (2020). Monsanto Roundup Trial Schedule. Available at: www.baumhedlundlaw.com/toxic-tort-law/monsanto-roundup-lawsuit/monsanto-court-papers/monsanto-roundup-trial-schedule.

Boody, G. (2020). Farming to Capture Carbon & Address Climate Change Through Building Soil Health: How Well-Managed Grazing & Continuous Living Cover Benefit the Climate, Our Waters, Farmers & Taxpayers Through Improved Soil Health [White Paper]. Land Stewardship Project. Available at: https://landstewa rdshipproject.org/repository/1/3143/grazing_white_paper_1_23_20.pdf.

CB Insights. (2020). Plant-Based Meat Industry: Global Meat Market's Meatless Future [Research Brief]. CB Insights Research. Available at: www.cbinsights.com/research/ future-of-meat-industrial-farming.

Davidson, E. (2009). The contribution of manure and fertilizer nitrogen to atmospheric nitrous oxide since 1860. *Nature Geosci*, 2, 659–662. doi:10.1038/ngeo608.

Farm and Dairy. (2012). Codex approves international standards for ractopamine. Farm and Dairy. Available at: www.farmanddairy.com/news/codex-approves-international- standards-for-ractopamine/39119.html.

Food and Agricultural Policy Research Institute. (2020). Baseline Update for U.S. Agricultural Markets. June 2020. Available at: www.fapri.missouri.edu/wp-content/ uploads/2020/06/2020-June-Update.pdf.

Impossible Foods. (n.d.). FAQ: Is it Halal or Kosher? Impossible Foods. Retrieved August 6, 2020. Available at: http://faq.impossiblefoods.com/hc/en-us/articles/ 360034765814.

Intergovernmental Panel on Climate Change. (2019). Summary for Policymakers. In P. R. Shukla*et al.* (Eds.). *Climate Change and Land: An IPCC Special Report on climate change, desertification, land degradation, sustainable land management, food security, and greenhouse gas fluxes in terrestrial ecosystems*. Intergovernmental Panel on Climate Change. Available at: www.ipcc.ch/site/assets/uploads/sites/4/2020/02/SPM_Updated- Jan20.pdf.

Kromhout, D., Spaaij, C.J.K., de Goede, J., & Weggemans, R.M. (2016). The 2015 Dutch food-based dietary guidelines. *European Journal of Clinical Nutrition*, 70(8), 869– 878. doi:10.1038/ejcn.2016.52.

Mbow, C., Rosenzweig, C., Barioni, L.G., Benton, T.G., Herrero, M., Krishnapillai, M., Liwenga, E., Pradhan, P. Rivera-Ferre, M.G., Sapkota, T., Tubiello, F.N., & Xu, Y. (2019). Food security. In P.R. Shukla*et al.* (Eds.), *Climate Change and Land: An IPCC Special Report on climate change, desertification, land degradation, sustainable land management, food security, and greenhouse gas fluxes in terrestrial ecosystems*. Available at: www.ipcc.ch/site/assets/uploads/sites/4/2020/02/SRCCL-Chapter-5.pdf.

Plant Based Foods Association. (2020). Retail Sales Data. Plant Based Foods Association. Available at: https://plantbasedfoods.org/marketplace/retail-sales-data/.

Reynolds, M. (2018, March 20). The clean meat industry is racing to ditch its reliance on foetal blood. *Wired UK*, March 20. Avaiable at: www.wired.co.uk/article/scaling- clean- meat-serum-just-finless-foods-mosa-meat.

Ritchie, H., Reay, D.S., & Higgins, P. (2018). Potential of Meat Substitutes for Climate Change Mitigation and Improved Human Health in High-Income Markets. *Frontiers in* Sustainable Food Systems, 2. https://doi.org/10.3389/fsufs.2018.00016.

Schieber, J. (2019). Novameat has a platform for 3D-printing steaks and has new money to take it to market. *TechCrunch*, September 5. Available at: https://social.techcrunch. com/2019/09/05/novameat-has-a-platform-for-3d-printing-steaks-and-has-new-money- to-take-it-to-market.

Sitzer, C. (2019). US milk sales drop by more than $1 billion as plant-based alternatives take off. *World Economic Forum*, April 2. Available at: www.weforum.org/agenda/ 2019/04/milk-sales-drop-by-more-than-1-billion-as-plant-based-alternatives-take-off.

Smetana, S., Mathys, A., Knoch, A., & Heinz, V. (2015). Meat alternatives: Life cycle assessment of most known meat substitutes. *The International Journal of Life Cycle Assessment*, 20(9), 1254–1267. doi:10.1007/s11367-015-0931-6.

Stephens, N., Di Silvio, L., Dunsford, I., Ellis, M., Glencross, A., & Sexton, A. (2018). Bringing cultured meat to market: Technical, socio-political, and regulatory challenges in cellular agriculture. *Trends in Food Science & Technology*, 78, 155–166. doi:10.1016/j.tifs.2018.04.010.

The National Academies of Sciences Engineering Medicine. (2019). Negative Emissions Technologies and Reliable Sequestration: A Research Agenda. The National Academies Press. doi:10.17226/25259.

Tilman, D. & Clark, M. (2014). Global diets link environmental sustainability and human health. *Nature*, 515(7528), 518–522. doi:10.1038/nature13959.

United Nations Framework Convention on Climate Change. (2019). Winners of the 2019 UN Global Climate Action Awards Announced [Press release]. United Nations Climate Change, September 26, 2019. Available at: https://unfccc.int/news/winners-of-the-2019-un-global-climate-action-awards-announced.

United States Food and Drug Administration Center for Veterinary Medicine. (2019). 2018 Summary Report On Antimicrobials Sold or Distributed for Use in Food-Producing Animals (Table 4, pp. 19, pp. 49). Food and Drug Administration. Available at: www.fda.gov/media/133411/download.

World Health Organization International Agency for Research on Cancer. (2017). Some organophosphate insecticides and herbicides (IARC Monographs No. 112; Monographs on the Evaluation of Carcinogenic Risks to Humans). Available at: www.ncbi.nlm.nih.gov/books/NBK436774/.

17 Dining Out

The True Cost of Poor Wages

Saru Jayaraman and Julia Sebastian

Introduction and Background

One of the greatest misunderstood "true costs" of meals in restaurants is labor. This is largely because the service sector is one of few industries in which the majority of the labor cost is not reflected in the meal, but instead paid in tips. For an industry that is disproportionately composed of low wage workers, workers of color and women, the coronavirus disease (COVID-19) crisis and national uprising for racial justice has exposed the untenability of this system of compensation and, in general, the deep structural inequities of the service sector. Although the COVID-19 pandemic has created an acute reality of economic peril for restaurant workers, the current situation is simply a heightened reflection of the precarities that perpetually underlie the restaurant industry. Although this chapter shines a light on the situation facing US restaurant workers during the pandemic of 2020, it also points toward how ill-prepared the industry is to provide for the basic necessities of its workers in times of economic crisis.

Prior to the pandemic, there were more than 13 million restaurant workers and nearly 6 million tipped workers across the USA, including restaurant (who account for over 80% of tipped workers), car wash, nail salon, tech platform delivery, and other workers, according to the Bureau of Labor Statistics in 2019. The National Restaurant Association had long argued that, given customer tips, businesses should be able to pay their tipped employees a subminimum wage, today just $2.13 an hour federally. A legacy of slavery, the subminimum wage for tipped workers today is also a gender equity issue. 70% of tipped workers are women, disproportionately women of color, who work in nail and hair salons and casual restaurants like IHOP and Denny's, live in poverty at three times the rate of the rest of the US workforce, and suffer from the worst sexual harassment of any industry because they are forced to tolerate inappropriate customer behavior in order to feed their families in tips ("Tipped Over the Edge", 2012). On top of this, research shows that workers of color earn less in tips than their white counterparts due to pervasive racial bias (Lynn *et al.*, 2008). Indeed, the voluntary nature of tips means that the true cost of labor is neither reflected in worker's wage nor in the cost of the meal, but rather is paid based on the whims and biases of customers.

Seven states—California, Oregon, Washington, Arkansas, Minnesota, Nevada, and Montana—have rejected this legacy of slavery and instead pay One Fair Wage: a full minimum wage with tips on top. According to a National Restaurant Association industry report, these states have comparable or higher restaurant sales per capita, job growth among tipped workers and the restaurant industry overall, and tipping averages than the 43 states with lower wages for tipped workers. They also claim half the rate of sexual harassment in the restaurant industry ("The Glass Floor", 2014).

The pandemic-induced economic collapse has affected few other industries more deeply than restaurants and food service. As mayors and governors across the country ordered shutdowns, and customers ceased dining out, restaurant owners shuttered their doors, and workers scrambled to join digital unemployment lines. Emerging national surveys of the restaurant sector show that four out of ten restaurants have closed their business, and the industry is predicting over $240 billion in financial losses ("Industry Research", 2020). By May 2020 the Bureau of Labor's State Occupational Employment and Wage Estimates reported nearly 6 million lost jobs across all food services and drinking places. According to the US Private Sector Jobs Quality Index, however, nearly 10 million low wage jobs in the restaurant and bar industries are at risk due to the COVID-19 fall-out ("Statement from JQI", 2020). Furthermore, it is workers of color and women, who disproportionately comprise the sector, who have been most acutely affected. The national unemployment rate for Black workers had risen to 16.7% from the outbreak, compared with 14.2% for white workers, according to the Bureau of Labor Statistics Current Population Survey. When we consider the impact on Black women particularly, the unemployment rate rises to 16.9% compared with only 12.8% for white men. Latina women, however, experienced the highest rate of unemployment, as nearly one in five are out of work (Gould, 2020).

It is the subminimum wage for tipped workers that has exacerbated the sheer destitution facing the millions of tipped workers who have lost their jobs during the COVID-19 crisis. During this unprecedented economic cliff, with unemployment rates surpassing those during the Great Depression, analysis conducted by the civil society organization One Fair Wage (founded by the author) shows that, on average, states are rejecting 44% of unemployment claims ("Locked out by Low Wages," 2020). However, surveys of tipped workers from this same research reveals that this statistic is closer to 60% for tipped service workers. This higher denial rate is in large part because workers are being told that their subminimum wage plus tips is too little to meet minimum income thresholds to qualify for benefits. In other words, these workers are being penalized because their employers paid them too little. Even among those who are eligible, unemployment insurance is being calculated based on the subminimum wage plus tips, and generally, this is an underevaluation of their tips.

Interviews with workers across the country are exposing the unjust intersection of subminimum wages, tipping, and the unemployment system. Charles

Almanza, a New York bartender and son of Nicaraguan immigrants, exemplifies the cruelty of getting caught in the unjust cross hairs of failed public policy. After the sudden closures of restaurants and nightclubs in New York, Charles filed for unemployment, only to discover that his W2 form stated that he had made only $5,000 over the seven months that he worked for his previous employer. Even though Charles knew his pay checks excluded any base wage, instead forcing him to live entirely off of tips (a common situation in states like New York where employer wages are negligible compared with tips) Charles did not know that his boss was also underreporting his tips. As a consequence, after six weeks of waiting for his unemployment check, it amounted to less than $300 a week. Charles's story is but one in a sea of emerging workers whose experiences have mobilized them to demand a more fair and dignified wage system. Millions of workers find themselves now unable to pay for rent, food for their children, or other bills. In fact, findings from One Fair Wage's research shows that 89% of nationally surveyed tipped workers report that they are either unable or unsure whether they can afford to make their rent or mortgage payment during this time. On top of this, 79% of surveyed service workers report being able to afford groceries for only up to two weeks or less. And now at a time when their family is most at risk, hundreds of thousands of tipped workers are being asked to return to work for the tipped workers' subminimum wage at a time when tips have dramatically declined—according to some employers, by as much as 75%.

Years of research demonstrating that workers of color earn less in tips owing to customer bias has now become painfully clear on a larger structural level—workers of color are disproportionately being denied unemployment insurance because they are more likely than white workers to have worked in casual restaurants where they received their tips in cash, and state unemployment insurance systems are automatically denying these workers because their incomes appear to be too low to meet the minimum threshold to qualify.

With tips drying up, workers are demanding a labor model in which the value of their labor is reflected in their wage, which would require employers and consumers to consider food service workers' labor like those of those of other workers—as part of the cost of the product, not as a separate, voluntary donation made by consumers.

Prior Initiatives for Change

Prior to the pandemic, a set of leading employers had worked voluntarily to move to One Fair Wage despite the fact that their state did not require it. These employers transitioned to a One Fair Wage compensation model through one of three ways.

First, these employers instituted a full minimum wage with tips on top and then shared tips among all non-management employees in the restaurant, allowing for a more equitable balance between back of house and front of house employees. Paying employees the full state minimum allows restaurant owners to redistribute tips both to kitchen and front of house staff even if the kitchen does not have direct contact with the customer. This model is contrary to one in which tipped workers receive a subminimum wage and thus legally must retain all tips in order to offset their low wages. In 2018 we worked with United States Congress Members to pass a rider to the Congressional budget bill that allowed employers who pay the full minimum wage to all workers the opportunity to permit tips to be shared among kitchen staff as well. Tip sharing with dining room staff has been customary in the seven One Fair Wage states for decades; the practice creates greater equity and unity between kitchen and dining staff and allows for cross-training between positions, allowing greater flexibility for the owner and mobility for workers.

A second initiative pursued by employers has been to move to a full minimum wage with additional income in the form of a service charge, which is also shared among all non-management employees. Finally, the third pathway involved employers moving to an entirely gratuity-free model, incorporating all tips and gratuities into workers' wages and thus into the cost of the meal.

Several employers who have implemented or contemplated these changes have found that, in many cases, by incorporating the true cost of food service labor into the cost of a meal, consumers have opted to dine at another restaurant that continues with the subminimum wage labor model. Especially for restaurants that chose a gratuity free model and thus the highest menu prices, they found that consumers could not understand that the labor cost typically paid out as a tip was now being incorporated into the actual menu and was thus costing the consumer the same overall amount. The fact that other restaurants were not incorporating the true cost of the labor into the cost of the meal meant unfair competition. This occurs, of course, in the context where consumers remain undereducated about the true cost of labor and tipping, as well as the negative externalities of a subminimum wage model that is a legacy of slavery and a source of discrimination and harassment for millions of workers of color and women nationwide.

One of the major challenges has been demonstrating to employers a change in consumer understanding and increased consumer support for employers willing to change their practices. It has thus been historically challenging to convince more employers to move away from the subminimum wage for tipped workers without being able to demonstrate a change in consumer understanding.

The Pandemic as a Portal for Change

Now with the COVID-19 pandemic and uprisings for racial justice, there is an opportunity for workers and employers to transform their industry so that labor costs are better incorporated into the cost of a meal, and so that consumers are informed and willing to pay the true cost of food service labor. The moment has provided opportunities to pilot new solutions which have shown that we can simultaneously support workers and ensure that responsible restaurant owners who care about their workers survive the crisis—and thus reshape the service sector, going forward. Significant economic and cultural shifts have brought a new set of restaurant owners who previously opposed or were hesitant about One Fair Wage forward, who are now showing willingness to commit to One Fair Wage and increased equity. For some, their eyes have been opened to the unsustainability of the system; for others, the moment has allowed them to break free from the confines of an old business model. Some are even working with us to design model restaurants.

Dan Simons, co-owner of the Farmers Restaurant Group, has seized upon the unimaginable shifts in the restaurant industry to work with One Fair Wage and its network of restaurants who lead in ethical labor practices.

Box 17.1

Before COVID-19, tipped employees at all nine full-service Founding Farmers' restaurant locations (based in Washington DC, Maryland, Virginia, and Pennsylvania) had been paid the subminimum wage for tipped workers in their state. However, the pandemic forced Simons to close dine-in service at all locations, resulting in the layoff of nearly 1,100 workers. The closures spurred Simons to build out a new market and grocery business model in addition to the restaurant take-out business. To operate his new business, Simons decided to rehire employees at the full minimum in each state.

As employees' positions changed from server to curbside deliverer, busser to grocery bagger, everyone became unified into a single team, all paid the same base wage. Simons is using a contribution from consumers that is similar to a gratuity charge and sharing it among all non-management employees on top of a full minimum wage. As he moves toward a full re-opening, Simons is testing a model that would include the previous menu price and a detailed break-down of the how the service charge will be used to cover employees costs such as for personal protective equipment (PPE), health insurance premiums, and additional employee benefits and safety supplies to address pandemic safety protocols. As Simons explains, "It's about building new compensation structures and new business models for the world we are in. For example, of course we need to provide employees PPE while certainly not making it a cost to the employee. Perhaps we can include both a fixed service charge and fixed COVID charge, which allows you to use that money as the business needs to protect our employees." Simons knows that it is critical to educate customers about what portion of the additional charge is going to pay a full living wage, to provide PPE or going to pay employee health insurance. He wants his customers to know the true cost of a meal and support the societal benefits that it brings to the essential service workers who are feeding us in times of crisis.

In this time of political opening, One Fair Wage has also partnered with state and local legislators to innovate new solutions towards simultaneously meeting the needs of workers, employers, and consumers. Based on conversations with restaurant leaders like Simons and others, we have developed a partnership with New York City and California governments to launch High Road Kitchens—a program in which restaurants provide meals on a sliding scale to low-wage workers, health care workers, and first responders, while also receiving financial support towards restaurant workers and responsible restaurant owners. Participating restaurants voluntarily commit to move to One Fair Wage and institute greater race and gender equity policies and practices by next year. In exchange for joining High Road Kitchens, restaurants will receive public and private dollars to re-hire their workers and re-purpose themselves as community kitchens to provide free meals to those who need them. The program is now likely to be replicated in Massachusetts and Michigan. Such a program seeks to provide both relief to struggling independent restaurant owners, free meals to workers and others in need, and most importantly, re-shaping the sector toward equity.

In this time of reflection of the impact of the pandemic and its impact on restaurant workers, it may be possible to leverage moments in which the greater public is gaining awareness of the true value of the workers who feed us in order to push forward more sustainable and equitable business practices. There is a new understanding among consumers about the "essential" nature of these workers and the ways in which these workers' health and well-being is directly connected to consumer health and well-being. Indeed, networks of restaurants around the country are beginning to coalesce to educate consumers about the dual benefits of consumer and worker health. Good Works Austin (GWA) provides one such example as a collaborative of around 30 restaurants in Austin, Texas that has collectively designed and committed to a series of protocols for how to safely and ethically reopen after COVID-19. Dedicated to worker and consumer health and safety, all restaurants that abide by these guidelines will receive promotional materials, as well as be a certified member of the GWA network. This project works both to provide consumers with a safe dining alternative while also educating them about the real need for worker health and safety.

As a result of the national uprisings for racial equity, there is also a new appreciation of the need to end historical legacies of slavery and address structural racism in every facet of American society, including in the ways in which workers are paid and treated. Restaurants around the country are newly reaching out to One Fair Wage's Restaurants Advancing Industry Standards in Employment (RAISE) network to receive coaching on how to transition to more racially equitable models around wage compensation models, as well as recruitment, hiring and promotions. It is incumbent upon consumers to support this shift as restaurants restructure their menu pricing to reflect the real cost of producing a meal. There is thus a moment of opportunity to build upon that new consumer awareness to educate consumers and engage them in supporting restaurants that are committed to change.

The pandemic and the global reckoning with race is both the gravest crisis in the service sector's history in the U.S. and also the greatest moment for transformation – for building power among workers and change among employers toward a sustainable future of equity and collective prosperity in which everyone understands and appreciates that the true cost of dining out must include the value of the skilled labor that produces and serves our meals.

References

The Glass Floor: Sexual Harassment in the Restaurant Industry. (2014). *The Restaurant Opportunities Centers United*. Available at: https://chapters.rocunited.org/wp-content/uploads/2014/10/REPORT_The-Glass-Floor-Sexual-Harassment-in-the-Restaurant-Industry2.pdf.

Gould, E. & Wilson, V. (2020). Black workers face two of the most lethal preexisting conditions for coronavirus—racism and economic inequality. Economic Policy Institute, June 1. Available at: www.epi.org/publication/black-workers-covid.

Industry Research. (n.d.). National Restaurant Association. Retrieved May 25, 2020. Available at: https://restaurant.org/Manage-My-Restaurant/Business-Operations/Covid19/Research/Industry-Research.

Locked Out by Low Wages. (2020). One Fair Wage. Available at: https://onefairwage.com/wp-content/uploads/2020/05/OFW_LockedOut_UI_COVID-19_-FINALUPDATE.pdf.

Lynn, M., Sturman, M., Ganley, C., Adams, E., Douglas, M., & McNeil, J. (2008). Consumer Racial Discrimination in Tipping: A Replication and Extension. *Journal of Applied Social Psychology*, 38(4), 1045–1060. doi:10.1111/j.1559-1816.2008.00338.x.

National Restaurant Association Restaurant Industry Outlook. (2017). National Restaurant Association. Available at: www.restaurant.org/Downloads/ PDFs/News-Research/2017_Restaurant_outlook_summary-FINAL.pdf.

Statement from the US Private Sector Job Quality Index ("JQI") Team on Vulnerabilities of Jobs in Certain Sectors to the Covid-19 Economic Shutdown. (n.d.). US Private Sector Job Quality Index. Retrieved June 15, 2020.Available at: https://d3n8a8pro7vhmx.cloudfront.net/prosperousamerica/pages/5561/attachments/original/1587123672/JQI_Team_Statement_One_on_COVID-19_Economic_Shutdown_Job_Impact_031920.pdf?1587123672.

Tipped Over the Edge: Gender Inequity in the Restaurant Industry. (2012). The Restaurant Opportunities Centers United. Available at: https://chapters.rocunited.org/wp-content/uploads/2012/02/ROC_GenderInequity_F1-1.pdf.

18 True Price Store

Guiding Consumers

Adrian de Groot Ruiz

Introduction

Currently, the production of food almost unavoidably involves hidden true costs such as climate change, biodiversity loss, and poverty. A kilogram (kg) of cocoa beans from West Africa, for example, is responsible for about 5kg of CO_2-eq emissions. In addition, farmers would need to receive $3.00 extra per kg of cocoa to earn a living income (True Price, 2018a).

Imagine a store where people can make choices to rectify such damage. In the True Price Store, one can pay to take out the CO_2 emissions and counteract poverty. If one enters the store, which is actually located in one of Amsterdam's main shopping streets, one sees a coffee corner where one can order drinks, as well as a pyramid of blue crates featuring a diverse range of products, such as cider from the organic fruit cooperative Fruitmotor. On Saturdays, one can buy fresh bread from Bakery Van Vessem, which optimizes its recipe to minimize its environmental damage. The windows show the best sellers: colorful chocolate bars of Tony's Chocolonely, one of the largest Dutch chocolate brands, which has been managing its true costs (with slave-free cocoa) since 2013.

If True Costs are a Problem, True Prices are a Solution

The True Price Store was founded on three insights. The first is that true costs are a major societal challenge. Owing to the external costs of global production and consumption, our economic system greatly damages the natural, social, and human capital that underpin society. Climate change is perhaps the best-known externality, which, in economic terms alone, could reduce global Gross Domestic Product by a quarter by the end of the century (Network for Greening the Financial System, 2020). However, there are many other serious externalities. On the environmental side, for example, about a million species are currently threatened with extinction (Intergovernmental Science-Policy Platform on Biodiversity and Ecosystem Services, 2019). On the social side, a fifth of the global working population and their families are (extremely or "moderately") poor, and in Africa the majority of workers live in poverty

(International Labour Office, 2019). Our current global economy enslaves more people for the sake of food production than most would realize (Hodal, 2019).

The external costs of the food system are estimated in the order of $12 trillion per year (Nature Editorial, 2019). Up to 37% of global greenhouse-gas emissions can be attributed to the food system (Science Advice for Policy by European Academies, 2020), a majority of the global working poor are employed in agriculture (World Bank, 2020a) and over 70% of children forced into child labor are linked to food production (International Labour Organization, 2017b). While 690 million people were undernourished in 2019 (Food and Agriculture Organization of the United Nations, 2020), excessive fat and salt content in food leads annually to 11 million deaths and 255 million healthy life years lost (Global Burden of Disease 2017 Diet Collaborators, 2017).

The second insight is that, if true costs are the problem, true prices could well be our best chance at a solution. An economy with external costs ignores and at times even rewards damage to society: products that externalize costs to others are on the whole more profitable to the producer and cheaper to the consumer. True prices are prices that reflect the true costs: if a product has a true price, then the external costs are transparent, paid for, and repaired. As a result, with true prices there is no unresolved damage to people or the planet. True prices additionally remove the perverse incentive that bad products are cheaper than good products.

In fact, if all products had a true price, the global economy would arguably be sustainable. If no product imposes harm on workers, consumers, or the environment, then nature, at the macro-level, is conserved, the climate does not warm up, human and labor rights are respected, and every worker earns enough to give her and her family a good life.

Mission Impossible?

This sounds too good to be true. If true prices would solve so many of the world's problems, why has nobody done this before? In fact, true pricing is a form of pricing externalities that economists have long understood to be the solution to internalizing externalities. A group of British economists—Pigou, Sigdwick, and Marshall—formulated the concept of externalities and proposed to price them through corrective "Pigovian" taxes, a century ago (Pigou, 1920; Laffont, 2008).

In practice, however, it has proven to be a mission impossible for economists and policymakers to systematically price the externalities of products. For starters, establishing what these prices should be has proven elusive, owing to the difficulty of establishing "what is true" and the complexity of computing externalities. This is compounded by the typical, political view that saw pricing externalities as a tax hike that consumers would not be willing to pay. As a result, pricing externalities has been, until recently, an economists' pipe dream.

True Price

The organization True Price was founded in 2012 on the belief that social and technological innovation made possible what was impossible a century ago. Its vision is that true pricing is the way to realize a sustainable global economy and its mission is to make it happen.

An important aim of the organization is to establish a global standard to determine true prices and advocate for true pricing. In True Price's theory of change, the most effective way to get businesses and governments to adopt true pricing is to lead by example. True Price has thus been working with businesses in food and agriculture from the start. It also holds that consumers and citizens need to be involved. So, as soon as sufficient businesses were on board, in 2019 True Price opened a store to bring true pricing to the consumer.

The store is a true pricing microcosm. Consumers who visit can see the true prices of various food products and pay for them, whereas the businesses who place their products in the store actively work to minimize the external costs of their products.

Roadmap

The remainder of this chapter will focus on three questions. First, *what* exactly is a true price and true pricing? Second, *why* try to realize true pricing if it has never worked before? And, finally, *how* can true pricing be implemented in practice?

What is True Pricing?

What is a True Price?

The true price of a product is the market price plus the *true price gap*. The true price gap consists of external costs or, colloquially, "hidden costs." More precisely, the true price gap is defined as the costs to remediate the harm resulting from the externalities of production and consumption that breach basic rights.[1] The true price gap reflects the costs of the actions that need to be taken to restore these harms. In the case of CO_2 emissions, it reflects the costs to take CO_2 out of the air; or, in the case of child labor, the costs to provide missed education to children, offer required medical and psychological support, and compensate children for the injustice suffered.

The bar of pure chocolate that can be bought in the True Price Store has a market price of $3.12. The true price gap is $0.99. This includes environmental costs like carbon emissions, deforestation, and pollution, as well as social costs like underpayment of farmers, child labor, and forced labor. The true price is thus $4.11.

It is important to note that one cannot realize a true price by just increasing the price. The extra money needs to be used to repair the damages done.

Hence, a product only has a true price if no external costs occur, or if all external costs are repaired.

What is True Pricing?

Next, what is *true pricing* in practice? Is it calculating true prices? Or taxing them? Or is it the ideal state where all products have a true price?

True Price defines true pricing as taking action to transition to a sustainable economy with true prices through *transparency* about true prices, *transformation* of products to prevent external costs, *transactions* to pay and remediate external costs, *taxation* of external costs, and *taking out* unacceptable external costs by prohibition. Hence, true pricing is something one can do, right now, and aims to solve the problem of an unsustainable economy.

Why Try?

When True Price was founded in 2012, there was little support for it among experts, who considered it to be a mission impossible. The perceived barriers can be summarized by three objections: i) it is impossible to establish what is "true;" ii) externalities cannot be calculated; and iii) people will not want to pay higher prices. We will discuss each challenge and explain how social and technological innovation has allowed them to be overcome.

What is True?

The first key challenge is a question posed by philosophers and consumers alike when they first hear about true pricing: "but what is true?" They wonder how social and environmental effects are monetizable and whether *all* things can be monetized, including child labor and biodiversity. And beyond that, how is it possible to come to a single price if people value things differently? They ask how to trace the infinite number of consequences of production, and whether monetizing harmful actions like slavery enables their commodification and justifies them through the profit, or pleasure, that they enable.

Economists have traditionally tackled the pricing of externalities through *shadow prices*. They assume that the perfect *shadow price* should take the form of a tax that factors in all positive and negative externalities, perfectly balances all (internal and external) costs and leads to a market equilibrium that benefits all parties. This has its origin in nineteenth-century (British) utilitarian and naturalistic conceptions of society, which still underpin neoclassical economics. In this paradigm, market outcomes are believed to represent an almost natural equilibrium of market forces that, ideally, should maximize the sum of the utility experienced by all individuals.

Calculating this perfect shadow price runs into all the aforementioned problems and has proven elusive for over a century. True Price found a solution:

find truth in the rights of people. A price is considered true if, in producing and consuming a product, all basic rights are respected.

This *rights-based approach* builds on the social innovation represented by the postwar consensus that the social order should be based on universal rights. After the Second World War, a global understanding that people have human rights grew. The United Nations subsequently began to recognize labor and environmental rights and the twenty-first century saw the recognition of the responsibility of market players to respect rights (United Nations, 2011). The set of universal rights is evolving, its interpretation varies per country, and adherence to them is highly imperfect. Still, universal rights have become a global consensus: all countries have come to adopt the Universal Declaration of Human Rights, just as all adopted the 2030 Agenda outlining the Sustainable Development Goals (United Nations General Assembly, 2015).

True Price argues that the implications of globally accepted rights and responsibilities are that, if market players cause negative externalities that breach a human right, they have the responsibility to remediate this harm. As a result, it is not necessary to measure all positive and negative externalities to calculate the true price. Nor does the true price gap reflect the intrinsic value of damages, such as child labor or climate change or an "exchange rate" to off-set these harms. Rather, the true price specifies what buyers ought to pay if they want to meet their responsibilities toward their fellow people in the marketplace.

Based on above principles, True Price developed a framework to establish true prices (True Price Foundation, 2020a). This framework has been successfully applied to calculate the true price of the food products found in the Store.

Too Complex to Compute and Account

The second barrier to establishing true prices is the theoretical complexity of computing externalities and accounting for them in practice. This has first been made possible by relatively recent theoretical advances in scientific fields such as environmental and ecological economics and environmental and social life cycle analysis, resulting in a new True Cost Accounting discipline. For example, a recent United Nations–The Economics of Ecosystems and Biodiversity study represented a milestone in the economic analysis of ecosystem services and biodiversity but only began in 2007.

Accounting for externalities in practice has been made possible by the recent information revolution. The cost of storing, communicating, and processing information has dramatically declined, unlocking data at an unprecedented rate. This makes it possible to gather, account, aggregate, and verify the necessary data. Whereas accounting for externalities is still immature, it was either impossible or prohibitively expensive just two decades ago.

At a modest scale, the businesses that provide products in the True Price are current examples of the possibility of computing and accounting true prices using the latest information.

Too Expensive

The final perceived barrier is that consumers and voters will be reluctant to support true pricing, as it increases their cost of living, albeit to the detriment of others, such as poor farmers or future generations.

In the end, this is an empirical question. The latest science suggests an increasing willingness on the part of consumers to participate in true cost purchasing. The selfishness of people is a fundamental tenet of classical economics, but based on armchair speculation. Actual research conducted by behavioral economists in this century suggests that the majority of individuals are willing to sacrifice material wealth for the sake of others, if others do so as well (Fehr and Fischbacher, 2003). Recent research suggests that 37%-54% of consumers are willing to pay more for sustainable food (PwC, 2019), for example. Anecdotal evidence from the True Price Store suggests that a majority of customers are willing to pay the true price, including the many unsuspecting customers—like tourists—who come to buy cool chocolate bars and have never heard of true prices.

Even if many people would not accept higher prices, this is not a showstopper. A strong argument can be made that true prices can drop significantly. Preventing externalities is typically much cheaper than remediation. Currently, there is no pressure to reduce unknown true costs. True pricing would unleash the power of markets to decrease external costs by leveraging innovation, competition, and entrepreneurship. Finally, if governments are smart, they will tax external costs and decrease the price of sustainable and healthy food with the revenues.

The picture above shows a pyramid of blue crates in the True Price Shop in February 2020. Each crate contains a product for which the true price is known or will soon be known. In white the retail price is shown—the price at which the product is typically sold for in stores. In blue, the true price of the product is shown. For example, one crate contains bananas that typically costs 1.52 per kilogram and reveals that their true price is €1.86. Another crate contains a pair of jeans with a typical retail price of €40 and a true price of €73. True Price aims to place such blue crates with true priced products in stores and restaurants of other organizations as part of the "blue crate movement."

How?

The previous sections presented the case that true prices are an effective and feasible solution to external costs. This leaves the question: *how* can true pricing be implemented? True Price envisions a five-step implementation (True Price Foundation, 2019):

1   The provision of *Transparency* about true prices of products by businesses and the use of this transparency by consumers.
2   The *Transformation* of production by businesses to prevent external costs.

Figure 18.1 True Price Store display

3   The *Transaction* by consumers to pay for repairing external costs that cannot be prevented.
4   The *Taxation* of external costs and the subsidization of sustainable food by governments to incentivize businesses to produce sustainable products and enable consumers to buy them.
5   The *Taking out* of externalities by regulation where it is feasible and remediation is undesirable.

These '5Ts' have a logical order. In practice they can occur in parallel or in different order.

Transparency

The starting point is transparency. This requires producers to compute and disclose their true prices, providing consumers and other buyers with the information needed to make sustainable decisions. Transparency also provides the information required for the other steps.

True prices are computed in five phases. In the *scoping phase*, all relevant processes of a product's lifecycle are determined, together with relevant negative externalities per process. In the *measurement phase*, the externalities are quantified, providing footprints like tons of hectares of land used or full-time equivalent hours of child labor. Measurement requires data collection. Ideally all data is primary data, collected at all the production sites across the globe. In practice, one has to work with a combination of primary data, estimates from product-specific lifecycle models, and data from macroeconomic input-output models. In the *monetization phase*, the footprints are monetized by estimating the remediation costs, using local factors where possible. In the *aggregation phase*, all remediation costs are summed to come to the true price gap. Finally, in the *validation phase*, results are validated.

Consider a pure chocolate bar of Tony Chocolonely's. It was the first company to calculate their true price and supply bars in the Store. The key parts of its lifecycle are farmers growing cocoa beans, chocolate processors using beans to make cocoa liquor and butter, sugar plantations, and the chocolate factory making the bars. Other parts include the production of lecithin, aluminum, and paper, as well as transportation and retail.

Tony's was founded with the mission to create a slave-free chocolate sector. To maximize its impact on the sector, Tony's built a value chain that others can emulate. Therefore, it sources cocoa from smallholder farmers in Ghana and Côte d'Ivoire, who produce the majority of the world's cocoa (Ceres, 2020). Similarly, Barry Callebaut, the world's largest chocolate processor, processes its beans. For each step, the potentially relevant external costs were established based on previous research.

Through its *bean to bar* program, Tony's knows the cooperatives that it sources from. This greatly facilitates the measurement phase, as primary research can be done on the main ingredient. Data from most other ingredients come from secondary sources. The analysis then results in footprints. For example, the total emissions per bar are 0.66 $kgCO_2$-eq.

Based on a monetization factor of $0.13/$kgCO_2$-eq, this implies remediation costs of $0.09 per bar. By similarly monetizing and aggregating all remediation costs, a true price gap of $0.99 is established. $0.95 of the remediation costs are related to cocoa cultivation and $0.03 to sugar cultivation. The largest environmental costs are land use (10% of the gap), climate change (9%), and soil pollution (7%). The largest social costs were underearning of farmers and underpayment of workers (29%), child labor (14%), and harassment (10%).

The validation phase showed that the calculated results were justified, although at this stage footprints and remediation costs involve uncertainty.

After computing the true price, it can be shown to customers. Business clients can use this information to reduce the true price gap of the products that they sell. Consumers can use it to be as sustainable as possible (select the product with the lowest true price gap) or otherwise search for products with a lower true price gap and affordable price.

Tony's, Van Vessem, and others show their true price in the True Price Store. As this is just one store, it is more significant that they use this information in their own communications. Because true prices are not widely available, brands typically provide a benchmark to give their consumers the context that they need. For example, Van Vessem uses this information to show consumers that its bread is twice as sustainable as the average bread in the Netherlands (True Price, 2018b).

Transformation

The second step is the *transformation* of production to realize (more) sustainable products. By changing the product, the ingredients and the ways of production, businesses can reduce their true costs.

For example, Van Vessem—a baker with seven stores—uses its data on true prices to develop recipes that lower the true price gap of its bread (ibid.).

Tony's also uses true price data to inform its interventions. When Tony's first calculated its true prices, the external costs of their cocoa were around $9.30 per kilogram (True Price, 2018a). On the one hand, that was better than the $16.60 of external costs of the average cocoa from Western Africa. On the other hand, it was not fully sustainable. It took various steps, including calculating the price that farmers would need to receive to realize a living income, better monitoring of child and forced labor, and measuring their carbon footprint. Later, Tony's started to pay above the market price to close the living income gap. Tony's managed to reduce its external costs from $9.30 in 2013 to $5.30 in 2018 (ibid.). To be able to pay farmers more, it needed to increase its price and explained this to consumers. Despite this, Tony's has been commercially successful, becoming one of the largest chocolate brands in Dutch supermarkets, surpassing traditional chocolate giants.

Transaction

The third step is *transaction*. In the short run, it is impossible for consumers (and businesses) to only purchase products without external costs, as these simply do not exist. Hence, to meet their responsibility, buyers need to be given the opportunity to pay the true price and remediate harms in the best way possible. Remediation is just starting to become available. It requires the availability of organizations that provide remediation in a highly reliable and effective manner.

Currently, remediation for two externalities can be provided in the Store. Hence, consumers can currently see the true price, but pay a "truer price." The externality with most remediation providers is climate change. In the True Price Store, consumers can pay to remediate the carbon emissions from their products. This is provided to a company that plants forests in deserts and provides real-time data on trees planted. Consumers can also pay to remediate underpayment to workers. However, owing to the difficulty of reaching individual workers in the value chain, at the moment this is given to a non-governmental organization that gives verifiable direct payments to people living in poverty. While an imperfect implementation of true pricing, the Store is working on a better system with the businesses involved.

Taxation

The fourth step is *taxation*. Transparency, transformation, and transaction enable market participants to buy and sell sustainable food if they want to. In addition, they create an incentive for businesses to make products more sustainable. Still, they do not resolve the perverse incentive that less sustainable food is cheaper than more sustainable food. Nor do they alleviate the problem that for low-income families it can be a real problem to pay more for food. This means that taxation is an important step in true pricing: governments can make value-added tax proportional to the true price gap by, for example, making more sustainable products cheaper and less sustainable products more expensive. This closes the incentive and affordability problems.

Taking-Out

The final step is governments *taking out* products that have unacceptable external costs. For various externalities, remediation is a perfectly acceptable manner to deal with external costs, and taxation is a suitable form of government intervention. For example, for CO_2 it does not matter whether it is avoided or taken out of the air quickly. Other externalities, such as forced labor, ought to be prevented. Hence, in such cases the prohibition of these externalities forms the final step of true pricing.

In practice, prohibitions are problematic. First, they often exist but are not enforced effectively. Second, governments have no jurisdiction to prohibit or enforce prohibition in other countries. Third, consumers and businesses have no way at all to prohibit or enforce prohibitions. Hence, until there is an effectively enforced global prohibition, transformation, transaction, and taxation are needed for such externalities.

Conclusion

Currently, it is possible to calculate the true price of a product and show it to consumers. Various businesses are applying it and there is a store where consumers can see and pay the true price. All these things were inconceivable less than a decade ago. This means that pricing externalities is no longer a pipe dream. However, optimism is still required to see true pricing taking over the global economy. The fact that there is at least one store where prices are a bit truer, however, could warrant a healthy dose of such optimism.

Note

1 An important question is if animals and nature are included in social and environmental rights held by humans or if they have rights in themselves.

References

Ceres. (2020). Engage the Chain. An Investor Brief on Impacts that Drive Business Risks – Cocoa. Available at: https://engagethechain.org/sites/default/files/commodity/Cocoa%20Brief%20Engage%20the%20Chain%20Oct2020.pdf.

Fehr, E. & Fischbacher, U. (2003). The nature of human altruism. *Nature*, 425, 785–791. doi:10.1038/nature02043.

Food and Agriculture Organization of the United Nations, International Fund for Agricultural Development, UNICEF, World Food Programme, and World Health Organization. (2020). The state of food security and nutrition in the world 2020. Transforming food systems for affordable healthy diets. Rome: Food and Agriculture Organization of the United Nations. doi:10.4060/ca9692en.

Global Burden of Disease 2017 Diet Collaborators. (2017). Health effects of dietary risks in 195 countries, 1990–2017: a systematic analysis for the Global Burden of Disease Study 2017. *The Lancet* 393(10184), 1958–1972. doi:10.1016/S0140-6736(19)30041-8.

Hodal, K. (2019). The briefing. One in 200 people is a slave. Why? *The Guardian*, February 25, 2019. Available at: www.theguardian.com/news/2019/feb/25/modern-slavery-trafficking-persons-one-in-200.

Intergovernmental Science-Policy Platform on Biodiversity and Ecosystem Services. (2019). *Global assessment report on biodiversity and ecosystem services of the Intergovernmental Science-Policy Platform on Biodiversity and Ecosystem Services.* E.S. Brondizio, J. Settele, S. Díaz, & H.T. Ngo (Eds.).

International Labour Organization. (2017b). Global estimates of Child Labour: Results and trends, 2012–2016. Available at: www.ilo.org/wcmsp5/groups/public/@dgreports/@dcomm/documents/publication/wcms_575499.pdf.

Laffont, J.J. (2008). Externalities. In *The New Palgrave Dictionary of Economics*. London: Palgrave Macmillan. doi:10.1057/978-1-349-95121-5_126-2.

Nature Editorial. (2019). Counting the hidden $12-trillion cost of a broken food system. *Nature*, 574(296). doi:10.1038/d41586-019-03117-y.

Network for Greening the Financial System. (2020). NGFS Climate Scenarios for central banks and supervisors. Available at: www.ngfs.net/sites/default/files/medias/documents/ngfs_climate_scenarios_final.pdf.

Pigou, A.C. (1920). *The Economics of Welfare*. London: Macmillan.

PwC. (2019). 2019 Global Consumer Insights Survey Sustainability. Retrieved October 11, 2020. Available at: www.pwc.nl/en/insights-and-publications/services-and-industries/retail-and-consumer-goods/2019-consumer-insights-survey/sustainability.html.

Science Advice for Policy by European Academies. (2020). *A sustainable food system for the European Union*. Berlin. doi:10.26356/sustainablefood.

True Price. (2018a). The True Cost of Cocoa. Tony's Chocolonely 2018 progress report. https://trueprice.org/wp-content/uploads/2018/11/The-True-Price-of-Cocoa.-Progress-Tonys-Chocolonely-2018.pdf.

True Price. (2018b). The First Ever True Price of Bread. Retrieved July 18, 2020. Available at: https://trueprice.org/consumer/bread-bakker-van-vessem.

True Price Foundation (2019). A roadmap for True Pricing. Available at: https://trueprice.org/a-roadmap-for-true-pricing.

True Price Foundation. (2020a). Principles for True Pricing. Available at: https://trueprice.org/principles-for-true-pricing.

True Price Foundation. (2020b). Monetisation Factors for True Pricing. Available at: https://trueprice.org/monetisation-factors-for-true-pricing.

United Nations. (2011). Guiding principles on business and human rights: Implementing the United Nations "Protect, Respect and Remedy" framework. Available at: www.ohchr.org/Documents/Publications/GuidingPrinciplesBusinessHR_EN.pdf.

United Nations General Assembly. (2015). Transforming our world: the 2030 Agenda for Sustainable Development. Available at: www.un.org/ga/search/view_doc.asp?symbol= A/RES/70/1&Lang=E.pdf.

World Bank. (2020a). Understanding Poverty. Overview. Retrieved July 13, 2020. Available at: www.worldbank.org/en/topic/poverty/overview.

Conclusion

Mobilizing the Power and Potential of True Cost Accounting

Nadia El-Hage Scialabba, Carl Obst, Kathleen A. Merrigan and Alexander Müller

There is an increasing public and scientific debate about the potential for True Cost Accounting (TCA) and the need for TCA to play an important role in the policies and decisions of all agri-food system stakeholders, including those of governments, businesses, communities, and every citizen. In recent decades, the recognition of the need for a new and encompassing accounting system that takes into account the hidden environmental costs of production has started to change the economic thinking far beyond conservation circles. The appreciation of the negative (and sometimes positive) impacts of production on the environment has become common, together with the recognition that economic reporting does not adequately consider the impacts of activities on the natural resource base, or on social wellbeing and human health. However, there is a wide gap between the multitude of colorful Corporate Social Responsibilities reports and actual company impacts on natural, human, and social resources, precisely because the mainstream international standards of economic accounting and reporting exclude externalities. With the current awareness of the true (or full) costs of economic activities, it is time to go beyond discussion and design of TCA approaches and move towards implementation. A range of opportunities is explored in this chapter, as well as likely challenges.

From a theory of change perspective, much is being done by the TCA community of practice, but less attention is paid to who needs to do what differently for TCA to succeed. Scientific and methodological breakthroughs will keep emerging and offering new opportunities to improve TCA measurements. However, tangible effects on policy and decision-making are essentially related to socio-political processes. It is only through social processes that lead to a consensus on an agreed set of processes and overall framework that trust will be built for making choices that establish sustainable food systems. Thus, it is the mobilization of governments and multi-stakeholder community networks that will be crucial to the effective realization of TCA's potential.

True Cost Accounting (TCA) cannot be a panacea, and nor can TCA advocates assume that wide adoption of the process will magically change the current way of doing business and making policy. As highlighted through advancing the United Nations (UN) Sustainable Development Goals (SDGs), mindsets and institutional structures are far from the trumpeted integrated, transdisciplinary approaches that cut across all human and natural spheres. Moving towards holistic approaches is not easy, but it is encouraging to see that TCA has already heightened public awareness on food system externalities. TCA is an important tool to advance a global transition to sustainable food systems, but each societal actor has a role to play in making change happen.

Where We Came From and Are Going To

TCA has successfully changed mindsets. The Food and Agriculture Organization of the United Nations launch of the Food Wastage Footprint in 2014 marked a sudden shift in public awareness about the environmental and social impacts of food loss and waste. The mantra "if food wastage was a country, it would represent the third largest emitting country in the world" went global within days. For the first time, food system externalities were quantified, and people woke-up to reality. It did not really matter if the emissions were 3.5 Gt or 4.4 Gt of CO_2 equivalents per year (depending on the year of the dataset used), or which emission factor or carbon price was used to quantify the social cost of carbon at $394 billion per year. The huge hidden costs of food wastage were made visible. Donor funds, which were scarce for investment in reducing post-harvest losses, rapidly became available, thanks to allocations made by environmental (rather than agricultural) budgets.

Similarly, efforts to quantify the climate impacts of agricultural practices that accelerate soil erosion have opened new dialogue about the need for public support and market mechanisms to support soil-enhancing practices. Nowadays, the link between food and agriculture systems, climate change, antibiotic resistance, and noncommunicable diseases is clear to all, even if the interaction pathways are not fully established. Looking back, it can confidently be stated that TCA has played a significant role in changing political debates and public mindsets, beyond the dollar values that one can assign to individual TCA assessments.

Gross Domestic Product (GDP) for successful economies. The scientific effort and political debate to define the "true costs" of food must be placed within the successful measurement of the economy that perceives annual GDP growth as the world's most powerful statistical indicator (Lepenies, 2016). GDP is not only the measure of a country's economic output; it also is understood to describe, in a single number, the success of the overall development of a country. GDP is not a general law of nature expressed in statistical calculations, but rather the result of a long process of attempts to measure the economic reality of a country and express it as a single statistical indicator. As such, GDP is a "social construct" created by people and accepted by society. GDP measures the total economic output of a country based on monetary values; the fact that the value of goods and services is based only on their market value automatically excludes whatever has no market value. Thus, the value of biodiversity and fertile soils, which have no market price, do not influence GDP, at least in the short run. TCA, however, by considering natural, social, human, and produced capitals involved in food and agriculture systems (The Economics of Ecosystems and Biodiversity, 2018), provides a social construct that reconsiders the basic concept of how all countries in the world measure their development. Adopting and implementing TCA for food and agriculture systems is therefore bound to change the overarching perception of economic success and its actual expression in annual GDP growth.

TCA is a tool. Experts are continuing to refine TCA approaches by structuring accounts and assigning values that speak to the wonderful complexity of issues and relationships that constitute our lives. The nascent TCA toolbox is currently in an adolescent stage, actively exploring possible futures and confident in its genuine capacity to change the world. However, any tool, even the most mature and well developed one, is a lifeless instrument unless people engage in using it. The ultimate responsibility for responding to the implications highlighted by using the tool rests with the user. Thus, the social and political process surrounding TCA's development and implementation, as well as actors' accountability, are of crucial importance for a transparent and effective food system transformation.

Towards informed decision-making. While acknowledging the unavoidable gap between scientific evidence and policy processes, TCA seeks to provide evidence for decision-makers to consciously manage complexity. Complexity is defined as a network of multiple interacting factors and unknowns that cannot be addressed in a piecemeal approach. TCA's broad lens aims to offer a high-resolution snapshot of our agri-food ecosystems, by giving a meaningful place to the variety of mineral, plant, animal, human, and produced goods and services, and hence providing a richer picture of the dynamic canvas of life. Developing this richer picture also supports better recognition and understanding of clouds on the horizon that indicate unknowns, risks, or patterns that deserve attention. By providing a clear picture, policymakers, investors, producers, and communities can better evaluate what to support (or not) for the future of food. When TCA is eventually embedded in standard reporting systems of enterprises, measuring and valuing all positive and negative externalities will provide a very different picture of the interaction of businesses with nature, society, and individuals. Currently, several frameworks try to capture the complex reality of a defined eco-agri-food system; an inventory of methodological frameworks, resources, databases, and case studies provides an overview of where we stand today (Bandel *et al.*, 2020).

Where Do We Stand?

The richness of material that this book has drawn together under the banner of TCA is impressive. The richness speaks to the desire for new and more encompassing approaches to assessing and analyzing food systems; to the breadth of the skills and experience that can and must be applied; and to the momentum that is building for change. This chapter draws out some key insights from considering the chapters as a reflection, on the part of the authors, of the status of TCA. It provides suggestions for taking TCA forward so that it can positively influence the sustainability of our food systems around the world.

Seven insights emerge from stepping back and considering the book chapters as a whole.

Complex systems. The first is that there are many "pieces" in the TCA puzzle. Joining together material on the health consequences of diets, with the need for the conservation of natural resources, the growing of crops and breeding livestock, the supply chain risks of major food conglomerates, and the precarious nature of work of those employed in the processing and dining sector is both magnificent and overwhelming. How can these all possibly be fit together by a long-standing systems thinker, let alone a short-term financial analyst, a policy specialist, a politician, a farmer, or a voter? There thus remains a significant challenge to demonstrate how all of the pieces that legitimately fall under the TCA banner can be brought together, such that food and agricultural systems can be assessed holistically and results can be presented in simple terms.

System boundaries and responsibilities. Second, food supply chain boundaries extend very far, upstream and downstream, with sustainability impacts on the environment and communities that become less visible as the spatial coverage increases. Studies have so far set TCA assessment boundaries according to data, resources, and time available for individual projects. Excluding or including a geographic impact area yields results that are bound to remain incomplete and potentially unfair to affected populations. While, ideally, TCA assessments should set boundaries within the realm of control or influence of financial and operating policies and practices, the "system" impacts are often planetary. This interconnectedness points to the need for a greater understanding of the responsibilities and accountabilities of all societal actors, at community, national, and international scales. Furthermore, it calls for the development of meaningful legal and institutional frameworks that are conducive to TCA implementation and adoption.

Incorporating the social dimension. Third, notwithstanding the broad coverage of topics in this book, there are important areas poorly reflected in the chapters that should, ideally, be the heart of the conversation. These include social capital, particularly in terms of individual and culturally important connections, and the wider suite of ecosystem services beyond the inputs to food production on which farm management and related supply chains can have significant influence. This is not to say that these topics are not mentioned across the chapters, but rather that these distinctly "non-market" aspects of food systems do not appear to receive the level of discussion that most people supportive of TCA would agree is needed. Social issues are difficult to quantify, and creating science-based targets for worker welfare or racial justice is not value-free. However, addressing deeply rooted systemic inequalities requires particular efforts to measure and communicate: 2020 is a turning point, and we need to completely rethink how we approach social issues.

Risks and thresholds. Fourth, and building on the previous point, because of the common interest in using TCA to "amend the bottom line" and move away from financial profit as the sole measure of success, there is a tendency to focus on applying standard economic pricing approaches in a more holistic way. Put differently, a general flavor of the chapters is how to adjust or extend current marginal pricing approaches to production decisions and applying standard approaches to the pricing of externalities. For many, this is a general understanding of the intent of TCA. However, what is missing in this application is a broader appreciation of systemic and non-marginal risks and the extent to which we are approaching, or passing, ecological or societal thresholds. While in theory, prices should rise in order to reflect scarcity, history reveals that humanity regularly ignores any such signals or finds substitutes. Moreover, when there are no prices for non-marketed goods that are present in the prevailing institutional framing (i.e., there are externalities) there will be no price signals. In this context, the importance of applying other aspects of economic theory (and accounting) around wealth and balance sheets becomes fundamental. Understanding risks and thresholds in terms of the available natural, produced, human, and social capital is a central thesis of the UN Environment TEEBAgriFood framework. This is not a perspective that is well developed in the chapters. What is required is a stronger focus on the stocks of capitals themselves and their condition/quality, in addition to consideration of the benefits (or loss of benefits) associated with their use. A focus on stocks of capital directly facilitates measurement of thresholds and non-linearities and provides a basis for establishing informed targets and benchmarks. TCA on its own cannot determine the target thresholds, but it can structure the discussion. However, to do so, TCA requires not only a profit and loss statement but also a rich and comprehensive balance sheet.

Post COVID-19 narrative. Fifth, while only one paper tackles the challenges raised by the coronavirus disease (COVID-19), there is an opportunity for TCA to contribute further to the discussion in this space. Of course, the challenges facing agriculture and food systems have been both long-standing and will, unfortunately, continue to be faced beyond (hopefully) the time horizon in which solutions to the COVID-19 can be found. In that sense, the contexts for the papers are commonly focused on long standing environmental, social, and health challenges that are attributable to our current food systems. Nonetheless, it is also clear that COVID-19 has starkly highlighted many systemic concerns, but the policy responses have often been framed as choices between health and economics rather than in terms of integrated solutions. Indeed, COVID-19 has fueled two contrasting narratives: the need for local, resilient food production and the need for more international food trade in times of social distancing and lockdowns. Seen through a TCA lens, poor food and agricultural practices (e.g., deforestation, confined animals, wet markets) can be held responsible for the global pandemic. Perhaps this points to a key challenge for implementing TCA. If TCA approaches had been standard practice, then we might have readily reached shared conclusions about preventing and dealing with the global and immediate impacts of the pandemic in different parts of the world, rather than battling between the economic and health-focused solutions. TCA could provide advice on future health risks by assessing growing externalities, such as antibiotic resistances coming from the (over-)use of pharmaceuticals in industrialized livestock systems. We are more than capable, at least theoretically, of dealing with the complexity of balancing these objectives, but reaching that point will require a paradigm shift.

Government role. Sixth, if a paradigm shift is required and it needs to happen globally, the collected papers suggest that this will be either at local scale—farmers, True Price shops, communities—or from international processes. Both are undoubtedly required, but there is little discussion of the role of national governments in driving change. Perhaps it is failure at this level that motivates the search for solutions at other scales, but it seems difficult to imagine a pathway to the implementation of holistic food and agricultural systems that does not also involve the active engagement of national level jurisdictions. Undoubtedly, a prerequisite for national government-level TCA action is the standardization and harmonization of language, definitions, methods, and tools around TCA. While a few chapters speak to this—particularly Chapter 4 on methods and frameworks—the chapters as a whole reveal quite broad and relatively loose understandings of TCA. This is excellent for building a community but will be insufficient if large-scale adoption of TCA is the ambition. One possible pathway to greater government engagement is through substitution of TCA for cost-benefit analysis, as argued in Chapter 12 ("Embedding TCA Within US Regulatory Decision-Making"). To do so, it is necessary to understand the inner workings of governments in order to strategically embed TCA within existing processes. Among the many compelling arguments for national government adoption of TCA, two ideas seem particularly important. Given that governments are responsible for public goods, TCA would provide not only information on the value of these public goods but also make flows visible, leading to a different perception of public goods, the investments needed to maintain these goods, and the benefits that are derived from those investments. Second, the potential to introduce TCA into the taxation system to trigger a reconceptualization of the definition of assets could have far-reaching consequences.

Tool versus process. Finally, speaking to the ambition of TCA, many of the chapters point to the need to define success, that is, the purpose of establishing sustainable food systems. Chapter 15 ("Investing in the True Value of Sustainable Food Systems") notes that in considering TCA approaches, it inevitably leads to questioning fundamental choices and goals of business, society, and government. The UN SDGs provide a powerful basis for making these choices at farm, community, national, and global levels, but a challenge remains to establish TCA as the tool of choice to evaluate progress towards these goals. The chapters reveal clearly that TCA can be applied—this is a tremendous step forward. However, as Chapter 1 ("From Practice to Policy: New Metrics for the 21st Century") highlights, TCA is a technical tool—developing and implementing the process around using it must be the next focus.

Where Can We Go (and How Do We Get There)?

Communities, including food and agricultural practitioners and civil society organizations, have advanced scattered but widely diffused efforts for internalizing environmental and social externalities in market goods, such as witnessed by the organic agriculture and fair trade standards. A coalition of what so far has been considered an alternative movement, including environmental and human rights non-governmental organizations (NGOs), is starting to consolidate with initiatives such as Organic 3.0 (International Federation of Organic Agriculture Movement, 2016). Considering decades of practices with environmental and social Key Performance Indicators (KPIs) and the heightened awareness that any thematic focus is unlikely to succeed alone, a community of the willing is needed to identify and develop common TCA-KPIs, based on what can be achieved while keeping producers in business. In line with their respective mandates, NGOs already facilitate agri-food producers' recognition of externalities; this is evident in compliance with organic standards that reveal farmers' unintended environmental impacts. Most importantly, a push from the field and farming communities is the only way to blend important traditional knowledge of agri-ecosystems, the richness of communities' culture, and potential government regulation for TCA. The Global Alliance for the Future of Food Community of Practice for TCA has started to pool expertise to advance TCA, but it needs to extend its partnerships with farmer organizations, producer associations, standard-setting owners, and government representatives.

Businesses, including private companies, investors, and insurers, have been progressing fairly well with the idea of TCA, as a means to hedge against risk, as seen by the numerous initiatives of the Capitals Coalition (https://capitalscoalition.org/). In fact, in the face of supply disruption, companies have been leading change with Integrated Profit and Loss accounting. Tangible financial terms are being integrated in annual accounts and company valuations, as well as in credit ratings and insurance policies. Increasingly, due diligence tools are crafted to improve investors' decisions around capital allocation and portfolio goal setting. However, history teaches us that unless harmonized accounting standards are developed, TCA will follow the same fate as sustainability reporting where, depending on individual benchmark setting, all businesses will soon be flaunting successful operations in various shades of green. For TCA not to become a greenwashing highway, it must be integrated within a new accounting standard, together with the integration of clear thresholds within financial balance sheets. The Capitals Coalition, which united in January 2020 the Natural Capital Coalition and the Social & Human Capital Coalition, is a major effort of global collaboration of over 350 businesses and accountancies to bring nature and people into the heart of business decisions. Building on the Natural Capital Protocol, and on the Social and Human Capital Protocol, a variety of guidance documents (e.g., Biodiversity Guidance, September 2020) are being developed as companion decision-making frameworks. In addition, a small group of European food companies is taking the first steps to measure all capitals in their respective companies, with a view to implement integrated reporting guidelines for the production and consumption of food. This initiative, called "True Cost – From Costs to Benefits in Food and Farming" (https://tca2f.org/) (TMG and Soil & More Impacts, 2020) aims to provide standardized guidance to make hidden costs and benefits visible along the entire value chain, providing a complete picture of the interaction of a company with people, society, and the environment. The US Sustainable Accounting Standards Board has been developing standards for the food and beverage sector that consider key issues and accounting metrics including environment, social capital, human capital, business model and innovation, and leadership and governance. The provisional Agricultural Products Sustainability Accounting Standard published in June 2015 (Sustainability Accounting Standards Board, 2015) could be joined, for instance, by the Capitals Coalition, TCA2F, and others, and collectively taken forward to reflect issues of global concern and consequent harmonious application for the whole business community. With a common baseline, internal and external reporting of companies and risk assessments would allow decisions-makers to create and develop long-term value, instead of focusing on short-term profits.

Governments have so far been virtually absent from the TCA landscape. Although they have agreed on the SDGs for national development, moving towards the Goals remains trapped within old-fashioned institutional structures. As demonstrated by the organic agriculture sector prior to the establishment of organic regulations, markets alone cannot trigger or scale-up change; worldwide, consumers' demand for organic products largely exceeds supply, owing to a lack of policies for supporting organic producers. Most importantly, the public good can only be guaranteed by government rules and enforcement. Indeed, COVID-19 has pushed governments back into the center of the arena for the security of humanity. With contributions from civil society and businesses, governments need to advance TCA on three fronts:

- Establishing the legal framework for a TCA standard, such as is done for corporate accounting standards, in order to secure a fair playing field for all, prevent fraudulent practices, and reduce the cost of supporting multiple approaches.
- Adopting TCA as an administrative process for the elaboration of policy incentives (positive and negative), that orient all stakeholders (smallholder farmers, private multinationals and line ministries) to opt for the appropriate decisions. In particular, TCA should substitute the classical cost-benefit analysis to ensure that, to the greatest extent possible, distortion can be resolved once the externalities are evaluated, and the true-cost of various actions are transparent to policymakers; and
- TCA implies actions far broader than the food and agriculture system *per se*. With the current state of affairs, power and inequity are two obstacles to progress. Currently, cheap food policies are used as social safety nets.

Further, and most importantly, the power exerted from the highly concentrated agri-food input and retail sector often works against addressing externalities. In this context, regulations requiring TCA might work to dis-incentivize natural and human resources exploitation while, at the same time, opening the pathway for adopting alternative competition and anti-trust policies to address the agricultural input-machinery-insurance and food market oligopolies.

Inter-governmental institutions, including the UN system, Bretton Woods institutions, CGIAR research institutions and regional commissions, have been developing and practicing TCA, including: the World Bank project on mainstreaming Wealth Accounting and the Valuation of Ecosystem Services (WAVES) in national economies; the "beyond GDP" UN System of Environmental-Economic Accounting framework (United Nations *et al.*, 2014) that standardizes and classifies countries' statistics and accounts for environmental data; and the UN Environment TEEBAgriFood framework for better understanding, managing, and valuing the impacts of food and agriculture systems. Inter-governmental institutions are precious entry points for governments in order to progress along three main fronts:

- To explore the implications of TCA and eventually develop a TCA Index that would complement—and eventually replace—Gross Domestic Product (GDP) or Human Development Index (HDI). In fact, GDP is a post-World Wars index focused on reconstruction and economic production capacity. The 1990 United Nations Development Programme's HDI better reflects well-being by considering health, education, and living standards. In our globalized era of climate change and pandemics, we need an index that better reflects our modern issues, in particular one that encompasses environmental thresholds. Chapter One "From Practice to Policy: New Metrics for the 21st Century" introduces such a TCA Index, as a means to simplify complexity for decision-making, while moving away from actual monetization. It is interesting to note that SDG 17.19 hints to such an index: *"by 2030, build on existing initiatives to develop measurements of progress on sustainable development that complement gross domestic product, and support statistical capacity-building in developing countries."*
- Through the UN statistical system, adopt universally accepted concepts and definitions for data across all dimensions of sustainability. Common data standards can form the basis for the development of a universal TCA standard and establishing relevant sustainability thresholds. This is the reality for economic measurement and has been for decades. The theory is in place for the other dimensions but it needs the institutional process in order to be driven forward.
- In the longer term, TCA practice and implementation could assist countries negotiating trade reforms that assess national stock flows through international trade, with trade rules accounting for virtual water, virtual

land, virtual pollution, and unsuitable labor conditions. The World Trade Organization (WTO) trade rules favor the lowest cost, that strongly lock-in negative externalities within national boundaries. Although the WTO allows countries to adopt trade measures regulating "product characteristics or their related Production and Processing Methods," this concept remains controversial from a conceptual and policy point of view. Currently, the free flow of capital and labor flattens countries' comparative advantage and we are witnessing a race to the bottom towards the lowest production cost possible. Thus far, the trade of certified organic products has been facilitated by the existence of international standards, as requested by Sanitary and Phytosanitary Measures (the SPS Agreement), because environmental requirements (e.g., no pesticides) are perceived as health and safety requirements. This highlights the importance of an eventual common international TCA reference standard. This could follow the blueprint of the European Union Organic Regulation that is in line with the international standard laid out by the Codex Alimentarius Guidelines; provides the basis for individual country regulations and conformity assessment procedures; and is open enough to private standards that may be more stringent than the national rule (e.g., Soil Association, Demeter).

Clearly, the different stakeholder group initiatives ought to progress in harmony. The current push from the base is changing the narrative in an effective way. Networks are forming but they need to link up with other networks and scale-up their efforts. Suppliers, clients, employees, companies, investors, communities, governments, and conservationists will have different scopes for TCA assessments, but the agreement of all parties on the TCA baseline is crucial.

This book reveals the extent to which TCA has, and can continue, to drive a broadening of mindsets in achieving the sustainability of our food and agricultural systems. This chapter has highlighted areas where more can be done and areas where increased collaboration is required. Fundamentally, the opportunities that exist for TCA are immense. The chance to build on changing mindsets is real and action is needed now. TCA's history proves its potential; its future can drive us towards sustainable solutions.

References

Bandel, T., Sotomayor, M.C., Kayatz, B., Müller, A., Riemer, O., & Wollesen, G. (2020). *True Cost Accounting, Inventory Report.* Global Alliance Future of Food. Available at: https://futureoffood.org/wp-content/uploads/2020/07/TCA-Inventory-Report.pdf.

Food and Agriculture Organization of the United Nations. (2014). *Food Wastage Footprint: Full-Cost Accounting. Final Report.*

International Federation of Organic Agriculture Movement. (2016). *Organic 3.0 for Truly Sustainable Farming and Consumption.* (2nd Ed.). IFOAM-Organics International & The Sustainable Organic Agriculture Action Network.

Lepenies, P. (2016). *The Power of a Single Number, A Political History of GDP*. New York, NY: Columbia University Press.

Sustainability Accounting Standards Board. (2015). *Agricultural Products Sustainability Accounting Standard*. *Consumption 1 Sector*. Sustainability Accounting Standards Board Provisional Standards.

The Economics of Ecosystems and Biodiversity. (2018). *Measuring What Matters in Agriculture and Food Systems*. United Nations Environment Programme.

TMG and Soil & More Impacts. (2020). True Cost – From Costs to Benefits in Food and Farming. Available at: https://tca2f.org/wp-content/uploads/2020/09/TCA_one-two-pagers.pdf.

United Nations, European Commission, Food and Agriculture Organization of the United Nations, International Monetary Fund, Organisation for Economic Co-operation and Development, & World Bank. (2014). System of Environmental-Economic Accounting 2012 Central Framework. New York, NY: United Nations. Available at: https://seea.un.org/sites/seea.un.org/files/seea_cf_final_en.pdf.

Index